SUPERFOOD AND FUNCTIONAL FOOD - THE DEVELOPMENT OF SUPERFOODS AND THEIR ROLES AS MEDICINE

Edited by **Naofumi Shiomi**
and **Viduranga Waisundara**

Superfood and Functional Food - The Development of Superfoods and Their Roles as Medicine

http://dx.doi.org/10.5772/65088
Edited by Naofumi Shiomi and Viduranga Waisundara

Contributors

Toshiyuki Takahashi, Luis Felipe Quiroz, Carolina Rosas, Claudia Stange, Marcial Cariqueo, Alberto Leguina-Ruzzi, Anca Corina Farcas, Maria Tofana, Dan Cristian Vodnar, Sonia Ancuta Socaci, Sevda Inan, Cenk Aydin, Nilay Seyidoglu, Daniela Peña, Claudio Inostroza-Blancheteau, Alejandra Ribera-Fonseca, Marjorie Reyes, Caroline Nunes, Franz Burini, Roberto Burini, Beatrice Ekesa, Eduardo Farinaro, Francesco Cacciatore, Roberto Marcantonio, Saverio Stranges, Maurizio Trevisan, Elisabetta Della Valle, Duangporn Werawatganon, Marius Emil Rusu, Daniela-Saveta Popa, Ines Drenjančević

CBS, Edition **2018**

Published by InTech

Janeza Trdine 9, 51000 Rijeka, Croatia

Publishing Process Manager Romina Rovan

Technical Editor SPi Global

Cover Designer InTech Design Team

Additional hard copies can be obtained from orders@intechopen.com

Superfood and Functional Food - The Development of Superfoods and Their Roles as Medicine, Edited by Naofumi Shiomi and Viduranga Waisundara
p. cm.
Print ISBN 978-953-51-2941-7
Online ISBN 978-953-51-2942-4

Contents

Preface

Based on the concept of "medicine and food are the same," which means that people who eat suitable foods will become healthy, many foods that are effective for improving health have been discovered and investigated in Japan. However, the effects of these foods were not always sufficient to treat or prevent disease. To improve these effects, artificial foods have been created in which effective compounds have been added. Such foods are referred to as "functional food." In Japan, many products containing purified dietary fiber, polyphenols, calcium, etc., are commercially used as functional foods to maintain health.

The use of natural food containing large amounts of effective compounds is another way to improve the insufficient effects. Such foods are called "superfoods"—famous superfoods include fruits and vegetables, such as acai, coconut, and hemp. The idea of superfoods was discussed by doctors and researchers studying diet therapy in Canada and the USA. The roles of superfoods in medicine have been reported in many countries; thus, there is great interest in the potential for healthy food to protect against diseases.

Nowadays, the market has increased to more than several hundred billion dollars, and new sources and functions of superfoods have been developed.

This book and the topic-related book *Superfood and Functional Food- An Overview of Their Processing and Utilization* introduce recent advances in the fields of superfood and functional food.

I recommend looking at the book *Superfood and Functional Food- An Overview of Their Processing and Utilization* in addition to this book because it provides an overview of the many kinds of superfoods and functional foods and provides the readers with important information about the characteristics of these two types of food.

This book focuses on "the production of superfoods and their role as medicine." In the early chapters, prominent researchers introduce the role and production of spirulina or microalgae, the production of functional fruits through metabolic engineering, and the use of food waste. Moreover, effective methods for preparing and consuming superfoods and functional foods are introduced. These chapters will provide improvement to the readers' understanding on the development of superfood. In the latter chapters, other prominent researchers introduce the medical effects of superfoods and functional foods. The roles of a Mediterranean diet in health are introduced first. In the subsequent chapters, the protective effects of isoflavones, anthocyanins, curcumin, glutamine, and unsaturated fatty acids—which are contained in superfoods and functional foods—against several diseases and their therapeutic use are discussed. These chapters will suggest that superfoods and functional foods are very effective in the prevention and treatment of many diseases. I believe that this book will be useful for researchers and students who are studying or developing new functional foods and superfoods.

Finally, I would like to thank Ms. Romina Rovan and the publishing process managers of InTech Publisher for their great support and assistance throughout the writing and publication process of this book.

Naofumi Shiomi
Kobe College,
Japan

Chapter 1

A Prominent Superfood: *Spirulina platensis*

Nilay Seyidoglu, Sevda Inan and Cenk Aydin

Additional information is available at the end of the chapter

http://dx.doi.org/10.5772/66118

Abstract

Our planet's resources have been declining, as you know. The life qualities of humans have also changed a little because of their economy, nutrition, sports, and family life. Therefore, more alternative resources are being sought after by humans. Also, in the food supply for animals, scientists have been researching different and alternative supplements for growth performance, immunity, reproduction, and metabolism. *Spirulina platensis* and its contents have been linked to a nutritional component in both human and animal health and welfare. Growth and immunomodulation properties of this supplement have been widely studied in animals and humans, recently. Nowadays, nutritional specifics of *S. platensis* are a main focus for researchers. *S. platensis* is a single-cell protein due to its rich components, such as protein, essential amino acids, fatty acids, antioxidant pigments, carotenoids, beta-carotene, and phycocyanin. Today, researchers study the nutritional quality and investigate the effects of *S. platensis* on growth, immunity, antioxidant, antitoxicologic, anticancerogenic, cholesterol and glucose metabolism, and fertility. For these reasons, *S. platensis* may be acceptable as an alternative and/or superfood for the next generation. So, we review this information regarding *S. platensis* using historical background, literature reviews, qualitative studies, and microscopic appearances in this chapter.

Keywords: super food, *Spirulina platensis*, microalgae

1. Introduction

Population growth, depletion of food resources, and balanced diets require the usage of new food sources. For many years, there have been antibiotics, hormones, or drugs used for improving health and immunity, and to fight against disease. Today, antibiotic resistance has become a reality, and using a more natural approach to additives in both humans and animals has become a more acceptable alternative. The natural additives are using a protein source to replace the use of the antibiotics, hormones, and drugs. Natural additives are contained in a big scientific family, but mostly they come from plant derivatives and extracts. The nutritional status of these supplements is important for use as a food additive. Among these additives, microalgae are prevalent throughout history. The utilization of these algae as a protein source has been observed by researchers for many years. There are several types of microalgae, but especially Spirulina, namely *S. platensis*, has been studied more than others due to its rich components, positive effects, and being a supplement that is nontoxic.

S. platensis is a filamentous cyanobacterium known as blue-green algae, which is often used as a single cell protein. This microalgae contain essential amino acids, proteins, fatty acids, antioxidant pigments, carotenoids, beta-carotene, and phycocyanin. It has been designated as a health food by the World Health Organization (WHO), and it has the potential to become one of the best alternative treatments in the twenty-first century. Also, according to the National Institutes of Health (NIH), *S. platensis* can be used as a treatment for the nervous system and metabolism, including weight loss, diabetes, and high cholesterol. And today, well-known scientific sources say that Spirulina is a "superfood" and a "miracle from the sea."

1.1. Classification

S. platensis is a member of Phormidiaceae family. It is a filamentous and multicellular cyanobacterium which is figured like a cylindrical filament [1]. Also, it is a photosynthetic bacterium, and according to Bergey's Manual of Determinative Bacteriology (1974) it is considered to be in eukaryotic organisms [2]. Actually, there have only been one more algae in this family, named Arthrospira, which was confirmed by Gomont in 1989 [3]. He explained that Spirulina and Arthrospira are different due to their features such as helix type, cell wall, visibility under microscopy, diameters, and filaments (**Figure 1**). According to Botanics, the name of *S. platensis* was called *Arthrospira platensis* [4] at first because of its oxygenic photosynthetic feature, but today, the worldwide researchers use the term "Spirulina" for this microalgae.

1.2. Historical perspective

S. platensis was first isolated from Lake Texcoco by the Aztecs in the sixteenth century and they called it "tecuitlatl" [5]. Later, Dangeard happened upon the Kanembu tribe which had been harvesting these excellent microalgae from Lake Chad in Africa [6]. He then coined the name "dihe" for *S. platensis* which had been used for bread, meals, and cakes in the 1940s. *S. platensis* was analyzed chemically and it quickly prompted research in 1964 [7]. During that year, studies began on this microalgae by botanists, microbiologists, and scientists, and also reviewed by some researchers [8, 9].

Early in the 1990s, NASA studied the cultivation of *S. platensis* as a food source for long-term outer-space programs. They modified the growth process using environmental factors and suggested that this microalgae could be used as palatable diet [10]. Also, in 1967, *S. platensis* was touted as a "wonderful future food source" by the International Association of Applied Microbiology [11].

The World Health Organization reported that *S. platensis* has no risk and is a good food supplement for health [12]. Included in this issue, in 2003, the Intergovernmental Institution studied this microalgae for malnutrition (IIMSAM) and developed a charter with the United Nations Economic and Social Council (UNECOSOC). They agreed that Spirulina should be used against malnutrition for humans, especially in developing countries.

In 2011, the National Institutes of Health proposed that *S. platensis* could be used in human research, but they requested further studies on the effects of Spirulina [13]. *S. platensis* was suggested as a safe dietary supplement by The Food and Drug Administration (FDA) in 2012 [14]. They recommended a 3–10-g daily dose of this microalgae for human health. Notably, according to the European Food Safety Authority (EFSA), *S. platensis* also helps to control the blood sugar level for glycemic health [15].

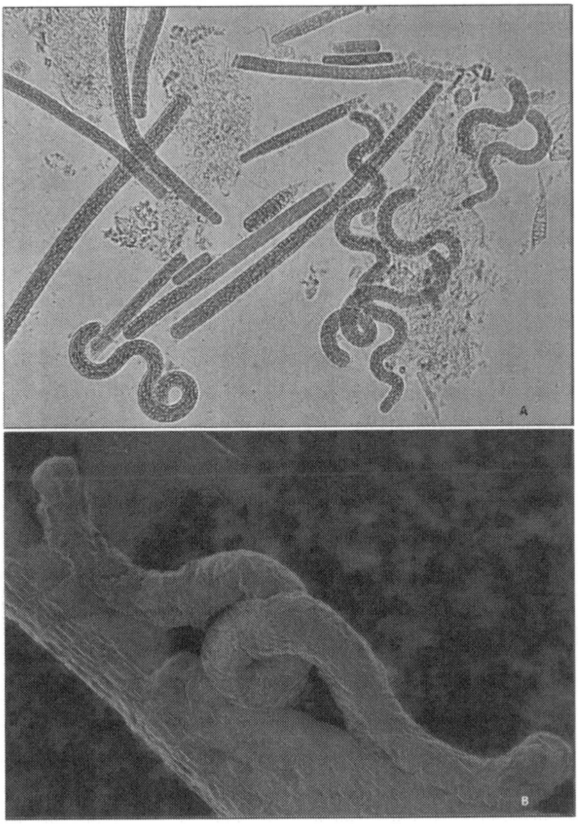

Figure 1. (A) Microscopic view of microalgae *Spirulina platensis* and (B) scanning electron micrograph of *Spirulina platensis*. Photograph by N. Seyidoglu.

1.3. Nutritional composition

The superfood *S. platensis* includes bioactive components such as proteins, amino acids, minerals, vitamins, pigments, nucleic acids, carbohydrates, and lipids, shown in **Tables 1–6**.

Food protein origin	Protein (%)
Spirulina powder	60–70
Whole dried egg	47
Beer yeast	45
Skimmed powdered milk	36
Whole soybean flour	36
Parmesan cheese	36
Wheat germ	27
Peanuts	26
Chicken	19–24
Fish	19.2–20.6
Beef meat	17.4

Table 1. Quantity of *Spirulina platensis* proteins and other foods [33].

Protein and amino acids	g/100 g
Protein	57.47
Tryptophan	0.929
Threonine	2.97
Isoleucine	3.209
Leucine	4.947
Lysine	3.025
Methionine	1.149
Cystine	0.662
Phenylalanine	2.777
Tyrosine	2.584
Valine	3.512
Arginine	4.147
Histidine	1.085
Alanine	4.515
Aspartic acid	5.793
Glutamic acid	8.386
Glycine	3.099
Proline	2.382
Serine	2.998

National Nutrient Database for Standard Reference, Release 28 slightly revised May, 2016. Available from: https://ndb.nal.usda.gov/ndb/foods/show/3306?fgcd=&manu=&lfacet=&format=Full&count=&max=50&offset=&sort=default&order=asc&qlookup=11667&ds

Table 2. Protein and amino acids in *Spirulina platensis* powder (nutritional value per 100 g).

Vitamins	mg/100g
Provitamin A equiv.	2.330 × 103 IU/kg
Vitamin E d-a-tocopherol	5
Thiamin B1	3.5
Riboflavin B2	4.0
Niacin B3	14.0
Vitamin B6 pyridoxine	0.8
Vitamin B12 cobalamin	0.32
Folic acid	0.01
Biotin	0.005
Phantothenic acid	0.1
Vitamin K	2.2

Table 3. Vitamins in *Spirulina platensis* powder [133].

Fatty acids	(%)
Myristic acid	0.23
Palmitic acid	46.07
Palmitoleic acid	1.26
Oleic acid	5.26
Linoleic acid	17.43
Gamma-Linolenic acid	8.87
Others	20.88

Table 4. Fatty acid composition of *Spirulina platensis* powder [134].

Mineral	mg/100g
Calcium	700
Chromium	0.28
Copper	1.2
Iron	100
Magnesium	400
Manganese	5.0
Phosphorus	800
Potassium	1400
Sodium	900
Zinc	3.0

Table 5. Minerals in *Spirulina platensis* powder [133].

Pigments	mg/100g
Carotenoids	370
Chlorophyll a	1000
C-Phycocyanin	14,000

Table 6. Pigments in *Spirulina platensis* powder [133].

1.3.1. Protein and amino acids

S. platensis is the most useful microalgae for nutrition due to its components, especially protein. The nutritional level of protein is almost 70% of its dry weight and also has a high quantity and quality belonging to amino acids [1]. *S. platensis* contains all of the essential amino acids, as shown in **Table 2**. Researchers reported that although methionine and cysteine are found in a lower value, albumin and casein are found in a higher value, of animal proteins, respectively, in eggs and milk [8, 16]. *S. platensis* contains biliproteins, especially C-phycocyanin which is 20% of all protein fractions. C-Phycocyanin molecule has an antioxidant feature, which regulates immunity and protects the organism against diseases [17].

1.3.2. Vitamins

S. platensis has the richest vitamin source of vitamin A (beta-carotene), vitamin E, thiamin (vitamin B1), biotin (vitamin B7), and inositol (vitamin B8) in food. Beta-carotene is in a biotransformed state which can be absorbed by humans, and is also important for antioxidant processes in organisms [18]. On the other hand, there is a conflict of cobalamin (vitamin B12) content in *S. platensis*. Some researchers reported that *S. platensis* has no reliable vitamin B12. They explain that it is a pseudovitamin B12 which is inactive and in a form that the human organism cannot uptake at a cellular level [19, 20]. However, other researchers claimed that *S. platensis* has a great amount of B12 compared to other sea algae and they indicated that vitamin B12 in this microalgae is important for vegetable nutrition, especially for humans who do not eat meat [21, 22].

1.3.3. Minerals

S. platensis contains many minerals such as potassium, calcium, chromium, copper, iron, magnesium, manganese, phosphorus, selenium, sodium, and zinc. This microalgae is a good component due to its iron, calcium, and phosphorus contents. The ferrous component in this microalgae can be easily digested and bioactive in an organism easily which is important for pregnant adult females [23]. The utilization of calcium and phosphorus contents of *S. platensis* has an important impact on bone calcification and improves bone health [24].

1.3.4. Lipids

Lipid contents of *S. platensis* are only 4–7%, but it has important essential fatty acids for humans: gamma-linolenic acid and linolenic acid. These components are also mediators of immune and cardiovascular system due to their precursor effects of prostaglandins and leukotrienes [25].

http://dx.doi.org/10.5772/66118

The *S. platensis'* other lipids are stearidonic acid, eicosapentaenoic acid, docosahexaenoic acid, and arachidonic acid, respectively.

1.3.5. Carbohydrates

S. platensis contains 13.6% carbohydrates, which are glucose, mannose, galactose, and xylose. Nevertheless, it does not contain cellulose, which cannot be absorbed by humans and thereby this feature makes *S. platensis* easily digestible and a safe nutrient for human consumption. It is significant for people who have intestinal malabsorption and for the elderly [24]. Likewise, there is also a polysaccharide molecule, isolated from *S. platensis*, which has a huge molecular weight. This polysaccharide has an immunomodulator effect called "immulina" by scientific authorities [26, 27].

1.3.6. Nucleic acids

Nucleic acids play a role in uric acid metabolism. They catabolize the uric acid to adenine and guanine which causes gout and cardiovascular diseases [28]. So, the World Health Organization recommends about 80-g daily dosage of *S. platensis*.

1.3.7. Pigments

S. platensis has some natural pigments which color this microalgae, such as c-phycocyanin, chlorophyll, xanthophyle, beta-carotene, zeaxanthin, and allophycocyanin. The most important are phycocyanin, chlorophyll, and beta-carotene. C-Phycocyanin is the most important pigment, which includes iron, and contains 14% of its dry weight. Also, *S. platensis* is one of the best nutrients that contains the highest chlorophyll value (1%). Chlorophyll is known as a detoxifier and purifier phyto-nutrient. It improves the carbohydrate, protein, and lipid metabolism and affects reproduction positively. Carotenes constitute half of this microalgae, especially beta-carotene. The carotenes and xanthophyle in *S. platensis* are demonstrated in different metabolism pathways in the body, and also better influence the function of vitamins and minerals in an organism [29]. Nowadays, diets rich in carotenes are found to be important for human health due to its effects in reducing the risk of diseases [30, 31] (**Table 6**).

2. Utilization of *S. platensis* worldwide

2.1. Usage as food

Plants and plant extracts have been the focus for improved health in recent years. *S. platensis* is one of the most sought after natural alternatives for nutrition in both human and animal. *S. platensis* is a microalgae that has been consumed for centuries due to its high nutritional value and supposed health benefits. Apart from its easy production, *S. platensis* has a high nutritional ability. Its affects on growth, antioxidants and antiviral features, immunomodulator activity, and hypocholesterolemic influence have been proposed by researchers over the years. Likewise, it is indicated as a nontoxic supplement, and the World Health Organization has supported it as a health nutrient [32]. *S. platensis* is used in many countries, such as Mexico, United States, Japan, Taiwan, India, Singapore, Germany, Spain, Switzerland, Holland, and many others. It is added in food marketing such as candies, chewing gums, appetizers, sports tablets, and bread. As well as its many uses in food, it is a component in some cosmetics such as creams, masks, tonics, and shampoos [33].

Natural additives have also been added to animal feed for healthy animal growth in recent years. At the same time, in the farming sector, it is preferred as it is a natural and economical product, as well as healthy, and it is shown to have rapid growth performance. *S. platensis* is one of the most sought after ingredients for animal feed as compared to other nutrients due to its high protein contents and nourishing features. Its growth, antiviral, antidiaretic, antioxidant, probiotic, hypocholesterolemic, antiallergic, analgesic, anthelmintic, anticarcinogenic, antiparasitic, immune system activator, and cardiovascular protective effects for animals have been reported by researchers [34–39].

S. platensis grows naturally in shallow bodies of water and in the presence of an alkaline medium of high salinity [40, 41]. The primary component for growing this microalgae is sodium bicarbonate. The production systems for this microalgae are found in Thailand, United States, Africa, China, and Chile, mostly where the Pacific Ocean, fresh water, and deep oceans exist. On the other hand, in Turkey and Bulgaria, *S. platensis* has been cultivated experimentally and recently.

Clinical and experimental trials have shown that *S. platensis* can be utilized for both human and animal safety. There have also been many studies that can help explain the benefits of this interesting microalgae. Its high biological components are an interest for scientists in recent centuries. Although it has been reported as a nontoxic supplement, current studies have continued to test its safety.

S. platensis can be used for immune enhancement, growth, as a nutritional food source, protector of metabolism, and many other important benefits for both humans and animals. It is amazing that all of these different features exist in this one specific microalgae. This is why scientific evidences call this microalgae a "super food." Nevertheless, in that respect there is always a need for continued studies regarding natural additives such as *S. platensis* to explain the study of their effects on humans and animals.

2.2. Effect of *S. platensis* on the growth of bacteria and animals

S. platensis does not contain cellulose on its cell wall. Therefore, this microalgae can be absorbed in the intestinal mucosa and improve the intestinal function and mucosal digestion. Although *S. platensis* can repress the harmful microorganism such as Candida, it can help to increase the good microorganism such as *Lactobacillus* and *Bifidobacteria*. So, this increase of *Lactobacillus* population helps the absorption and digestion of food [42–44]. At the same time, the biological components in *S. platensis*, such as phycocyanin, polysaccharide, and gamma-linolenic acid, have an important role for improving overall body function. The Scientific Committee on Food (SCF) and the European Food Safety Authority (EFSA) also recommend 10 g of *S. platensis* as a supplement for daily intake in order to protect the health of humans, and research indicates that there is no risk with this microalgae use as a food [14].

The focus on *S. platensis* is due to its protein bioavailability, and that is the reason for this important microalgae to be compared to others. Its high protein content can improve growth performances of both humans and animals. The application of *S. platensis* for protein malnutrition has resulted in good weight gain, hematological responses, and positive nitrogen balance in metabolism with no side effects. Foods containing high protein are especially useful for malnutrition in humans, as malnutrition is a global problem. Studies, which estimate the effects, were performed in Africa, where malnutrition is prevalent, especially in children. The children and older people were separated according to their protein malnutrition first, and then rehabilitated with *S. platensis* for these studies [45–49]. The studies resulted in positive weight gain, normalized blood values, and optimized the health of human immunodeficiency virus (HIV)-negative children. The study of Simpore et al. [47] compared HIV-negative and -positive children, and showed a positive weight gain between 15 and 25 g/day with children given *S. platensis*. They reported that *S. platensis* is a good food source for malnutrition. On the other hand, Azabji Kenfazk et al. [49] studied HIV-infected and malnourished adults, using *S. platensis* for 12 weeks. At the end of that study, positive improvement in body composition and body weight was concluded.

There are many different studies that point out the growth performance of *S. platensis* in animals [50–57]. For example, Moreira et al. [50] studied the Wistar rat using *S. platensis* as an added nutrient at 8.8, 17.6, and 26.4% doses of forage. They reported that there was a significant increase in weight in the 17.6% group. Heidarpour [35] used 0-, 2-, 6-, and 25-g *S. platensis* for cattle, and noted weight gain every 15 days. He observed no statistical differences in growth performances when comparing all groups. On the other hand, although some researchers found positive effects of *S. platensis* as a supplement with fish [52], some of them reported no significant changes in growth performances in fish [53, 54]. Seyidoglu and Galip [51] tried to elucidate the effects of *S. platensis* on growth performance in rabbits. They indicated that there was a positive effect of supplementing *S. platensis* on growth performance due to dose, animals, and environmental changes.

When comparing all these studies, there were different results about the supplementing dose and effects of the *S. platensis*. So, there continues to be more studies which are necessary to determine dietary concentration and the effects of this interesting microalgae.

3. Utilization of *S. platensis* for health

3.1. Immune system and allergy

Hematopoietic system is important for repairing tissues, generating important body cells, and protecting healthy regulation. The immune system is one of the most important systems within the hematopoietic system. Together, they are all responsible for protecting the host. The immune system of the organism is classified as an innate immune system and adaptive immune system. The innate immune system is the first barrier to protect the organism against infections. This system includes macrophages, neutrophiles, natural killer cells, and lectins. On the other hand, providing a more specialized and active defense against diseases is called an adaptive immune system, in which there are antibodies, lymphocytes, and cytokines. These two immune systems are in a sensitive balance with each other.

S. platensis can produce high protein, amino acids, vitamins, beta-carotene, pigments, and polysaccharides as a bioactive agent. All these components have an enhanced effect on the production of antibodies and cytokines. Especially polysaccharide, in this microalgae, has an effect on macrophages and T- and B-cell proliferations, and so it is said that *S. platensis* can improve the resistance of the organism. However, the effects of *S. platensis* on the immune system have not yet been precisely determined. The first experimental study was performed on mice in 1994, and it investigated at the effects of supporting antibody production [58]. In that study, it was reported that C-phycocyanin and polysaccharide in *S. platensis* activated the proliferation of monocytes, erythrocytes, granulocytes, and fibroblastosis in the bone marrow, and thereby the hematopoietic and immune systems were activated. In the University of Mississippi, a polysaccharide that is called "Immulina" was extracted from *S. platensis* by researchers [26]. They measured the immunostimulatory activity on human monocyte cells in vitro, and reported positive monocyte activation due to the effect of polysaccharide. Some researchers demonstrated that *S. platensis* plays an important role in the balance of immune system cells [59–65]. All these researchers reported that polysaccharides and phycocyanin have a positive role in erythropoietin activity, which is based on improved T-lymphocytes, and triggered leukocytes and bone marrow growth. Moreover, Løbner et al. [60] observed the increased CD4+ cell proliferation in humans using Immulina. There are two studies which also used *S. platensis* supplement (Immulina) with healthy humans. They reported that hemoglobin levels, natural killer cell activity, and monocytes were increased [27, 61]. Although some of the studies did showed the immune stimulatory effect of this microalgae on adaptive immune system [62–64], some of the studies [65] found no effect on the immune system, which can be explained by mutation in protocols and strains, and also the ratio of *S. platensis*.

An allergic response is a reaction of the immune system against a harmless substance such as pollen, nutrition, house mites, or other substances. Today, it is an increasing problem in the world. The protection and treatment process of allergies is aided by natural foods, especially *S. platensis*. According to researchers [66], *S. platensis* can regulate T-helper cells (Th) in allergic rhinitis. In that study, which was the first human study investigating at allergies, the role of T-helper 2 cells (Th2) and IL-4, which induced the production of IgE, was inhibited by this microalgae. According to the results, *S. platensis* supplements had a positive effect on allergic patients. In another study about food allergies, the researchers investigated the immunoglobulins role (IgA, E, G1) in the protective effects of *S. platensis*. They suggested that *S. platensis* may enhance the IgA antibody, which worked as a blocking antibody toward IgE, and thereby had protective effects against allergic reactions [67].

The supplementing of *S. platensis* was also used for adolescent animals, which have an immature immune system, which has been shown to improve the immune system and living ratio [62, 68]. Some researchers studied this concept with poultry and reported that there was a positive immunomodulator effect of *S. platensis* through the decreasing of the nutrients in macrophages [69]. According to other studies in animals, there have been increases in hemoglobin, erythrocytes, natural killer value, T–lymphocytes, and cytokine activity with this microalgae [70–72]. Prompya and Chitmanat [53] studied fish over a 60-day duration using this microalgae and found a statistically significant increase in white and red blood cells. There was another research which studied newborn pigs, and the results found a significant increase in cytokines and interleukins [72].

For many years, *S. platensis* has been used as a food additive for both humans and animals. According to scientific findings, the components are sufficient for healthy nutrition, the protective activity of the body and disease therapies. Also, according to the Food and Drug Administration, *S. platensis* has been designated as a "safe food" [14] due to its natural properties for health therapies.

3.2. Anemia

Anemia refers to a decreased number of circulating red blood cells and is the most common blood disorder. Insufficient nutritional intake, toxic metals, and environmental contamination cause there to be a disruption in the red blood cell production pathways, and thereby anemia is the result. Also, iron deficiency is the most common cause of anemia in pregnant women, older people, and children [61]. In literature reviews, several studies have shown that several types of anemia have been treated by *S. platensis* due to its phycocyanin content [73–75]. The mechanism of C-phycocyanin is explained through the stimulation of the hematopoiesis and the endogenous erythropoietin (Epo). The Epo is known as an indicator for the proliferation and differentiation of erythrocytes. Along with this result, some research have also demonstrated that *S. platensis* has a positive impact on different types of anemia due to its rich components such as essential amino acids, folic acids, vitamin B12, and high iron which have an important role in erythropoiesis [48, 76, 77]. There are also some animal studies regarding anemia that have shown the beneficial effects of *S. platensis* on hemoglobin and serum iron levels [47, 86, 88].

3.3. Obesity

S. platensis has a hypocholesterolemic effect due to its C-phycocyanin component. It was reported that C-phycocyanin inhibits the reabsorption of bile acids in the ileum and also cholesterol in the jejunum [78–80]. In some studies, humans using *S. platensis* supplements showed lower results in cholesterol and triacylglycerol levels, and an increase in high-density lipoprotein levels. All of these effects indirectly reduced both diastolic and systolic blood pressure and gave a protective effect on the cardiovascular system [51, 81–83]. In another study [84], researchers treated hyperlipidemia nephrotic syndrome with *S. platensis* by applying 1-g *S. platensis* per day for 2 months and observed whether *S. platensis* decreased essential fatty acids and cholesterol values or not. They concluded that *S. platensis* consumption decreases lipid profile and helps to reduce the hyperlipidemia nephrotic syndrome. Also, all these researchers suggested that *S. platensis* is important to maintain a healthy cardiovascular system including blood lipid profile as well as treating precardiovascular disease. In vascular lesions such as coronary artery disease, the proteoglycan metabolism protecting cardiovascular cells is associated with exogen polysaccharides that are present in *S. platensis*. This pathway was studied by Sato et al. [85] and has been found to be an important element in coronary artery disease.

Cardiovascular diseases, obesity, and diabetes are linked with each other. The risk of cancer development is enhanced by these diseases in both humans and animals. On that point, some researchers point out the effects of *S. platensis* on obesity and diabetes [86–89]. During a 4-week study, *S. platensis* supplement (2.8 g) was taken by obese people, and the total body weight and biochemical values were determined. A reduction in body weight and lower cholesterol levels in obese humans was observed, in the lower significant level. Also, the other researchers observed the positive effects on diabetics using supplements of *S. platensis* [86, 89]. In these studies, obese humans with high blood sugar and lipid profiles were studied to determine the antidiabetic mechanism of this microalgae and have suggested that the gamma-linolenic acid in *S. platensis* may be attributed to the reduction in hyperglycemia.

S. platensis has been applied to animal feed and it has been reported that *S. platensis* plays a substantial part in lipid metabolism in animals, such as a decreased effect on total cholesterol, lipid profile, and glucose [5, 35, 90, 91]. They suggested that *S. platensis* could reduce serum cholesterol, and thereby have positive effects on lipid metabolism. In fact, cholesterol metabolism is significant in these creatures, especially in the milk production during lactation. The fatty acid profile of this microalgae is a prominent source and may stimulate milk production. The application of *S. platensis* to both humans and animals has been reviewed by The Dietary Supplements Information Expert Committee (DSI-EC) with experimental researches of animals, human clinical, and animal studies, and has reported that *S. platensis* does not have any risk for nutritional consumption. However, as there are quite limited studies in animals, especially in ruminants shown by researchers, more animal studies will be necessary to study this functional microalgae.

3.4. Healing and antibiotic effects

Wound healing is a process of repairing skin or tissue, and this process is also important for regulating hemostasis. During the healing process, bacteria and other pathogens are present at damaged areas where the pyretic situation occurs as a result of the inflammation. Natural pharmaceutical compounds are generally used to heal such wound areas. In addition, *S. platensis* or its extracts have been widely used in creams, solutions, raw juices, and ointments for skin health in recent times. Collagen fibrils, which is the plant constituent contained in microalgae, have attributes that have positive effects on wound closure during the healing process [92]. Rabadiya et al. [93] suggested that the antibiotic effects of *S. platensis* had inhibitive effects of bacteria and promoted skin healing, during the scarring process. Also, another study suggested that aqueous extract of *S. platensis* has a healing activity and it is an economical method for promoting skin, especially for diabetic wounds [94].

The anti-inflammatory effect of *S. platensis* is explained as an inhibitive effect of gamma-linolenic acid [95–97]. Gamma-linolenic acid is important to control inflammation and cell proliferation. The high value of gamma-linolenic acid inhibits the work of prostaglandin and the progression of inflammation. On the other hand, some researchers reported that *S. platensis* and its extract C-phycocyanin, can regulate the cytotoxicity and inflammation-associated factors such as ions, COX-2, tumor necrosis factor (TNF)-α, and IL-6 with BV-2 microglial cell during the inflammatory process [98].

Antibacterial activity of *S. platensis* is also caused by the activation of phagocytosis in mononuclear cells and this bacterial clearance is associated with liver health. The increase in T-cell and mononuclear phagocytes in liver by *S. platensis* has been reported [99].

S. platensis and its extracts, especially calcium, do not allow the viruses to attack and infect the cells. On that point, there are some written reports about the inhibition effect of viral replication and natural defenses [100]. Referring to the animal studies, *S. platensis* has been shown to be beneficial as an antiviral agent and lead to a limitation of foot and mouth disease [101]. The researchers studied the calcium extract of this microalgae in vitro, and indicated that the replication of viruses, such as herpes, measles, or mumps, was interfered by this extract. In some other studies, aqueous extracts of *S. platensis* diminished the HIV-1 virus and enterovirus replication in T-cells, Langerhans, and peripheral blood mononuclear cells due to the polysaccharides activity of this microalgae [102, 103].

Helminth infections contribute to diseases such as anemia, eosinophilia, and malnutrition. Studies about marine natural products, which are used for anthelmintic situation, were reviewed by Mayer et al. [104]; however, sufficient anthelmintic effect by *S. platensis* on the parasites was not observed.

3.5. Fertility

There are many factors that affect infertility in female humans and animals such as age, size and physical condition, reproductive history, and nutrition [105]. *S. platensis* is an amazing food for supporting fertility and pregnancy due to its contents. It was reported that high protein and essential amino acid components of *S. platensis* may have improved fertility by enhancing the gonad weight and gonadosomatic index, and thereby had positive effects on reproductive function [106]. Granaci et al. [107] studied with boars and found that *S. platensis* can increase the fertilizing ability of sperms. Some researchers suggested that *S. platensis* improves the sperm motility and tone due to lactate dehydrogenase (LDH) in spermatozoa, which is increased by this microalgae [108, 109]. Also, it is known that thyroid hormones (T3 and T4) are associated with increased testosterone stimulation [110], which in turn helps spermatogenesis, which were studied in rats supplemented with *S. platensis*. It was also described that these thyroid hormones regulated by this microalgae can show an improvement in rats, which have a testicular injury and dysfunction, due to its antioxidant components [111, 112].

3.6. Antioxidant, anticancer and antitoxicity effects

The natural antioxidants are vitamins (B1, B5, B6, and E6), minerals (zinc, manganese, and copper), amino acid (methionine), beta-carotene, and trace elements (selenium). *S. platensis*, which contains phenolic acids, beta-carotene, and tocopherols, is a very important natural source for the intake of antioxidants. The antioxidant effect has been examined in vivo and in vitro [113, 114]. *S. platensis* has antioxidant and immunomodulatory properties which appear in the mechanism of tumor destruction and also in cancer prevention [115]. Some researchers studied liver cancer and reported that lymphocyte activity and survival rate in cancer-stricken organisms can be increased by the supply of *S. platensis* [17] through C-phycocyanin activating the immune system and playing an important role to prevent the progress of local and oral cancer [116].

Beta-carotene contained in *S. platensis* at a high value protects the free radicals and tumors induced by chemicals and enhances the immunologic resistance of the body, also decreasing lung cancer [117, 118]. The inhibitory effects of *S. platensis* and its extracts on carcinogenesis for both humans and animals were reported in some studies [119–121]. Grawish et al. [119] showed the inhibition of dysplastic tumoral changes in cheek pouch mucosa in hamsters. In another study, the protective phyto-antioxidant functions of liver tumors were determined, by an increase of the Bax/Bcl-2 ratio, which is associated in the apoptosis mechanism of hepatocellular carcinoma cell line HepG2 [120]. Additionally, *S. platensis* and its contents have protective effects against drugs, chemicals, and xenobiotics on liver tissue [120, 122, 123]. Abdel Daim et al. [124] reported that the protective mechanism of *S. platensis* against Deltamethrin induced oxidative stress through the inhibition of lipid peroxidation and releasing of free radicals or enhancing of the activity superoxide dismutase. Related to all these studies, it has been suggested that *S. platensis* may have a positive effect on anticancerogenic and oxidative situations.

S. platensis consists of proteins, lipids, carbohydrates, elements, and vitamins such as β-carotene, riboflavin, cyanocobalamin, α-tocopherol, and α-lipoic acid [125]. As discussed, with all these substances, *S. platensis* has beneficial effects against nephrotoxicity and cardiotoxicity [125–127]. Mohan et al. [126] showed that *S. platensis* may protect against cisplatin-induced nephrotoxicity in rats. Also, Khan et al. [127] described the protective effect of *S. platensis* against doxorubicin-induced cardiotoxicity. In the world, there are some threats which are spreading dangerously such as arsenic and radiation in the water. The millions of people living in Bangladesh, India, Taiwan, and Chile are consuming high concentrations of arsenic-contaminated drinking water and thousands of them are exposed to chronic arsenic poisoning [128]. Specific treatment for this situation is unavailable. Misbahuddin et al. [128] showed that *S. platensis* extract plus zinc could be beneficial for the treatment of chronic arsenic poisoning with melanosis and keratosis. Likewise, in another study it was determined that *S. platensis* could protect the testes against mercury chloride-induced testicular damage by its rich antioxidants and antitoxicity activity [129]. An important example of radiation and *S. platensis* effects is the Chernobyl disaster. In Ukraine and Belarus, people live with radiation, which is in contaminated water, land, and nutrients. Due to this effect, poisoning, leukemia, cancer, birth defects, anemia, and thyroid disease have appeared. On that point, there is some unpublished work which talks about the effects of *S. platensis* on these symptoms and diseases [130]. Also, the protective effects of this microalgae and its extract polysaccharides and phycocyanin were shown by Belookaya et al., Wu et al., and Qishen et al [130–132]. They reported that *S. platensis* and its extracts decrease the radioactivities, and improve the bone marrow reproduction and immune system.

4. Conclusions

A prominent super food, *S. platensis*, has been known for its importance for health instead of medicine for centuries. Many studies have been performed on the effects of this interesting microalgae on both humans and animals. Today, studies observe at the nutritional quality and investigate the medicinal aspects of *S. platensis* on growth, hematopoietic system, immune system, allergy, anemia, cholesterol, obesity, diabetes, wound healing, fertility, viral and bacterial diseases, parasites, and helminth diseases. Besides these effects, anti-inflammatory, antibiotic, antipyretic, antioxidant, anticancer, and antitoxicity effects have also been determined by researchers. The potential effects have been addressed with in vivo and in vitro experiments, and contribute to the literature.

Acknowledgements

Special thanks are to Susan Korucubasi who assisted our chapter in proofreading.

Author details

Nilay Seyidoglu[1*], Sevda Inan[2] and Cenk Aydin[3]

*Address all correspondence to: nseyidoglu@nku.edu.tr

1 Department of Physiology, Veterinary Faculty, Namik Kemal University, Tekirdag, Turkey

2 Department of Pathology, Veterinary Faculty, Namik Kemal University, Tekirdag, Turkey

3 Department of Physiology, Veterinary Faculty, Uludag University, Bursa, Turkey

References

[1] Ciferri O. Spirulina, the edible microorganism. Microbiology Reviews. 1983. PMCID: PMC283708

[2] Sánchez M, Bernal-Castillo J, Rozo C, Rodríguez I. Spirulina (arthrospira): an edible microorganism: a review. Universitas Scientiarum. 2003;8(1):7–24. PMC283708

[3] Castenholz RW, Waterbury JB. Oxygenic photosynthetic bacteria. In: Staley JT, Bryant MP, Pfenning N, Holt JG., editors. Bergey's Manual of Systematic Bacteriology. Williams and Wilkins Co; Baltimore, 1989. p. 1710–1806. ISBN:0683041088 9780683041088 0683011189 9780683011180

[4] Gershwin ME, Belay A. Spirulina in Human Nutrition and Health. CRC Press; USA, 2007. DOI: 10.1007/s10811-009-9467-0.

[5] Habib M, Ahsan B, Parvin M, Huntington TC, Hasan MR. A review on culture, production and use of spirulina as food for humans and feeds for domestic animals and fish. Food and Agriculture Organization of The United Nations; 2008. Retrieved 2011. Available from: ftp://ftp.fao.org/docrep/fao/011/i0424e/i0424e00.pdf [Accessed 2016/08/18].

[6] Abdulqader G, Barsanti L, Tredici M. Harvest of *Arthrospira platensis* from Lake Kossorom (Chad) and its household usage among the Kanembu. Journal of Applied Phycology. 2000;12:493–498. DOI: 10.1023/A:1008177925799

[7] Zarrouk C. Contribution to the cyanophyceae study: influence various physical and chemical factors on growth and photosynthesis of *Spirulina maxima*. [thesis]. Faculty of Science, University of Paris; 1966.

[8] Vonshak A. *Spirulina platensis* (Arthrospira): Physiology, Cell-Biology and Biotechnology. Taylor & Francis; London, 1997. DOI: 10.1023/A:1008177925799

[9] Siva Kiran RR, Madhu GM, Satyanarayana SV. Spirulina in combating protein energy malnutrition (PEM) and protein energy wasting (PEW)—A review. Journal of Nutrition Research. 2015;3(1):62–79. DOI: 10.13140/RG.2.1.3149.0325

[10] Tadros MG. Normal, Al. Characterization of Spirulina biomass for CELSS diet potential. NASA Technical Reports Server. 1988; DOI: 19940009624

[11] Sasson A. Micro Biotechnologies: Recent Developments and Prospects for Developing Countries. Place de Fontenoy, Paris. France: United Nations Educational, Scientific and Cultural Organization (UNESCO), BIOTEC Publication 1/2542; 1997. p. 11–31.

[12] Intergovernmental Institution for the use of Micro-algae Spirulina Against Malnutrition. Preamble-Introductory Note [Internet]. 5 March 2003. [Updated: 2 July 2014]. Available from: www.iimsam.org/http://www.iimsam.org/images/IIMSAM-CHARTER.pdf [Accessed: 2016/08/18]

[13] Natural Medicines Comprehensive Database Consumer Version. "Blue-green algae" [Internet]. November 18, 2010. [Updated: Retrieved April 15, 2011]. Available from: https://medlineplus.gov/druginfo/natural/923.htlm [Accessed: 2016/08/18]

[14] Food Drug Administration. FDA Agency Response Letter GRAS Notice No. GRN 000127 [Internet]. 2003. Available from: http://www.fda.gov/Food/FoodIngredientsPackaging/GenerallyRecognizedasSafeGRAS/GRAS Listings/ucm153944.htm [Accessed: 2012/12/7]

[15] EFSA European Food Safety Authority (EFSA). Scientific Opinion on the substantiation of health claims related to various food(s)/food constituent(s) claiming maintenance of normal blood glucose concentrations. In: EFSA, editor. EFSA Panel on Dietetic Prod-

ucts, Nutrition and Allergies; Parma, Italy. EFSA Journal; 2010. p. 1490. DOI: 10.2903/j.efsa.2010.1490

[16] Fujisawa T, Narikawa R, Okamoto S, Ehira S, Yoshimura H, Suzuki I, Masuda T, Mochimaru M, Takaichi S, Awai K, Sekine M, Horikawa H, Yashiro I, Omata S, Takarada H, Katano Y, Kosugi H, Tanikawa S, Ohmori K, Sato N, Ikeuchi M, Fujita N, Ohmori M. Genomic structure of an economically important cyanobacterium, *Arthrospira (Spirulina) platensis* NIES-39. DNA Researches. 2010;17:85–103. DOI: 10.1093/dnares/dsq004

[17] Iıjıma N, Fugii I, Shimamatsu H, Katoh S. Anti-tumor agent and method of treatment therewith. U.S. Patent Pending. 1982;1150. Available from: https://docs.google.com/viewer?url=patentimages.storage.googleapis.com/pdfs/US8404650.pdf [Accessed: 2013/08/06].

[18] Kapoor R, Mehta U. Utilization of beta-carotene from *Spirulina platensis* by rats. Plants Foods for Human Nutrition. 1993;43(1):1–7. DOI: 8464841.

[19] Watanabe F. Vitamin B12 sources and bioavailability. Experimental Biological Medicine (Maywood). 2007;232(10):1266–1274. DOI: 10.3181/0703-MR-67.

[20] American Dietetic Association; Dietitians of Canada. Position of the American Dietetic Association and Dietitians of Canada: Vegetarian Diets. Journal of the American Dietetic Association. 2003;103(6):748–765. DOI: 10.1053/jada.2003.50142

[21] Khan Z, Bhadouria P, Bisen PS. Nutritional and therapeutic potential of Spirulina. Current Pharmaceutical Biotechnology. 2005;6:373–379. DOI: 16248810

[22] Becker EW. Handbook of microalgal mass culture. In: Richmond A, editor. Nutritional Properties of Microalgae: Potentials and Constraints. Boca Raton, Florida: CRC Press; 1986. p. 339–419. ISBN-10: 0849332400 ISBN-13: 978-0849332401

[23] Puyfoulhoux G, Rouanet JM, Besancon P, Baroux B, Baccou JC, Caporıccıo B. Iron availability from iron-fortified Spirulina by an in vitro digestion/Caco-2 cell culture model. Journal of Agriculture and Food Chemistry. 2001;49:1625–1629. DOI: 11312906

[24] Walter P. Effects of vegetarian diets on aging and longevity. Nutrition Reviews. 1997;55(1):61–68. DOI: 10.1111/j.1753-4887.1997.tb06106.x

[25] Kulshreshtha A, Zacharia AJ, Jarouliya U, Bhadauriya P, Prasad GB, Bisen PS. Spirulina in health care management. Current Pharmaceutical Biotechnology. 2008;9:400–405. DOI: 18855693

[26] Pugh N, Ross SA, Elsohly HN, Elsohly MA, Pasco DS. Isolation of three weight polysaccharide preparations with potent immunostimulatory activity from *Spirulina platensis*, Aphanizomenon flos-aguae and Chlorella pyrenoidosa. Planta Medica. 2001;67:737–742. DOI: 10.1055/s-2001-18358

[27] Nielsen CH, Balachandran P, Christensen O, Pugh ND, Tamta H, Sufka KJ, Wu X, Walsted A, Schjørring-Thyssen M, Enevold C, Pasco DS. Enhancement of natural killer

cell activity in healthy subjects by Immulina®, a Spirulina extract enriched for Braun-type lipoproteins. Planta Medica. 2010;76(16):1802–1808. DOI: 10.1055/s-0030-1250043.

[28] Seegmiller JE, Laster L, Howell RR. Biochemistry of uric acid and its relation to gout. The New England Journal of Medicine. 1963;268:764–773. DOI: 10.1056/NEJM196304042681406

[29] Babadzhanov AS, Abdusamatova N, Yusupova FM, Faizullaeva N, Mezhlumyan L. G, Malikova M KH. Chemical composition of *Spirulina platensis* cultivated in Uzbekistan. Chemistry of Natural Compounds. 2004;40:3. DOI: 10.1023/B:CONC.0000039141.98247.e8

[30] Stahl W, Sıed H. Bioactivity and protective effects of natural carotenoids. Biochimica et Biophysica Acta. 2005;1740(2):101–107. DOI: 10.1016/j.bbadis.2004.12.006

[31] Chew BP, Park JS. Carotenoid action on the immune response. Journal of Nutrition. 2004;134(1):257–261. DOI: 14704330

[32] Hasler C. Functional foods: Benefits, concerns and challenges—A position paper from the American Council on Science and Health. Journal of Nutrition. 2002;132:3772–3781. DOI: 12468622

[33] Henrikson R. "Superfood Spirulina microalgae future", Microalgae Spirulina, superalimento del futuro. 2nd ed. Ronore Enterprises; 1994. 222 p. Available from: http://es.slideshare.net/maguifley/microalga-spirulina-robert-henrikson-50-mb

[34] Armstrong DG. Gut-active growth promoters, control and manipulation of animal growth. In: Buttery PJ, LIndsay D, Haynes NB, editors. Control and Manipulation of Animal Growth. London: Butterworths; 1986. p. 21–37. Available from: https://books.google.com.tr/books?id=ha_YBAAAQBAJ&pg=PA21&lpg=PA21&dq=armstrong,+gut+active+growth&source=bl&ots=nI5Z8CbU0R&sig=herW-RY_CWIG8o7luNFj115FHB_I&hl=tr&sa=X&ved=0ahUKEwiPkbH7q9_OAhULshQKHXHTCSIQ6AEIHTAA#v=onepage&q=armstrong%2C%20gut%20active%20growth&f=false

[35] Heidarpour A, Fourouzandeh-Shahrakı AD, Eghbalsaied S. Effects of *Spirulina platensis* on performance, digestibility and serum biochemical parameters of Holstein calves. African Journal of Agricultural Research. 2011;6(22):5061–5065. DOI: 10.5897/AJAR11.1076

[36] Kay RA. Microalgae as food supplement. Critical Reviews in Food Science and Nutrition. 1991;30:555–573. DOI: 10.1080/10408399109527556

[37] Kahraman Z. "Herbal Feed Additive and Their Usage in Laying Hen Diets", Bitkisel yem katkı maddelerinin yumurta tavuğu yemlerinde kullanımı (in Turkish). Tavukçuluk Araştırma Dergisi. 2009;8(1):34–41. Available from: http://arastirma.tarim.gov.tr/tavukculuk/Belgeler/web%20English%20Doc/journal%20(Dergimiz)/Dergimiz%20Cilt%208%20Sayi%201/Cilt%208%20Sayi%201%20Makale%207%20Bitkisel

%20Yem%20Katk%C4%B1%20Maddelerinin%20Yumurta%20Tavu%C4%9Fu %20Yemlerinde%20Kullan%C4%B1m%C4%B1.pdf

[38] Halidou Doudou M, Degbey H, Daouda H, Leveque A, Donnen P, Hennart P, Dramaix-Wilmet M. The effect of spiruline during nutritional rehabilitation: systematic review. Revue d Épidémiologie et de Santé Publique. 2008;56(6):425–431. DOI: 10.1016/j.respe. 2008.08.004.

[39] Kocaoğlu Güçlü B, Kara K. "Use of Alternative Feed Additives in Ruminant Nutrition: 1. Probiotic, Prebiotic, and Enzyme", Ruminant Beslemede Alternatif Yem Katkı Maddelerinin Kullanımı: 1. Probiyotik, Prebiyotik ve Enzim. (In Turkish). Journal of Veterinary Faculty of University of Erciyes. 2009;6(1):65–75.

[40] Saxena PN, Ahmad MR, Shyan R, Amla DV. Cultivation of Spirulina in sewage for poultry feed. Experientia. 1983;39:1077–1083. DOI: 10.1007/BF01943117

[41] Vonshak A, Tomaselli L. Arthrospira (Spirulina): systematics and ecophysiology. In: Whıtton BA, Potts M, editors. Ecology of Cyanobacteria. Kluwer Academic Publishing; The Netherlands, 2000. p. 505–523. DOI: 10.1007/0-306-46855-7_18

[42] Fox RD. Spirulina production & potential. Aix-en-Provence: Edisud; 1996. p. 89. ISBN: 285744883X, 9782857448839

[43] Richmond A. Mass culture of cyanobacteria. In: Mann N, Carr N, editors. Photosynthetic Prokaryotes. 2nd ed. Plenum Press; New York, 1992. p. 181–2010. DOI: 10.1007/978-1-4757-1332-9_6

[44] Pulz O, Gross W. Valuable products from biotechnology of microalgae. Applied Microbiology and Biotechnology. 2004;65:635–48. DOI: 10.1007/s00253-004-1647-x

[45] Dogan M. "The Effect Mechanisms of Probiotic Bacteria in Gastrointestinal System", Probiyotik Bakterilerin Gastrointestinal Sistemdeki Etki Mekanizması. Electronic Journal of Food Technologies. 2012;7(1):20–27. e-ISSN:1306-7648.

[46] Belay A. The potential application of Spirulina (Arthrospira) as a nutritional and therapeutic supplement in health management. The Journal of the American Nutraceutical Association. 2002;5:27–48. Available from: http://www.macoc.fr/resources/The+Journal+of+the+American+Nutraceutical+Association+vol+5+spring+2002+Spiruline.pdf

[47] Simpore J, Zongo F, Kabore F, Dansou D, Bere A, Nikiema JB, Pignatelli S, Biondi DM, Ruberto G, Musumeci S. Nutrition rehabilitation of HIV-infected and HIV-negative undernourished children utilizing spirulina. Annals of Nutrition & Metabolism. 2005;49(6):373–380. DOI: 10.1159/000088889

[48] Yamani E, Kaba-Mebri J, Mouala C, Gresenguet G, Rey JL. Use of spirulina supplement for nutritional management of HIV-infected patients: Study in Bangui, Central African Republic. Medecine Tropicale. 2009;69(1):66–70. DOI: 19499738

[49] Azabji Kenfack M, Edie Dikosso S, Loni G, Onana A, Sobngwi E, Gbaguidi E, Ngougni Kana AL, Nguefack T, Von der Weid D, Njoya O, Ngogang J. Potential of *Spirulina platensis* as a nutritional supplement in malnourished HIV-infected adults in Sub-Saharan Africa: a randomised, single-blind study. Nutrition and Metabolic Insights. 2011;4:29–37. DOI: 10.4137/NMI.S5862

[50] Moreira LM, Rocha ASR, Ribeiro CLG, Rodrigues RS, Soares LS. Nutritional evaluation of single-cell protein produced by *Spirulina platensis*. African Journal of Food Science. 2011;5(15):799–805. DOI: 10.5897/AJFS11.184

[51] Seyidoglu N, Galip N. Effects of *Saccharomyces cerevisiae* and *Spirulina platensis* on Growth performances and biochemical parameters in rabbits. Journal of Kafkas Veterinary Faculty. 2014;20(3):331–336. DOI: 10.9775/kvfd.2013.9988

[52] Derkenbaşı S, Ünal H, Karayücel I, Aral O. Effect of dietary supplementation of different rates of Spirulina (*Spirulina platensis*) on growth and feed conversion in guppy (Poecilia reticulara Peters, 1860). Journal of Animal and Veterinary Advances. 2010;9(9):1395–1399. DOI: 10.3923/javaa.2010.1395.1399

[53] Promya, J, Chitmanat C. The effects of *Spirulina platensis* and *Cladophora* algae on the growth performance, meat quality and immunity stimulating capacity of the African sharptooth catfish (*Clarias gariepinus*). International Journal of Agriculture and Biology. 2011;13:77–82. ISSN Online: 1814–9596 10–272/DJZ/2011/13–1–77–82

[54] Ungsethaphand T, Peerapornpisal Y, Whangchai N and Sardsud U. Effect of feeding *Spirulina platensis* on growth and carcass composition of hybrid red tilapia (Oreochromis mossambicus × O. niloticus). Maejo International Journal of Science and Technology. 2010;4(02):331–336. Available online at www.mijst.mju.ac.th

[55] Peiretti PG, Meineri G. Effects of diets with increasing levels of *Spirulina platensis* on the performance and apparent digestibility in growing rabbits. Livestock Science. 2008;118:173–177. DOI: 10.1016/j.livsci.2008.04.017

[56] Grinstead GS, Tokach MD, Dritz SS, Goodband RD, Nelssen JL. Effects of *Spirulina platensis* on growth performance of weanling pigs. Animal Feed Science and Technology. 2000;83:237–247. DOI: 10.1016/S0377-8401(99)00130-3

[57] Araújo KGL, Facchinetti AD, Santos CP. Influence of intake of Spirulina biomass on body weight and feed intake in rats. Food Science and Technology. 2003;3(1): 6–9. Available from: http://www.scielo.br/scielo.php?script=sci_arttext&pid=S0101-20612003000100003 (Accessed 1999/07/06).

[58] Cheng-Wu Z, Chao-Tsi T, Zhen ZTY. The effects of polysaccharide and phycocyanin from *Spirulina platensis* on peripheral blood and hematopoietic system of bone marrow in mice. Proceedings of the Second Asia-Pacific Conference on Algal Biotechnology. National University of Singapore. 1994. p. 58.

[59] Mao TK, Van de Water J, Gershwin ME. Effect of Spirulina on the secretion of cytokines from peripheral blood mononuclear cells. Journal of Medicinal Food. 2000;13:135–140. DOI: 10.1089/jmf.2000.3.135

[60] Løbner M, Walsted A, Larsen R, Bendtzen K, Nielsen CH. Enhancement of human adaptive immune responses by administration of a high-molecular-weight polysaccharide extract from the cyanobacterium *Arthrospira platensis*. Journal of Medicinal Food. 2008;11(2):313–322. DOI: 10.1089/jmf.2007.564

[61] Selmi C, Leung PS, Fischer L, German B, Yang CY, Kenny TP, Cysewski GR, Gershwin ME. The effects of Spirulina on anemia and immune function in senior citizens. Cellular and Molecular Immunology. 2011;8:248–254. DOI: 10.1038/cmi.2010.76

[62] Qureshi MA, Garlich JD, Kidd MT. Dietary *Spirulina platensis* enhances humoral and cell-mediated immune functions in chickens. Immunopharmacology and Immunotoxicology. 1996;18:465–476. DOI: 10.3109/08923979609052748

[63] Baojiang G. Study on effect and mechanism of polysaccharides of *Spirulina platensis* on body immune functions improvement. Proceedings of 2nd Asia Pacific Conference on Algal Biotechnology, Garland Publishers; 1994. p. 24.

[64] Watanuki H, Ota K, Tassakka AC, Kato T, Sakai M. Immunostimulant effects of dietary *Spirulina platensis* on carp. Aquaculture. 2008;258:157–163. DOI:10.1016/j.aquaculture.2006.05.003

[65] Hirahashi T, Matsumoto M, Hazeki K, Saeki Y, Ui M, Seya T. Activation of the human innate immune system by Spirulina augmentation of interferon production and NK cytotoxicity by oral administration of hot water of *Spirulina platensis*. International Immunopharmacology. 2002;2:423–434. DOI: 11962722

[66] Mao TK, Van de Water J, Gershwin ME. Effects of a Spirulina based dietary supplement on cytokine production from allergic rhinitis patients. Journal of Medicinal Food. 2005;8(1):27–30. DOI: 10.1089/jmf.2005.8.27

[67] Hayashi O, Hirahashi T, Katıh T, Miyajima H, Hiranı T, Okuwaki Y. Class specific influence of dietary *Spirulina platensis* on antibody production in mice. Journal of Nutritional Science and Vitaminology. 1998;44 (6):841–845. DOI: 10197315

[68] Belay A, Toshimitsu K, Yoshımıchı O. Spirulina (Arthrospira): Potential application as an animal feed supplement. Journal of Applied Phycology. 1996;8:303–311. DOI: 10.1007/BF02178573

[69] Hamad AA, Saud IA, Ali A, Qureshi MA. Enhancement of chicken macrophage phagocytic function and nitrite production by dietary *Spirulina platensis*. Immunopharmacology and Immunotoxicology. 2001;23(2):281–289. DOI: 10.1081/IPH-100103866

[70] Simsek N, Karadenız A, Karaca T. Effects of the *Spirulina platensis* and Panaxginseng oral supplementation on peripheral blood cells in rats. Revue de Médecine

Vétérinaire. 2007;158(10):483–488. Available from: http://www.revmedvet.com/2007/RMV158_483_488.pdf

[71] Qureshi MA, Ali RA. *Spirulina platensis* exposure enhances macrophage phagocytic function in cats. Immunopharmacology and Immunotoxicology. 1996;18(3):457–463. DOI: 10.3109/08923979609052747

[72] Shen YB, Piao XS, Kim SW, Wang L, Liu P, Yoon I, Zhen YG. Effects of yeast culture supplementation on growth performance, intestinal health, and immune response of nursery pigs. Journal of Animal Science. 2000;87:2614–2624. DOI: 10.2527/jas.2008-1512

[73] Simsek N, Karadeniz A, Keles ON, Unal B. *Spirulina platensis* feeding inhibited the anemia- and leukopenia-induced lead and cadmium in rats. Journal of Hazardous Materials. 2009;164(2–3):1304–1309. DOI:10.1016/j.jhazmat.2008.09.041

[74] Zikic V, Stajn AS, Ognjanovic BI, Pavlovic SZ, Saicic ZS. Activities of superoxide dismutase and catalase in erythrocytes and transaminases in the plasma of carps (*Cyprinus carpio* L.) exposed to cadmium. Physiological Research. 1997;46(5):391–396. DOI: 9728486

[75] Kostic MM, Ognjanovic B, Dimitrijevic S, Zikic RV, Stajn A, Rosic GL, Zivkovic RV. Cadmium-induced changes of antioxidant and metabolic status in red blood cells of rats: in vivo effects. European Journal of Haematology. 1993;51(2):86–92. DOI: 10.1111/j.1600-0609.1993.tb01598.x

[76] Hug C, Von der Weid D. Spirulina in the fight against malnutrition. Geneva (Switzerland) 15 Foundation Antenna Technologies. Rue de Neuchâte. 2011; No.: 29-1201. 16. Available from: http://www.antenna.ch/medias/Spirulina-Assessment-and-Prospects.pdf (Accessed Feb 2011).

[77] Mani U, Sadliwala A, Iyer U, Parikh P. The effect of Spirulina supplementation on blood haemoglobin levels of anaemic adult girls. Journal of Food Science and Technology. 2000;37:642–644. ISSN 0022-1155

[78] Yeganeh S, Teimouri M, Amirkolaie AK. Dietary effects of *Spirulina platensis* on hematological and serum biochemical parameters of rainbow trout (*Oncorhynchus* mykiss). Research in Veterinary Science. 2015;101:84–88. DOI: 10.1016/j.rvsc.2015.06.002

[79] Nagaoka S, Shimizu K, Kaneko H, Shibayama F, Morikawa K, Kanamaru Y. A novel protein C-phycocyanin plays a crucial role in the hypocholesterolemic action of *Spirulina platensis* concentrate in rats. Journal of Nutrition. 2005;135(10):2425–2430. DOI: 16177207

[80] Bhat VB, Madyastha KM. Scavenging of peroxynitrite by phycocyanin and phycocyanobilin from *Spirulina platensis*: protection against oxidative damage to DNA. Biochemical and Biophysical Research Communication. 2001;285:262–266. DOI: 10.1006/bbrc.2001.5195u

[81] Juarez-Oropeza MA, Mascher D, Torres-Duran PV, Farias JM, Paredes-Carbajal MC. Effects of Spirulina on vascular reactivity. Journal of Medicinal Food. 2009;12(1):15–20. DOI: 10.1089/jmf.2007.0713

[82] Torres-Duran PV, Ferreira-Hermosillo A, Juarez-Oropeza MA. Antihyperlipidemic and antihypertensive effects of *Spirulina maxima* in an open sample of Mexican population: a preliminary report. Lipids in Health and Disease. 2007;26:6–33. DOI: 10.1186/1476-511X-6-33

[83] Samuels R, Mani UV, Iyer UM, Nayak US. Hypocholesterolemic effect of Spirulina in patients with hyperlipidemic nephrotic syndrome. Journal of Medicinal Food. 2002;5(2):91–96. DOI: 10.1089/109662002760178177

[84] Park HJ, Lee YJ, Ryu HK, Kim MH, Chung HW, Kim WY. A randomized double-blind, placebo-controlled study to establish the effects of Spirulina in elderly Koreans. Annals of Nutritional and Metabolism. 2008;52(4):322–328. DOI: 10.1159/000151486

[85] Sato T, Yamamoto C, Fujiwara Y, Kaji T. Biological activities of exogenous polysaccharides via controlling endogenous proteoglycan metabolism in vascular endothelial cells. Journal of Pharmaceutical Society of Japan (Yakugaku Zasshi). 2008;128(5):717–723. DOI: http://doi.org/10.1248/yakushi.128.717

[86] Becker EW, Jakober B, Luft D, Schmulling RM. Clinical and biochemical evaluations of Spirulina with regard to its application in the treatment of obesity. Inst. Chem. Pfanz. Nutrition Reports International. 1986;33(4):565. ISSN: 0029-6635

[87] Mani UV, Desai S, Iyer U. Studies on the long-term effect of Spirulina supplementation on serum lipid profile and glycated proteins in NIDDM patients. Journal of Nutraceuticals Functional & Medical Foods. 2000;2(3):25–32. DOI:10.1300/J133v02n03_03

[88] Hosoyamada Y, Takai Y, Toshimitsu K. Effect of water soluble and insoluble fractions of Spirulina over serum lipids and glucose resistance of rats. Journal of Japan Society of Nutrition and Food Science. 1991;44:273–277. DOI: http://doi.org/10.4327/jsnfs.44.273

[89] Parikh P, Mani U, Iyer U. Role of Spirulina in the control of glycemia and lipidemia in type 2 diabetes mellitus. Journal of Medicinal Food. 2001;4(4):193–199. DOI: 10.1089/10966200152744463

[90] Iwata K, Inayama T, Katoh T. Effect of *Spirulina platensis* on plasma lipoprotein lipase activity in fructose induced hyperlipidemia in rats, Journal of Nutritional Science and Vitaminology. 1990;36:165–171. DOI: 10.3177/jnsv.36.165.

[91] Simkus A, Oberauskas V, Laugalis J, Zelvyte R, Monkevicience I, Sederevicius A, Simkiene A, Pauliukas K. The effect of weed *Spirulina platensis* on the milk production in cows. Abstract, Veterinarija ir Zootechnika. 2007;38:60. ISSN 1392-2130

[92] Panigrahi BB, Panda PK, Patro VJ. Wound healing activity of spirulina extracts. International Journal of Pharmaceutical Sciences Review and Research. 2011;6(2):132–135. DOI: 10.17179/excli2014-697

[93] Rabadiya B, Patel P. Spirulina: Potential clinical therapeutic application (review). Journal of Pharmacy Research. 2010;3(8):1726–1732. ISSN: 0974-6943

[94] Syarina P, Karthivashan G, Abas F, Arulselvan P, Fakurazi S. Wound healing potential of *Spirulina platensis* extracts on human dermal fibroblast cells. EXCLI Journal. 2015;14:385–393. DOI: 10.17877/DE290R-7598

[95] Hudson BJ, Karlis IG. The lipids of the alga Spirulina. Journal of the Science of Food Agriculture. 1974;25:759–763. DOI: 10.1002/jsfa.2740250703

[96] Nichols B, Wood B. The occurrence and biosynthesis of gamma linolenic acid in *Spirulina platensis*. Lipids. 1986;3(1):46–50. DOI: 10.1007/BF02530968

[97] Remirez D, González R, Merino N, Rodriguez S, Ancheta O. Inhibitory effects of Spirulina in zymosan-induced arthritis in mice. Mediators of Inflammation. 2002;11(2): 75–79. DOI: 10.1080/09629350220131917

[98] Chen JC, Liu KS, Yang TJ, Hwang JH, Chan YC, Lee IT. Spirulina and C-phycocyanin reduce cytotoxicity and inflammation-related genes expression of microglial cells. Nutritional Neuroscience. 2012;15(6):252–256. DOI: 10.1179/1476830512Y.0000000020.

[99] Qureshi M, Kidd MT, Ali RA. Spirulina extract enhances chicken macrophage functions after in vitro exposure. Journal of Nutritional Immunology, 1996;3(4):35–45. DOI: 10.1300/J053v03n04_04

[100] Hayashi T, Hayashi K., Maeda M, Kojima I. Calcium Spirulan, an inhibitor of enveloped virus replication, from a blue-green alga *Spirulina platensis*. Journal of Natural Products. 1996; 59(1):83–87. DOI: 10.1021/np960017o

[101] Daoud HM, Soliman EM. Evaluation of *Spirulina platensis* extract as natural antivirus against foot and mouth disease virus strains (A, O, SAT2). Veterinary World. 2015;8(10): 1260–1265. DOI: 10.14202/vetworld.2015.1260-1265.

[102] Shih SR, Tsai KN, Li YS, Chueh CC, Chan EC. Inhibition of enterovirus 71-induced apoptosis by allophycocyanin isolated from a blue-green alga *Spirulina platensis*. Journal of Medical Virology. 2003;70(1):119–125. DOI: 10.1002/jmv.10363

[103] Ozdemir G, Karabay NU, Dalay MC, Pazarbasi B. Antibacterial activity of volatile component and various extracts of *Spirulina platensis*. Phytotherapy Research. 2004;18(9):754–757. DOI: 10.1002/ptr.1541

[104] Mayer AMS, Rodrigues AD, Berlinck RG, Hamann MT. Marine pharmacology in 2005–6: Marine compounds with anthelmintic, antibacterial, anticoagulant, antifungal, anti-inflammatory, antimalarial, antiprotozoal, anti tuberculosis, and antiviral activities; affecting the cardiovascular, immune and nervous systems, and other miscellaneous mechanisms of action. (review). Biochimica et Biophysica Acta. 2009;1790:283–308. DOI: 10.1016/j.bbagen.2009.03.011

[105] Woodhead AD. Nutrition and reproductive capacity in fish. Symposium of Proceedings of Zoology Society. London. 1960;19:23-28. Available from: http://

journals.cambridge.org/download.php?file=%2FPNS%2FPNS19_01%2FS0029665160000104a.pdf&code=61907413b7b8bb1b8eb4aa14d4bda045

[106] James R, Sampath K, Thangarathinam R, Vasudevan I. Effect of dietary Spirulina level on growth, fertility, coloration and leucocyte count in red swordtail, xiphophorus helleri. The Israeli Journal of Aquaculture. 2006;58(2):97–104. DOI: http://hdl.handle.net/10524/19166

[107] Granaci V. Achievements in the artificial insemination of swine. Bulletin of University of Agricultural Sciences and Veterinary Medicine Cluj-Napoca. Animal Science and Biotechnologies. 2007;382–386. Available from: http://journals.usamvcluj.ro/index.php/zootehnie/article/viewFile/2012/1937

[108] Kistanova E, Marchev Y, Nedeva R, Kacheva D, Shumkov K, Georgiev B, Shimkus A. Effect of the *Spirulina platensis* included in the main diet on the boar sperm quality. Biotechnology in Animal Husbandry. 2009;25(5–6):547–557. Available from: http://www.doiserbia.nb.rs/img/doi/1450-9156/2009/1450-91560906547K.pdf (Accessed 2009).

[109] Belay A, Ota Y, Miyakawa K, Shimamatsu H. Current knowledge on potential health benefit of Spirulina. Journal of Applied Phycology. 1993;5(2). DOI: 10.1007/BF00004024

[110] Dohle GR, Smit M, Weber RFA. Androgens and male fertility. World Journal of Urology. 2003;21(5):341–345. DOI: 10.1007/s00345-003-0365-9

[111] Yousef MI, Abdallah GA, Kamel KI. Effect of ascorbic acid and vitamin E supplementation on semen quality and biochemical parameters of male rabbits. Animal Reproduction Science. 2003;76(1–2):99–111. DOI: http://dx.doi.org/10.1016/S0378-4320(02)00226-9

[112] Rezvanfar MA, Sadrkhanlou RA, Ahmadi A. Shojaei-Sadee H, Rezvanfar M, Mohammadirad A, Salehnia A, Abdollahi M. Protection of cyclophosphamide-induced toxicity in reproductive tract histology, sperm characteristics, and DNA damage by an herbal source; evidence for role of free-radical toxic stress. Human and Experimental Toxicology. 2008;27(12):901–910. DOI: 10.1177/0960327108102046

[113] Falquet J. Nutritional Aspects of Spirulina. Antenna Technologies. Available from: http://www.antenna.ch/en/documents/AspectNut_UK.pdf

[114] Pratt DE. Natural antioxidants from plant material. In: Huang MT, Ho CT, Lee CY, editors. Phenolic Compounds in Food and their Effects on Health. II. American Chemical Society, Washington, 1992; p. 54–71. Chapter DOI: 10.1021/bk-1992-0507.ch005

[115] Karkos PD, Leong SC, Karkos CD, Sivaji N, Assimakopoulos DA. Spirulina in clinical practice: evidence-based human applications. Evidence Based Complementary and Alternative Medicine: eCAM. 2011;2011:531053 DOI: 10.1093/ecam/nen058.

[116] Li B, Zhang X, Gao M, Chu X. Effects of CD59 on antitumoral activities of phycocyanin from *Spirulina platensis*. Biomedicine & Pharmacotherapy. 2005;59(10):551–560. DOI: 10.1016/j.biopha.2005.06.012

[117] Suda D, Schwartz J, Shklar G. Inhibition of experimental oral carcinogenesis by topical beta carotene. Harvard School of Dental Medicinal Carcinogenesis.1986;7(5):711–715. DOI: 10.1093/carcin/7.5.711

[118] Schwartz J, Shklar G, Reid S, Trickler D. Prevention of experimental oral cancer by extracts of Spirulina-Dunaliella algae. Nutrition and Cancer. 1988;11(2):127–134. DOI: 10.1080/01635588809513979

[119] Grawish ME. Effects of *Spirulina platensis* extract on Syrian hamster cheek pouch mucosa painted with 7,12-dimethylbenz[a]anthracene. Oral Oncology. 2008;44(10): 956–962. DOI: 10.1016/j.oraloncology.2007.11.0141.

[120] Ismail MF, Ali DA, Fernando A, Abdraboh ME, Gaur RL, Ibrahim WM, Raj MH, Ouhtit A. Chemoprevention of rat liver toxicity and carcinogenesis by Spirulina. International Journal of Biological Science. 2009;2;5(4):377–387. PMID: 19521547 PMCID: PMC2695150.

[121] Mishima T, Murata J, Toyoshima M, Fujii H, Nakajima M, Hayashi T, Kato T, Saiki I. Inhibition of tumor invasion and metastasis by calcium spirulan (Ca-SP), a novel sulfated polysaccharide derived from a blue-green alga, *Spirulina platensis*. Clinical & Experimental Metastasis. 1998;16(6):541–550. PMID: 9872601.

[122] Karadeniz A, Cemek M, Simsek N. The effects of Panax ginseng and *Spirulina platensis* on hepatotoxicity induced by cadmium in rats. Ecotoxicology and Environmental Safety. 2009;72(1):231–235. DOI: 10.1016/j.ecoenv.2008.02.021.

[123] Vadiraja BB, Gaikwad NW, Madyastha KM. Hepatoprotective effect of C-phycocyanin: protection for carbon tetrachloride and R-(+)-pulegone-mediated hepatotoxicity in rats. Biochemical and Biophysical Research Communications. 1998;19;249(2):428–431. DOI: 10.1006/bbrc.1998.9149.

[124] Abdel-Daim MM, Abuzead SM, Halawa SM. Protective role of *Spirulina platensis* against acute deltamethrin-induced toxicity in rats. PLoS One. 2013;8(9):e72991. DOI: 10.1371/journal.pone.0072991

[125] Karadeniz A, Yildirim A, Simsek N, Kalkan Y, Celebi F. *Spirulina platensis* protects against gentamicin-induced nephrotoxicity in rats. Phytotherapy Research. 2008;22(11):1506–1510. DOI: 10.1002/ptr.2522.

[126] Mohan IK, Khan M, Shobha JC, Naidu MU, Prayag A, Kuppusamy P, Kutala VK. Protection against cisplatin-induced nephrotoxicity by Spirulina in rats. Cancer Chemotherapy and Pharmacology. 2006;58(6):802–808. DOI: 10.1007/s00280-006-0231-8.

[127] Khan M, Shobha JC, Mohan IK, Naidu MU, Sundaram C, Singh S, Kuppusamy P, Kutala VK. Protective effect of Spirulina against doxorubicin-induced cardiotoxicity. Phytotherapy Research: PTR. 2005;19(12):1030–1037. DOI: 10.1002/ptr.1783.

[128] Misbahuddin M, Islam AZ, Khandker S, Ifthaker-Al-Mahmud, Islam N, Anjumanara. Efficacy of spirulina extract plus zinc in patients of chronic arsenic poisoning: a randomized placebo-controlled study. Clinical Toxicology (Phila). 2006;44(2):135–141. PMID: 16615668.

[129] El-Desoky GE, Bashandy SA, Alhazza IM, Al-Othman ZA, Aboul-Soud MA, Yusuf K. Improvement of mercuric chloride-induced testis injuries and sperm quality deteriorations by *Spirulina platensis* in rats. PLoS One. 2013;8(3) DOI:10.1371/journal.pone.0059177.

[130] Belookaya TC. From Chairman of Byelorussian Committee "Children of Chernobyl" May 31, 1991. Available from: http://www.naturalpharmainternational.com/1/upload/reduces_effects_of_radiation_for_the_children_of_chernobyl.pdf

[131] Wu LC, Ho JA, Shieh MC, Lu IW. Antioxidant and antiproliferative activities of Spirulina and Chlorella water extracts. Journal of Agriculture and Food Chemistry. 2005;53(10):4207–4212. DOI: 10.1021/jf0479517

[132] Qishen P, Guo BJ, Kolman A. Radioprotective effect of extract from *Spirulina platensis* in mouse bone marrow cells studied by using the micronucleus test. Toxicology Letters. 1989;48(2):165–169. DOI: 2505406

[133] Belay A. Mass culture of Spirulina outdoors. The Earthrise Farms experience. In: Vonshak, A., Ed. Spirulina platensis (Arthrospira): Physiology, Cell-biology and Biotechnology. Taylor and Francis. London. 1997. pp. 131–158.

[134] Othes S, Pire R. Fatty acid composition of Chlorella and Spirulina microalgae species. Journal of AOAC International. 2001;84:1708–1714.

Chapter 2

Quality Assessment of Microalgae Exposed to Trace Metals Using Flow Cytometry

Toshiyuki Takahashi

Additional information is available at the end of the chapter

http://dx.doi.org/10.5772/65516

Abstract

Seaweed has long been an important kitchen ingredient and a functional food material. Microalgae have attracted the same attention as seaweed from food, pharmaceutical and cosmetic companies because several algae contain unique functional materials. Industry application of algae requires the selection of useful algal species, evaluation of their features and monitoring of their quality in culture. Taking *Chlorella* for example, this chapter presents a method using flow cytometry (FCM) to assess not only the number of algae but also algal quality. First, *Chlorella* was cultured in media containing eluate from steel slag as an experimental factor and trace metals. After the treatment of algae with eluate, the number and physiological features of algae were evaluated, respectively, using hemocytometry and FCM. Results show that eluate from slag induced neither lethality nor growth inhibition. Coupled with hemocytometry, FCM was used to estimate vigorous and aberrant algal status. Consequently, the eluate did not give rise to algae stresses. Interestingly, the addition of slag eluate increased the amounts of the carbonate species. The increase in the carbonate species actually triggered the potential increase in aqueous CO_2 for photosynthesis, eventually inducing algal proliferation. These analyses can support evaluation of algal features and maintenance of their quality for industry application.

Keywords: food science, *Chlorella*, Chlorophyll, flow cytometry, fluorescence spectroscopy, trace elements, steel slag

1. Introduction

Aqueous photosynthetic organisms such as algae are the foundation of aquatic ecosystems. The quantities of aqueous photosynthetic organisms as producers in the aqueous ecosystem support yields of both fish and shellfish. In addition to its role as a producer in aqueous ecosystems, seaweed has long been an important kitchen ingredient as a functional food material, providing functional nutrients such as carotenoids and anti-oxidative fucoxanthin. A biorefinery that can take advantage of biofunctions presents an alternative concept to that of conventional refineries that manufacture materials using fossil fuels. Especially, autotrophic algal biorefineries present great advantages over conventional microbial biorefineries such as those using fermentation. Some microalgae species such as *Chlorella* spp., *Dunaliella salina* and *Haematococcus pluvialis* are appreciated respectively as sources of β-1, 3-glucan, β-carotene and astaxanthin [1]. They have attracted attention equal to that devoted to seaweed products from several pharmaceutical, vitamin supplement, cosmetic and food companies [1, 2] because these algae have functional materials that are rare among land plants. Moreover, other microalgae have attracted attention for use as biofuel materials [2, 3] and bioremediation materials for environmental biodegradation [4, 5].

The industrial application of algae demands the selection of useful algal *sp.*, the evaluation of algal features and the assessment of their qualities in culture. Open pond culture systems, rather than closed systems, are the main type of culture system for the commercial scale culture of microalgae because of their relative low cost [6]. As commonly known, human activities have major impacts on the global and regional cycles of most of the trace elements including toxic heavy metals [7]. Atmospheric transport and deposition are potentially important processes for delivering a wide range of anthropogenic contaminations to aquatic environments [7, 8]. Microalgae are very sensitive to changes in their environment [9]. Their overall metabolisms are greatly affected by even trace amounts of various organic and inorganic pollutants including heavy metals [9]. Such fear factors might pose a threat to open culture systems of algae. Therefore, it is especially important to routinely control and manage algal qualities in culture.

Taking green algae *Chlorella* spp., for example, this chapter presents a method to assess algal quality using flow cytometry (FCM). *Chlorella* was cultured in media containing eluate from steel slag as an experimental factor and trace metals. After treatment of algae with eluate, the number and physiological features of algae were evaluated respectively using hemocytometry and FCM. These analyses are expected to contribute to the evaluation of algal features and to the maintenance of their quality for industry applications.

2. Algal characteristics using FCM

Over the last few decades, FCM has become widely used as a powerful and valuable tool for studies of cell biology, microbiology, protein engineering and healthcare. Several functions of FCM include several procedures such as cell counting, biomarker detection and cell sorting through assessment of cell optical information. **Figure 1** presents an outline of a flow cytometric instrument used for this study. This flow cytometer, which detects several optical properties, is equipped with a green laser operating at 532 nm. Forward scatter (FSC) signals were collected to ascertain the cell size. Red fluorescence is detected in the red fluorescence channel through a 680/30 nm band pass filter. Simultaneously, a yellow fluorescence channel through a 576/28 nm band pass filter is used [10–13]. Each fluorescence is converted into an electrical pulse. The electrical intensity is then quantified for each level of fluorescence intensity.

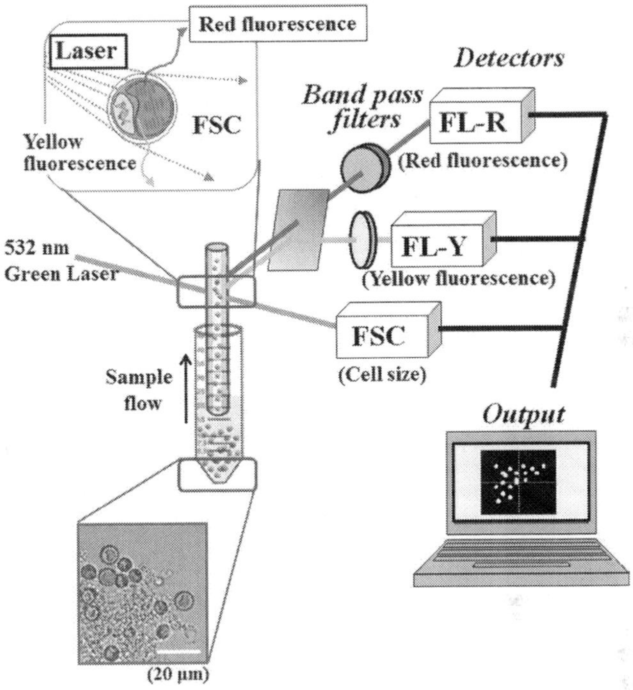

Figure 1. Overview of the flow cytometric system used for this study. Algae that had passed through a capillary were analysed. In addition to the red and yellow fluorescence derived from algae, FSC signals of algae were collected simultaneously as shown.

When heterotrophic cells, such as animal cells, are targeted for FCM measurement, fluorescence-labelling antibodies against certain biomolecules are used to detect and quantify the biomolecules. When using phototrophic cells, such as phytoplankton and plant cells, a photosynthetic pigment, chlorophyll, can also function as a biomarker similar to a fluorescence labelling antibody. When exposed to appropriate excitation light, chlorophyll in each cell irradiates red fluorescence (**Figure 2A** and **B**) [12]. **Figure 2C** depicts emission spectra of *Chlorella*-like algae [14, 16]. The wavelength of the maximal fluorescence of algal chlorophyll is approximately 680 nm (green curve in **Figure 2C**). Consequently, chlorophyll fluorescence is mainly detectable using the red fluorescence channel of the instrument used for this study. The cell size and chlorophyll content of algae are correlated strongly with the algal cell cycle [14–16].

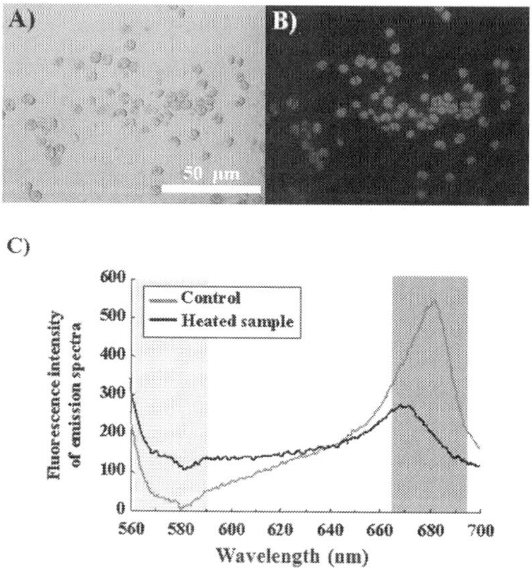

Figure 2. Fluorescence characteristics of algae and microphotographs of *Chlorella*-like algae isolated from protozoa *Paramecium bursaria*. Algal images obtained using bright field (A) and fluorescence microscopy (B) were referred from the literature [12]. Panel C presents fluorescence characteristics of algae obtained using fluorescence spectroscopy referred from the literature [10]. Emission spectra of algae are shown with (black line, heated algae) or without (green line, control algae). Yellow and pink areas, respectively, show detection ranges of yellow and red fluorescence channels used for FCM in this study.

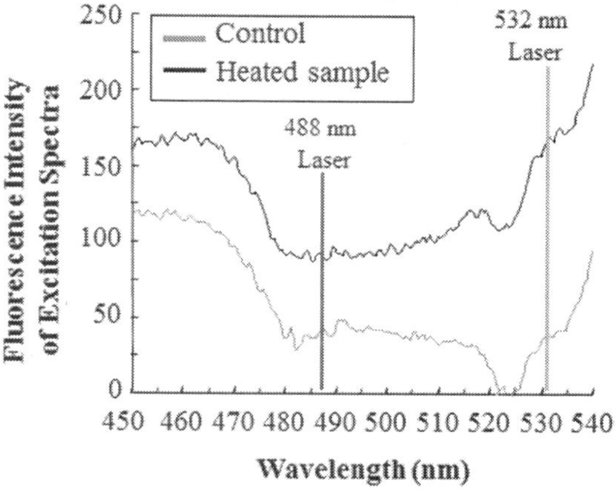

Figure 3. Excitation spectra of *Chlorella*-like algae with or without heat treatment referred from the literature [10]. The fluorescence intensities at 575 nm were measured to produce excitation spectra. Two vertical lines signifying 488 nm (blue) and 532 nm (green) are shown in the graph.

http://dx.doi.org/10.5772/65516

Chlorophyll is sensitive to physiological factors such as heat and acid. These physical factors eventually cause inactive chlorophyll because of degradation [11–13]. In fact, a previous study using *Chlorella*-like algae [10] demonstrated that algae without stress irradiated only red fluorescence derived from chlorophyll (green curve in **Figure 2C**). In contrast, dead algae, subjected to extraneous stress, tended to have less red fluorescence and more yellow fluorescence because of the biodegradation of chlorophyll (black curve in **Figure 2C**). Moreover, this instrument presents benefits for evaluation of algal status because the excitation efficiency of the green laser at 532 nm in the yellow fluorescence is higher than that of a conventionally used blue laser at 488 nm (**Figure 3**) [10]. It suggests that the red and yellow fluorescence intensities are regarded respectively as indices of vigorous algae and of variant algae when the green laser irradiates algae moving through the capillary of this flow cytometer [11, 13].

3. Features of steel slag used for this study

Iron and steel slags from blast furnace slag and steel making slag including converter slag and electric arc furnace (EAF) slag are produced as steel industrial by-products. All blast furnace slag can be recycled completely for the use of steel making slag base and cement as soil aggregate [17, 18], although several volumes of steel making slag, particularly EAF slag, ultimately end up in landfill sites [19]. New applications of slag, such as depurative and sand capping materials in aquatic environments [20] have been regarded both as decreasing the amounts of discarded slag and as reducing high costs of discarding slag. Several environmental pollution laws, however, have restricted slag use in aquatic environments because steel slag contains environmentally hazardous substances. The toxicity of eluate from EAF slag for aquatic organisms remains poorly understood [11, 13, 21], but the physiochemical properties and effects of converter slag on organisms have been documented often [22–27].

This study specifically examined stainless steel slag (designated as slag A) and common steel slag (slag B), exhausted respectively from oxidation processes of stainless and common steelmaking in EAF processes [11, 13]. **Table 1** presents compositions of EAF slags used for this study [11, 13, 21, 28, 29]. In brief, slag A contains more SiO_2, CaO, and Cr_2O_3 than slag B does, whereas slag A contains less FeO than slag B. All Fe and Cr compounds are described respectively as FeO or Cr_2O_3 because it is generally difficult to distinguish FeO and Cr_2O_3 formed form Fe and Cr in a suspended metal solution after alkali fusion of stainless steel slag [11, 13].

	FeO	SiO_2	CaO	Al_2O_3	MgO	MnO	Cr_2O_3	ZnO	NiO	CuO
Slag A	0.74	44.1	33	5.39	7.68	4.09	3.29	0.01	0.06	0.024
Slag B	35.1	19.2	20.8	15.2	4.1	5.1	0.43	0.071	0.028	0.025

Table 1. Chemical compositions of EAF steel slags used for this study (mass %).

4. Research methods

The author used *Chlorella* as the model organism representing algae in this study. Several methods used to examine algal behaviours have been established using *Chlorella* spp. The author used *Chlorella kessleri* (C-531 strain) which was obtained from the Institute of Applied Microbiology (IAM) culture collection at The University of Tokyo. The scientific name of *C. kessleri* was recently changed to *Parachlorella kessleri* because the taxonomy of *Chlorella* has been re-validated using multidisciplinary approaches based on combining classical and modern methods including molecular phylogeny and bioinformatics [30]. Before experiments, algae on the CA agar plates [31] were scratched with an inoculating needle and were suspended in CA liquid medium.

Steel slag was subjected to a leaching test based on JIS K0058-1: 2005 (Method for chemicals in slags Part 1: Leaching test) to elute metal components of slag with HCl [11, 13, 21, 28, 29, 32]. After elution, the solution was filtrated with a 0.45 μm pore filter to eliminate slag particles as described in previous reports. The filtrated eluate from the slag (designated respectively as eluate A and eluate B) was used for bioassay with *Chlorella* as a test solution including trace metals.

To assess the eluate effects on algal growth, *Chlorella* was cultured in CA medium containing an eluate from steel slag as an experimental factor and sources of trace metals [11, 13]. Compared with general culture media for algae, slag eluates used for this study contained insufficient nutrients for algal growth. To supplement nutrients for algal growth, the following assessments of algal growth were conducted with CA liquid medium at pH 7.2 [11, 13]. Nutrient amounts of CA medium containing eluates were the same as those of CA medium alone, but the concentrations of chemicals derived from each eluate differed from those of CA medium without eluate. Here, CA medium without eluate was designated as "control".

Algae (initial density of 1.0×10^4 cells/ml adjusted using hemocytometry) were cultured with CA medium containing eluate from each slag for 1 week in a plastic tube under an LD cycle (12 h light/12 h dark) at approximately 1100 lux of natural white fluorescent light and 23 ± 2°C as described in previous reports [11, 13]. After treatment of algae with eluate, the number and physiological features of algae were quantified respectively using hemocytometry and FCM. The algal proliferation ratio (average ± standard error) was expressed as a proportion of the number of algae treated with eluate to that of control without eluate [11, 13].

To investigate algal status using FCM, the algal status was analysed and estimated based on the corresponding fluorescence. In brief, the stress of each alga is portrayed as a two-dimensional graph (2D map) of red fluorescence intensity (665–695 nm) as the index of vigorous algae and yellow fluorescence intensity (562–590 nm) as the index of variant algae [11, 13]. To facilitate comparison of vigorous algae with stressed and dying algae, a reference standard of algae subjected to stress was prepared by treatment of algae with heat for 5 min at 100°C (designated as heated algae) [11, 13]. For FCM analyses, FSC signals detected only in the culture medium were removed as technical noise from FCM measurements as described in previous reports [11, 13]. The remaining signals were re-analysed as algal signals.

Aquatic CO_2 ($CO_2(aq)$) concentrations are related directly with photosynthesis and algal proliferation. $CO_2(aq)$, HCO_3^- and CO_3^{2-} are present as the carbonate species in solution, as presented in Eqs. (1)–(3) [11, 13].

$$CO_2\,(gas) \leftrightarrow CO_2\,(aq) \quad (1)$$

$$H_2O + CO_2\,(aq) \leftrightarrow H_2CO_3\,(aq) \leftrightarrow HCO_3^- + H^+ \quad (2)$$

$$HCO_3^- \leftrightarrow CO_3^{2-} + H^+ \quad (3)$$

For aqueous photosynthetic organisms, $CO_2(aq)$ of these carbonate species is particularly necessary to support photosynthesis. We ascertained the concentration of $CO_2(aq)$ in slag eluate using potentiometry with a diaphragm-type electrode to measure the $CO_2(aq)$ concentrations [11, 13]. Both HCO_3^- and CO_3^{2-} can be estimated as $CO_2(aq)$ in acidic conditions (≤ pH 4.0) as portrayed in **Figure 4** resulting from the following Henderson–Hasselbalch Eqs. (4) and (5) [11, 13].

$$pH = pK_1 + \log([HCO_3^-]/[CO_2\,(aq)]) \quad (4)$$

$$pH = pK_2 + \log([CO_3^{2-}]/[HCO_3^-]) \quad (5)$$

The respective pK values of $pK_1 = 6.35$ and $pK_2 = 10.33$ [33] were used for this study. The carbonate species aside from $CO_2(aq)$ were converted into $CO_2(aq)$ by adding a pH-adjustable solution. Then they were estimated as the amounts of total carbonate species. Each concentration of $CO_2(aq)$, HCO_3^-, and CO_3^{2-} was calculated from the amounts of total carbonate species and pH values using Eqs. (4) and (5) above. Here, the $[H_2CO_3(aq)]$ given by Eq. (2) was expressed as $[CO_2(aq)]$ in Eq. (4) because it was difficult to distinguish $CO_2(aq)$ from $H_2CO_3(aq)$ in solution, as described in previous reports [11, 13]. The result was expressed as the concentration of $CO_2(aq)$ (average ± standard error) under each condition [11, 13].

In general, the amounts of Ca^{2+} and Mg^{2+} are related to the water hardness and are highly reactive with carbonate species. To examine whether these elements contribute to the concentration of $CO_2(aq)$ in solution, the amounts of Ca^{2+} and Mg^{2+} were measured before and after treatment of algae with CA medium containing slag eluate [11, 13]. After treatment of algae with CA medium containing eluate, the culture tube including the algae was centrifuged. The supernatant, which no longer included algae, was collected and subjected to elemental analysis. Several organic compounds, such as biomolecules reportedly interfere with these measurements [34]. Therefore, this study applied colorimetric determination using specific chelate reagents to elucidate the amounts of Ca^{2+} and Mg^{2+} in the culture supernatant. In practice, the chlorophosphonazo-III method [35] and the xylidyl blue-I method [36] were used for the evaluation of the concentration of Ca^{2+} and that of Mg^{2+}. Moreover, elemental concentrations before treatment of algae with eluate were compared statistically with those after treatment using *t*-tests.

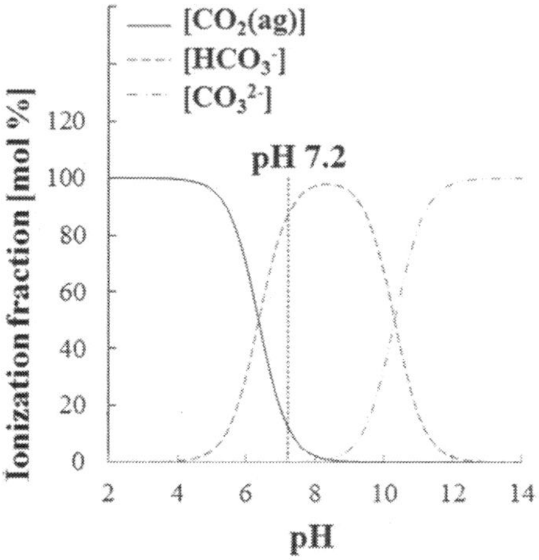

Figure 4. Concentrations of CO_2(aq), HCO_3^-, and CO_3^{2-} for each pH [mol%] referred from the literature [11, 13].

5. Results and discussion

Before evaluating the effects of slag eluates on algae, the concentrations of elements in each slag eluate were analysed and discussed (**Table 2**). In addition to the results of leaching tests for slag, the environmental quality standards (EQSs) for soil pollution and for marine and water pollution are shown as reference values in **Table 2** [13]. In brief, concentrations of Ca, Mg and Si eluted from the slag samples were high because these slags contained large amounts of those materials (**Table 1**). In contrast to the elements above, the eluted concentrations of Al and Fe were quite low in spite of their high concentrations in the slag particles. An earlier report [32] explained this contradictory phenomenon as a difference of these elements in terms of solubility. Leaching tests revealed that concentrations of components eluted from two slags used for this study were almost all lower than the respective EQSs, except for selenium (Se) in eluate from the slag A (eluate A), as described in earlier reports [11–13].

This study examined the effects of slag eluates as an experimentally stress factor on algal growth, particularly that of *Chlorella* spp. [11–13]. **Figure 5** shows the relation between the *Chlorella* proliferation ratio and concentrations of the slag eluate in the test solution. Here, all nutrient amounts derived from CA medium, other than elements derived from each slag eluate, were constant with each experimental condition. A detailed account of the results showed that the number of algae increased according to the concentration of each eluate up to 30 vol%. Subsequent comparison of the algal proliferation ratios in eluate A and eluate B showed that these ratios were almost equal at concentrations lower than 50 vol% of the respective eluates. However, the 70 vol% of eluate B showed a slightly more algal proliferation than that of eluate A, as described in previous reports [11–13]. After explaining the results from FCM analysis, we subsequently discussed the difference in the algal proliferation ratio between eluate A and eluate B.

Origin of slag		Eluate of EAF stainless steel oxidation slag (Slag A)	Eluate of EAF normal steel oxidation slag (Slag B)	Environmental quality standards				
				Soil pollution	Marine pollution	Water pollutant	Effluent standard	Drinking water standard
Regulated substances	Total As	ND[1] (RDL[2]: 0.001)	ND[3]	0.01	0.1	0.01	0.1	0.01
	Total B	0.16	0.28[3]	1		1[4]	10[4], 230[5]	1
	Total Be	ND (RDL: 0.0005)	ND[3]		2.5			
	Total Cd	ND (RDL: 0.0001)	ND[3]	0.01	0.1	0.01	0.03[6]	0.003
	Chromium (VI)	ND (RDL: 0.005)	ND[3]	0.05	0.5	0.05	0.5	0.05
	Total Cu	0.003	ND	0.001	3		3	1
	Total Pb	ND (RDL: 0.0005)	ND[3]	0.01	0.1	0.01	0.1	0.01
	Hg	ND (RDL: 0.0001)	ND[3]	0.0005	0.005	0.0005	0.005	0.0005
	Total Ni	0.001	ND	0.001	1.2			0.02
	Total Se	0.012	0.003[3]	0.01	0.1	0.01	0.1	0.01
	Total V	ND (RDL: 0.001)	0.01[3]		1.5			
	Total Zn	0.099	0.014[3]		2	0.03[7], 0.02[9], 0.01[9]	2	1
	F^-	ND (RDL: 0.1)	0.5[3]	0.8	15	0.8[4]	8[4], 15[5]	0.8

Origin of slag		Eluate of EAF stainless steel oxidation slag (Slag A)	Eluate of EAF normal steel oxidation slag (Slag B)	Environmental quality standards				
				Soil pollution	Marine pollution	Water pollutant	Effluent standard	Drinking water standard
Substances out of regulation	Total Al	ND	1.8					
	Total Ca	9.3	10.1					300[10]
	Total Fe	ND	0.23				10	0.3
	Total Mg	0.9	1.1					300[10]
	Total Mn	0.028	ND				10	0.05
	Total Si	1.8	1.9					
	Total N	0.4	0.362[3]			0.1–1[11] 0.2–1[8]	100	0.04[12], 10[13]
	Total P	ND (RDL: 0.1)	ND[3]			0.005–1[11] 0.02–0.09[8]	16	

[1]Not detected.

[2]Reportable detection limit.

[3]These data from a previous study reported by Takahashi et al. [17].

[4]Standard value is not applied to coastal waters.

[5]Standard value is applied to coastal waters.

[6]The Cd value has changed from 0.1 to 0.03 mg/L since December 2014.

[7]Habitable river or lake for aquatic life.

[8]Habitable coastal water for aquatic life.

[9]Habitable coastal water that requires conservation in particular for nidus and nursery ground.

[10]Total concentrations of both calcium and magnesium are limited for water hardness.

[11]Habitable lake for aquatic life.

[12]Total N contents derived from nitrite nitrogen.

[13]Total N contents derived from both nitrite nitrogen and nitrate nitrogen.

Table 2. Environmental quality standards regarding pollutions and others for effluent and drinking water, and concentrations of elements of each eluate (mg/L) quoted with permission from Ref. [9].

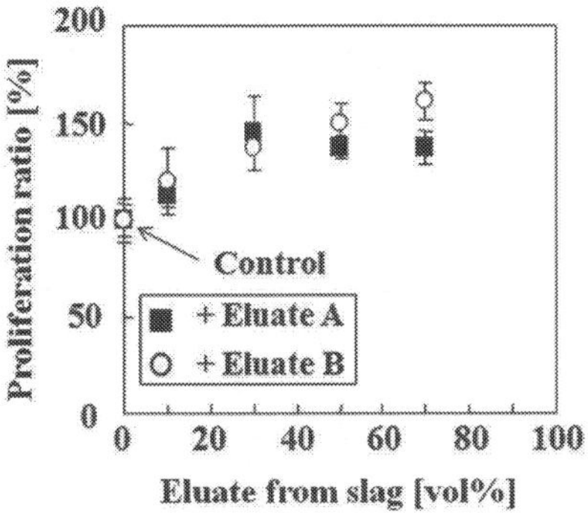

Figure 5. Effects of respective eluates on algal growth referred from the literature [11, 13].

In addition to algal population estimation using hemocytometry, estimation of the cellular status of algae using FCM was conducted. The results are presented two-dimensionally in **Figure 6** [11–13]. Here, each single dot on the 2D map represents optical information of a single alga. To compare the status of vigorous algae with that of stressed and dying algae, algae treated with heat (heated sample) were prepared as a reference standard of algae subjected to stress [11–13]. Results show that the 2D map of the red versus the yellow fluorescence intensity for control algae differs clearly from that for the heated algae (**Figure 6**). The 2D map of the red versus yellow fluorescence intensity for control algae showed respectively 10^2–10^3 on the red channel and 10^1–10^2 on the yellow, whereas that for the heated algae did 10^1–10^2 on the red channel and 10^1–10^3 on the yellow. It is particularly interesting that the dot distribution of algae treated with each slag eluate closely resembled that of control, although that with each eluate shifted slightly upward relative to that of control algae [11–13].

Quantitative analysis of algal distribution patterns (**Table 3**) was conducted along with qualitative analysis of those patterns (**Figure 6**). Each graph in **Figure 6** is divisible into four subareas (regions I–IV) based on algal viability [11–13]: region I represents an area for vigorous algae; region II includes dead and variant algae such as heated algae; region III includes algae with low red fluorescence intensity; and region IV includes data from which algae are virtually absent. Algal distribution patterning revealed clear differences between the algal distribution in control samples and those in heated algae. The signals of algae treated with eluate were also distributed almost entirely to region I. The ratio was 96.81 ± 2.60% in control, 98.15 ± 0.31% in eluate A, and 98.13 ± 0.24% in eluate B. Ratio analysis shows that the percentages of algae treated with slag eluates were slightly higher than those of control algae. Components dissolved from slags did not apparently give rise to algae stress because the quantities of algae in media containing the respective eluates were equal to or greater than those in media with no eluate (**Figure 5**). The tested slags contain metals such as copper, zinc and aluminium (**Table 2**). Aluminium, which is also not contained in the CA medium, has been particularly reported as inhibiting plant growth [37, 38]. Although the culture medium containing aluminium and other metal elements was predicted to affect algal growth and status, they caused no effect on algal growth directly. The data demonstrated that components eluted from slag showed no marked toxicity to algae. This assessment system using FCM, which estimates chlorophyll fluorescence of photosynthetic pigments, might be applicable to other algae, other phytoplankton and aquatic plants with chlorophyll, although this report presents data only for *Chlorella* spp. as a model organism. This technique can contribute to evaluation of algal features and monitoring their qualities in culture for industry application of algae.

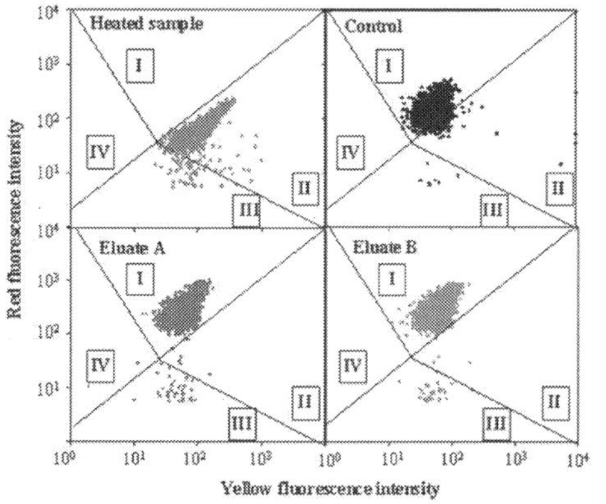

Figure 6. Distribution of *Chlorella* using FCM referred from the literature [11–13]. The red fluorescence intensity of algae is shown versus the yellow fluorescence intensity. The heated sample is the heat treatment sample of algae. Eluate A and eluate B denote solutions with respective concentrations of eluate A and eluate B of 50 vol%.

	I	II	III	IV
Control	96.81 ± 2.60	2.40 ± 2.91	0.73 ± 0.31	0.07 ± 0.08
Heated sample	0.28 ± 0.36	97.27 ± 0.81	2.59 ± 0.42	0.02 ± 0.03
Eluate A	98.15 ± 0.31	0.17 ± 0.08	1.67 ± 0.21	0.02 ± 0.03
Eluate B	98.13 ± 0.24	0.52 ± 0.51	1.29 ± 0.33	0.05 ± 0.05

Table 3. Distribution of untreated *Chlorella* and treated with heat or eluate from slag referred from earlier reports [7, 9].

It remains unclear why algae in media containing eluate proliferated more than algae in media without eluate (**Figure 5**). In general, the growth and proliferation of photosynthetic organisms, such as land plants and algae, depend strongly on photosynthetic efficiency. Photosynthesis is divisible mainly into two metabolizing systems: light-dependent reactions, which harvest light energy from sunlight and which perform electron transport; and the Calvin cycle, which performs CO_2 fixation to synthesize glucose. This study particularly examined the concentrations of CO_2, which are related to the Calvin cycle, because all experiments in this study were conducted under constant light conditions [11, 13]. This CO_2(aq) can be detected as infrared absorption near 2350 cm^{-1} using FT-IR [11, 13], which is identical to the infrared absorption attributable to anti-symmetric stretching of CO_2 [39]. To examine the relation between algal proliferation and the concentration of CO_2(aq) under treatment of algae with slag eluate, this study directly evaluated CO_2(aq) in medium containing eluate using FT-IR [11, 13]. The result shows that both eluates had higher infrared absorption identical to CO_2 than that of controls without slag eluate [11, 13]. Next, concentrations of CO_2(aq) under each test condition were quantified from the amounts of total carbonate species using a diaphragm-type electrode to measure CO_2(aq) and from calculation using the Henderson–Hasselbalch equations (**Figure 7**) [11, 13]. The result also showed that concentrations of CO_2(aq) in media containing eluate were higher than those of control samples. Moreover, the concentration in the medium containing eluate B had higher concentrations than that in eluate A. Speculating based on these obtained data, the addition of slag eluate appears to improve aqueous environments for photosynthetic organisms. It might facilitate algal photosynthesis more than CA medium alone. Consequently, increasing concentrations of CO_2(aq) by adding slag eluates induced greater algal proliferation than that in the control sample (**Figure 5**). **Figure 5** shows that this study also stumbled on the fact that the addition of eluate B to the culture medium induced greater proliferation of algae than that of eluate A. Accounting for the different concentrations of CO_2(aq) between eluate A and eluate B, the greater effects of eluate B than those of eluate A on algal proliferation might also be explained.

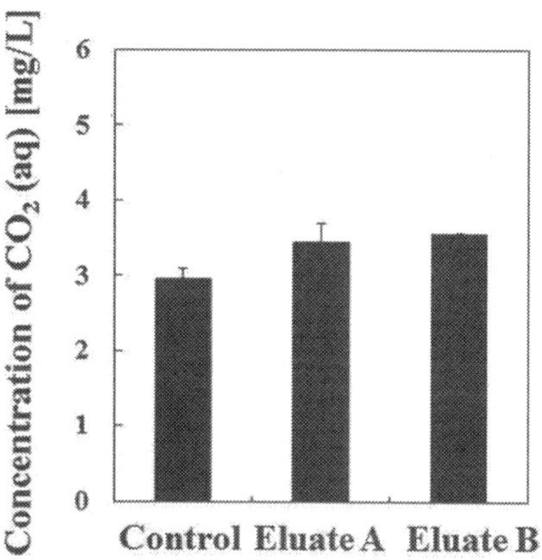

Figure 7. Concentrations of CO_2(aq) in each solution modified from the literature [11, 13]. Eluate A and eluate B, respectively, denote solutions in which concentrations of eluate A and eluate B were 70 vol%.

Formation of the greater concentrations of CO_2(aq) by addition of slag eluate than the control condition has been discussed in the literature [11, 13]. Ca^{2+} and Mg^{2+} are highly reactive substrates with the carbonate species. These elements conveniently existed in higher contents of eluates than other elements (see **Table 2**). As portrayed in **Figure 4**, the fraction of HCO_3^- is the highest content of the carbonate species at pH 7.2, which were experimental conditions used for this study. Therefore, Ca^{2+} in eluate, for instance, might be presumably reacted with HCO_3^- as described in the following Eq. (6).

$$Ca^{2+} + 2HCO_3^- \rightarrow Ca(HCO_3)_2(aq) \tag{6}$$

$Ca(HCO_3)_2$ can interfere in the chemical equilibrium of the carbonate species as HCO_3^- because $Ca(HCO_3)_2$ is ionized completely. Here, the ratios of concentrations of the respective carbonate species must be constant in solution. Consequently, the increase in HCO_3^- concentrations in solution prompts Eq. (2) to proceed leftward, resulting in increasing concentrations of $CO_2(aq)$. Increased $CO_2(aq)$ might be consumed by algae as a raw material of photosynthesis. Assessment of the transitional change of concentrations of Ca^{2+} and Mg^{2+} revealed no significant difference between the concentrations of these elements before and after incubation of algae with eluates, even at 7 days after incubation (**Table 4**). These obtained data support the hypothesis that the addition of slag eluate, particularly Ca^{2+} in eluate, increases the amounts of the total carbonate species and that the increase in the total amounts of the carbonate species by adding slag eluates triggers the potential increase of $CO_2(aq)$, eventually inducing algal proliferation.

	Before incubation	**After incubation**	***p* value (%)**
Concentration of Ca^{2+} (mg/L)			
Eluate A	9.904	8.945 ± 0.917	$p > 0.05$
Eluate B	10.464	9.763 ± 1.056	$p > 0.05$
Concentration of Mg^{2+} (mg/L)			
Eluate A	2.602	3.362 ± 0.381	$p > 0.05$
Eluate B	2.742	2.931 ± 0.075	$p > 0.05$

Table 4. Concentration of alkarin earth elements before and after incubation of *Chlorella* with each eluate, referred from references [7, 9].

The biochemical importance of $CO_2(aq)$ increased by slag eluates was discussed as described in earlier reports [11, 13]. In general, the increase of $CO_2(aq)$ can promote carbon dioxide assimilation by photosynthesis on an algal cellular level. However, the present concentrations of $CO_2(gas)$ in air determine the concentrations of $CO_2(aq)$ that can be dissolved in water. Consequently, the concentrations of $CO_2(gas)$ are regarded as a rate-determining factor of photosynthesis [40]. Its action as the rate-determining factor of $CO_2(gas)$ is common not only to land plants using $CO_2(gas)$ directly but also to aquatic organisms such as phytoplankton using $CO_2(aq)$ dissolved into water. This study indicates that slag components in solution did not cause toxicity to *Chlorella* and that the eluates were able to increase concentrations of $CO_2(aq)$, which functions as the rate-determining factor of photosynthetic organisms in the aqueous environment. This feature of slags in aqueous environment is regarded as beneficial for aqueous photosynthetic organisms including algae. This study was performed using *Chlorella* spp. as the model organism of algae and aqueous photosynthetic organisms. To present the usefulness of slag eluate and their components in algal culture more precisely, additional experiments must be done using photosynthetic organisms other than *Chlorella*.

6. Conclusion

For autotrophic algal biorefineries, biofuel materials, and bioremediation materials, it is important to evaluate features of interesting algae and their qualities in culture. In this study, *Chlorella* was cultured with CA medium containing an eluate from steel slag as an experimental factor and sources of trace metals. Results obtained from this study can be summarized as the following: (1) Slag eluates used for this study met the EQSs for soil pollution, effluent and drinking water, except for Se in eluate A. (2) Analyses of algae treated with the eluate revealed that the eluate from used slag induced neither lethality nor growth inhibition. (3) In addition to cell counting using hemocytometry, bioassay using FCM was able to estimate vigorous and aberrant algal growth simultaneously and graphically. (4) In contrast to comparison of control algae with the heat stress, the distribution of algae treated with the eluate was appropriately similar to that of control, suggesting that the eluate from slags did not give rise to algae stresses. (5) The addition of eluates to the medium increased the concentrations of CO_2(aq). The increased CO_2(aq), which was found to be related to the presence of Ca^{2+} in eluates, improved the rates of photosynthesis and algal proliferation.

Acknowledgements

This research was mainly supported by a Grant for Young Scientists from the Iron and Steel Institute of Japan and partly by a Grant-in-Aid for Young Scientists (B) (Grant No. 26870820).

Author details

Toshiyuki Takahashi

Address all correspondence to: mttaka@cc.miyakonojo-nct.ac.jp

Department of Chemical Science and Engineering, National Institute of Technology, Miyakonojo College, Miyakonojo, Miyazaki, Japan

References

[1] Spolaore P, Joannis-Cassan C, Duran E, Isambert A. Commercial applications of microalgae. J Biosci Bioeng. 2006;101:87–96. DOI: 10.1263/jbb.101.87

[2] Arashida R. Characteristics of the microalgae Euglena and its applications in foods and ecological fields. Japan Soc. Photosynthesis Res. 2012;22:33–38.

[3] Chisti Y. Biodeisel from microalgae. Biotechnol. Adv. 2007;25:294–306. DOI: 10.1016/j.biotechadv.2007.02.001 DOI:10.1016/j.biotechadv.2007.02.001#doilink

[4] Mallick N. Biotechnological potential of immobilized algae for wastewater N, P and metal removal: A review. BioMetals. 2002;15:377–390. DOI: 10.1023/A:1020238520948

[5] Hameed MSA, Ebrahim OH. Biotechnological potential uses of immobilized algae. Int. J. Agric. Biol. 2007;9:183–192.

[6] Borowitzka MA, Moheimani NR. Open Pond Culture Systems. In: Borowitzka MA, Moheimani NR, editors. Algae for Biofuels and Energy. Springer: Netherlands; 2012. pp. 133–152. DOI: 10.1007/978-94-007-5479-9_8

[7] Nriagu JO, Pacyna JM. Quantitative assessment of worldwide contamination of air, water and soils by trace metals. Nature. 1988;333:134–139. DOI: 10.1038/333134a0

[8] Cochran JK, Hirschberg DJ, Wang J, Dere C. Atmospheric deposition of metals to coastal waters (Long Island Sound, New York U.S.A): Evidence from saltmarsh deposits. Estuar. Coast. Shelf Sci. 1998;46: 503–522. DOI: 10.1006/ecss.1997.0299; DOI: 10.1006/ecss.1997.0299#_blank#Persistent link using digital object identifier

[9] Kaplan D. Absorption and adsorption of heavy metals by microalgae. In: Richmond A, Hu Q, editors. Handbook of Microalgal Cultre: Applied Phycology and Biotechnology, Second Edition, John Wiley & Sons, Ltd. Oxford, UK; 2013. pp. 602–611. DOI: 10.1002/9781118567166.ch32

[10] Takahashi T. Direct evaluation of endosymbiotic status in *Paramecium bursaria* using capillary flow cytometry. Cytometry Part A. 2014;85:911–914. DOI: 10.1002/cyto.a.22562

[11] Takahashi T, Yokoyama S. Bioassay of components eluted from electric arc furnace steel slag using microalgae *Chlorella*. (in Japanese) Tetsu-to-Hagané. 2015;101:506–514. DOI: 10.2355/tetsutohagane.TETSU-2014-130

[12] Takahashi T. Application of phytoplankton. In: Kanematsu H, Barry DM, editors. Corrosion Control and Surface Finishing – Environmentally Friendly Approaches. Springer: Japan; 2016. pp. 213–224. DOI: 10.1007/978-4-431-55957-3_19

[13] Takahashi T, Yokoyama S. Bioassay of components eluted from electric arc furnace steel slag using microalgae *Chlorella*. ISIJ Int. 2016;56:1495–1503. DOI: 10.2355/isijinternational.ISIJINT-2015-539

[14] Gerashchenko BI, Kosaka T, Hosoya H. Optical compartmentation of vegetating algae species as a basis for their growth-specific characterization. Cytometry. 2002;48:153–158. DOI: 10.1002/cyto.10120

[15] Gerashchenko BI, Takahashi T, Kosaka T, Hosoya H. Life cycle of unicellular algae. In: Current Protocols in Cytometry. John Wiley & Sons, Inc. New Jersey; 2010;52:11.19.1–11.19.6. DOI: 10.1002/0471142956.cy1119s52

[16] Takahashi T. Simultaneous evaluation of life cycle dynamics between a host paramecium and the endosymbionts of *Paramecium bursaria* using capillary flow cytometry. Sci. Rep. 2016;6:31638. DOI: 10.1038/srep31638

[17] Shinozaki H, Takahashi K. Iron and steel making slag and its applications for road constructions. Bull. Iron Steel Inst. Jpn. (Ferrum). 2010;15:177–182.

[18] Takahashi K, Yabuta K. Road temperature mitigation effect of "Road Gool" a water-retentive material using blast furnace slag. JFE Technical Rep. 2009;(19):58–62.

[19] Nippon Slag Association. Iron and steel slag. Tekko Sulagu Toukeinenpou. 2012;2012:2–10.

[20] Miyata Y, Sato Y, Shimizu S, Oyamada K. Environmental improvement in the sea bottom by steelmaking slag. JFE Technical Rep. 2009;13:41–45.

[21] Takahashi T, Ogura Y, Ogawa A, Kanematsu H, Yokoyama S. An effective and economic strategy to restore acidified freshwater ecosystems with steel industrial byproducts. J. Wat. Environ. Technol. 2012;10:347–362. DOI: 10.2965/jwet.2012.347

[22] Chlou CS, Chang CF, Chang CT, Shle JL, Chen YH. Mineralization of reactive black 5 in aqueous solution by basic oxygen furnace slag in the presence of hydrogen peroxide. Chemosphere. 2006;62:788–795. DOI: 10.1016/j.chemosphere.2005.04.072

[23] Kang WH, Hwang I, Park JY. Dechlorination of trichloroethylene by a steel converter slag amended with Fe (II). Chemosphere. 2006;62:285–293. DOI: 10.1016/j.chemosphere.2005.05.011

[24] Xue Y, Wu S, Hou H, Zha J. Experimental investigation of basic oxygen furnace slag used as aggregate in asphalt mixture. J. Haz. Mat. 2006;16:261–268. DOI: 10.1016/j.jhazmat.2006.02.073

[25] Chaurand P, Rose J, Briois V, Olivi L, Hazemann JL, Proux O, Domas J, Bottero JY. Environmental impacts of steel slag reused in road construction: A crystallographic and molecular (XANES) approach. J. Haz. Mat. 2007;139:537–542. DOI: 10.1016/j.jhazmat.2006.02.060

[26] Stimson J, Chae GT, Ptacek CJ, Emelko MB, Mesquita MM, Hirata RA, Blowes DW. Basic oxygen furnace slag as a treatment material for pathogens: Contribution of inactivation and attachment in virus attenuation. Wat. Res. 2010;44:1150–1157. DOI: 10.1016/j.watres.2009.11.054

[27] Miki O, Kato T, Kusui T. Bioassay test on steelmaking slag leachate using *Vibrio fischeri* and *Tigriopus japonicus*. J. Jap. Soc. Wat. Environ. 2010;33:141–146. DOI: org/10.2965/jswe.33.141

[28] Yokoyama S, Suzuki A, NIK Hisyamudin BMN, Kanematsu H, Ogawa A, Takahashi T, Izaki M, Umemoto M. Serial batch elution of electric arc furnace oxidizing slag

discharged from normal steelmaking process into fresh water. ISIJ Int. 2010;50:630–638. DOI: org/10.2355/isijinternational.50.630

[29] Yokoyama S, Shimomura T, Hisyamudin MNN, Takahashi T, Izaki M. Influence of amount of oxidizing slag discharged from stainless steelmaking process of electric arc furnace on elution behavior into fresh water. J. Phys. Conf. Ser. 2012;352:012051. DOI: 10.1088/1742-6596/352/1/012051

[30] Krienitz L, Huss VAR, Bock C. *Chlorella*: 125 years of the green survivalist. Trends Plant Sci. 2015;20:67–69. DOI: 10.1016/j.tplants.2014.11.005

[31] Nishihara N, Horiike S, Takahashi T, Kosaka T, Shigenaka Y, Hosoya H. Cloning and characterization of endosymbiotic algae isolated from *Paramecium bursaria*. Protoplasma. 1998;203:91–99. DOI: 10.1007/BF01280591

[32] Yokoyama S, Suzuki A, Izaki M, Umemoto M. Elution behavior of electronic arc furnace oxidizing slag into fresh water. Tetsu-to-Hagané. 2009;95:434–443. DOI: org/10.2355/tetsutohagane.95.434

[33] Jencksand WP, Regenstein J. Ionization constants of acids and bases. In: Lundblad RL, MacDonald FM, editors. Handbook of Biochemistry Molecular Biology. Boca Raton, Florida, CRC Press; 2010. pp. 305–351. DOI: 10.1201/b10501-74

[34] Fujimori E, Inagaki K, Haraguchi H. Multielement determination of major-to-ultratrace elements in biological samples by analytical plasma spectrometry. Bunseki Kagaku (Anal. Chem.). 1999;48: 57–67.

[35] Hokazono E, Osawa S, Nakano T, Kawamoto Y, Oguchi Y, Hotta T, Kayamori Y, Kang D, Cho Y, Shiba K. Development of a new measurement method for serum calcium with chlorophosponazo-III. Ann. Clin. Biochem. 2009;46: 296–301.

[36] Badyal A, Sodhi KS, Pandey R, Singh J. Serum magnesium levels: A key issue for diabetes mellitus. JK Sci. 2011;13:132–134.

[37] Bose J, Babourina O, Rengel Z. Role of magnesium in alleviation of aluminium toxicity in plants. J. Exp. Bot. 2011;62:2251–2264. DOI: 10.1093/jxb/erq456

[38] Zheng L, Lan P, Shen RF, Li WF. Proteomics of aluminium tolerance in plants. Proteomics. 2014;14:566–578. DOI: 10.1002/pmic.201300252

[39] Arikawa Y, Oguma K, Tsunoda K, editors. Basic Master Series, Analytical Chemistry (in Japanese). Ohmsha Ltd. Tokyo, Japan; 2013; pp. 193.

[40] Yoda K. Effects of terrestrial ecosystems on the carbon dioxide concentration in the atmosphere. (in Japanese) Chikyukagaku (Geochemistry). 1982;16:78–85.

Chapter 3

Functional Fruits Through Metabolic Engineering

Luis Quiroz-Iturra, Carolina Rosas-Saavedra and
Claudia Stange Klein

Additional information is available at the end of the chapter

http://dx.doi.org/10.5772/67219

Abstract

Metabolic engineering is a main focus of many plant biotechnology programs that look for the production of novel plant varieties with improved human health benefits. Among the most interesting goals are those that are focused in the production of functional fruits. A fruit can be considered as functional if it produces additional benefits to human health and well-being, beyond nutrition. Fruits that present higher levels of beneficial compounds such as essential vitamins, antioxidants, and phytochemicals can be considered as functional as those compounds have long-term benefits in reducing the occurrence of certain diseases. Through the expression, silencing, or mutagenesis strategies, many functional fruit crops have been produced during the last 40 years. Novel plants produce higher amount of carotenoids, antocyanins, and folic acid in their fruits, as well as higher color, sweetness, flavor, and aroma. The improvement of postharvest and resistance to biotic and abiotic stress in commercial plants has been also enhanced as it can led to a better fruit production. Taken together, this chapter will present a revision of the main fruits that have been improved by means of metabolic engineering within the framework of functional foods and super foods.

Keywords: biotechnology, functional food, fruits, metabolic engineering

1. Introduction

Today, the quality of life is becoming one of the pivotal reasons for people to be concerned about its health. Also, the steady increase in life expectancy accompanied by the growing cost of health care and the increasing rate of metabolic disorders (heart disease, obesity, diabetes, and arthritis) are the factors to consider in terms of life quality. On the other hand, vast scientific evidence determines a pivotal link between diet and human health, showing the crucial role of nutrition in the prevention of chronic diseases, such as coronary heart diseases, cancer, neurodegenerative, and respiratory disease, along with aging. Therefore, this scenario had led to the development of functional foods as a recognized category of foods to be part of an international strategy to overcome diseases related to human diet and life style.

The first conceptual approach developed by the European Commission Concerted Action on Functional Food Science in Europe (FUFOSE), coordinated by the International Life Sciences Institute (ILSI) Europe established that "A food can be regarded as 'functional' if it is satisfactorily demonstrated to affect beneficially one or more target functions in the body, beyond adequate nutritional effects, in a way that is relevant to either an improved state of health and well-being and/or reduction of risk of disease. Functional foods must remain foods and they must demonstrate their effects in amounts that can normally be expected to be consumed in the diet: they are not pills or capsules but part of a normal food pattern" [1, 2]. Today, an universal accepted definition for the term functional food is not accorded yet [1, 3]. In fact, this concept has been defined several times [4], although in most countries there is not a legislative definition [3].

From a practical point of view, a whole fruit that contains sufficient quantities of beneficial components represents the simplest example of functional food and therefore may be considered as functional fruit. The main characteristic of functional fruits is their high content of bioactive compounds, which promote a state of health and well-being and/or reduce the risk of some diseases and may even be used to cure some illnesses. However, there is a slight difference between conventional and functional fruits, even for experts such as nutritionists, because many if not most fruits contain natural components that provide benefits beyond basic nutrition, such as lycopene in tomatoes [5] or anthocyanins in pomegranate [6]. Moreover, not all fruits can be considered as functional because the health benefits strongly depend on the absorption and transformation of nutrients (bioaccesibility) during gastrointestinal digestion [7]. The bioaccesibility of these compounds permits to have a clear idea of their potential bioavailability, term that involves the biological activity of these compounds.

To meet future demand for functional foods, the food industry must address critical challenges such as developing strategies to increase the yield of healthy compounds in fruits or to add nutritional value with new components to improve the quality standards aiming to maintain the well-being. The beneficial components of functional fruits can be enhanced through special growing conditions [8–10], through breeding techniques [11] or through metabolic engineering for delivering truly unique health benefits [12]. In this context, metabolic engineering plays a key role in developing functional fruits, in terms of being defined as "the direct improvement of production, formation, or cellular properties through the modification of specific biochemical reactions or the introduction of new ones with the use of recombinant DNA technology" [13, 14]. In plants, metabolic engineering can be used to develop functional foods and super foods by handling the flow of primary and secondary metabolic pathways, allowing the redirection of carbon flow toward products of interest or the synthesis of new products.

Nowadays, metabolic engineering has been used to produce new plant varieties with higher levels of valuable compounds such as pro-vitamin A, antocyanins, folic acid, antioxidants, as well as higher color, sweetness, flavor, and aroma. Many novel functional fruit crops have been generated during the last 40 years by using molecular strategies such as the over-expression, silencing, or mutagenesis of specific genes. The improvement of postharvest and resistance to biotic and abiotic stress in commercial plants has been also enhanced as it can lead to a better fruit production.

Taken together, in the following sections, we will review the progress in biotechnological approaches in developing functional fruits by describing strategies employed in metabolic engineering, and the characteristics that have been improved in several agronomic traits to insert novel functional fruits into the market.

2. Metabolic engineering strategies

In plants, metabolic engineering has been developed through different approaches. Within them, the most widely used are the gain-of-function by transgenesis (including cisgenia and intragenia) and the loss-of-function carried out through gene silencing and mutagenesis. These strategies are achieved by *Agrobacterium tumefaciens* or biobalistics transformation systems.

2.1. Production of transgenic plants that express a functional gene

Transgenesis is the process by which one (or more) exogenous gene (called transgene) is inserted into a living organism, giving them a new feature that has to be stable over next generations. Cisgenia and intragenia terms are used when the exogenous genes belong to the same or related plants, respectively [15]. The functionality of the transgene in the new plant is accomplished due to the fact that the genetic code of all living organisms is exactly the same. This means that a specific DNA sequence may encode the same protein in any living organism. Transgenesis itself is a process that occurs in nature without human intervention. The better example is taken from plants that are infected by the bacteria *A. tumefaciens*. This bacteria produces the disease known as "grown call" in which *Agrobacterium* induces the formation of a tumor in the stem of more than 140 species of Eudicotyledoneae. The symptoms are caused by inserting a DNA segment in a semirandom manner in the plant genome. This DNA segment (known as T-DNA, 'transfer DNA') codifies for plant hormones that induce the generation of tumors when produced together in high levels in plant cells [16]. This feature has allowed scientists to use a modified strain of *A. tumefaciens* for inserting genes of interest instead of the phytohormone-inducing-tumor genes [17]. For instances, several commercial transgenic plant varieties produced through Agrobacterium transformation are cultivated and consumed in around 25 countries in the world. Most of them are transgenic varieties of maize, soybean, cotton, and canola that are tolerant to herbicides and resistant to insects, among others. Actually, fruits have also been improved through this technique and will be described later.

2.2. Gene silencing of specific genes in plants

In the case of plant gene silencing, also termed RNA interference (RNAi) or post-transcriptional gene silencing (PTGS), a small fragment of 100–400 pb of the gene in antisense orientation is introduced into the plant, causing the degradation of the target RNA, diminishing thus the amount of mRNA and of the protein from 40 to 99%. The process for gene silencing has been described extensively and is stated through the degradation of a double-stranded RNA molecule (dsRNA) in the cell coming from the hybridization of the antisense RNA and the endogenous sense RNA of the target gene, which triggers the RNAi pathway [18, 19]. The fragments generated include small interfering RNAs (siRNA) of about 21–23 nucleotides in length that through the host RNA-induced silencing complex (RISC) allows the systemic degradation of the target mRNA. It is believed that siRNA system has evolved as a cellular defense mechanism against RNA viruses or to combat the proliferation of transposons within the cell [19]. Currently, siRNAs are now widely used to suppress the expression of specific genes and to assess gene function. Gene silencing is considered a mechanism of gene knockdown, where the expression of a gene is reduced at least 99% but not completely [20].

2.3. Plant mutagenesis

Several standardized procedures that induce mutations have been used for the production of new crops varieties of commercial interest [21]. This technique uses physical or chemical mutagenic agents. The chemical procedure, by using EMS (ethyl-methane sulfonate), is simpler to achieve and has demonstrated to be one of the most reliable inducers of mutagenesis [21]. EMS tends to generate random changes in nucleotides, generating single-nucleotide polymorphisms (SNPs) or deletions (indel) that affect the functionality of some gene(s) in the genome [22]. This is translated in the modification of phenotypes and/or physiological characters. Mutations are inherit events in any alive species and naturally and randomly happen around 50,000 times per year which genetically change from normal cells to mutated cells every day [23]. Many of those alterations are repaired, but when not, a mutation persists being a key piece in the evolution of the species. Therefore, chemical mutagenesis has been used for more than 60 years in breeding programs in the world. Plants generated by mutagenesis do not require long evaluation processes as transgenic or silenced plants and are accepted and introduced to the market more efficiently. Approximately 2965 induced mutagenesis cultivars such as crops that include wheat, barley, and rice, and among others have been generated and released during the last 40 years. Novel varieties of fruits, which include Kiwifruits, produced through EMS mutagenesis, have also been approved for commercialization [21–24].

One of the most innovative system, which has gained great impact few years ago in the field of metabolic engineering, is the genome editing system CRISPR/Cas. This is a versatile and effective tool for editing genomes in a site-specific manner [25, 26]. The CRISPR/Cas system is a natural defense mechanism in eubacteria and Aequeas against plasmids and viruses [27, 28]. For metabolic engineering, a chimeric guide RNA (gRNA) that contains 20 nucleotides must specifically bind to their target sequence in the DNA. The target sequence must also contain the protospacer adjacent motif (PAM) sequence that is recognized by Cas9, cutting 3–4 nucleotides upstream of the PAM sequence [25]. The most common editing events are small Indels (insertion or deletion) of 1–10 nt [29]. The CRISPR/Cas9 system has been effectively used as a tool for editing the genome of numerous plants including *Arabidopsis thaliana, Nicotiana tabacum, Oryza sativa, Zea mays, Glycine max, Triticum aestivum,* and citrus [29].

3. Metabolic engineering for fruit-trees improvement

There are several strategies for metabolic engineering of plants. Nevertheless, the nutritive fruits are normally produced from trees or from "recalcitrant" plants that are difficult to regenerate *in vitro* and to transform [30]. Usually, the success of genetic transformation depends on the success of the regeneration process for each plant species. This is influenced by several factors, such as the genotype of the variety, the source of explant, and the degree of tissue differentiation. Therefore, the tissue culture conditions must be optimized for each range of each crop independently [31]. Generally, two methods of tissue culture have been used for regeneration of transgenic plants: organogenesis and somatic embryogenesis. Organogenesis is the process in which the regeneration of a new seedling occurs directly from the explants while in somatic embryogenesis, the formation of embryos from somatic cells that are present in the explant tissue is produced [32]. Somatic embryogenesis has many advantages over organogenesis, including its potentially high rates of multiplication, the genetic uniformity among embryo clones, the potential for expansion in bioreactors, and cryopreserved through synthetic seeds [33]. Despite the above, somatic embryogenesis has not been standardized for most fruit species, and therefore, somatic organogenesis is normally carried out for transgenic plant regeneration.

3.1. Disease control in plants to improve fruit crops

Owing to the economic importance that represents fruit production in various countries, there have been many efforts to generate plants with increased resistance against diseases and pathogens, whether fungi, bacteria, or viruses. Today, only virus resistance is used commercially, which only represents a small proportion of all transgenic plants grown in the world. Most of the plant-infecting viruses are of single-stranded RNA+ type, which codify for a capsid protein, movement protein, and replicases, among others. Within the first cases of success in the generation of virus-resistant plants through metabolic engineering, there are those using viral capsid protein (CP) as the transgene [34]. In this case, plants expressing the CP gene confer viral resistance mediated by RNAi gene silencing when plants are infected by the specific virus.

In 2004, the first example of a genetically engineered horticultural crop that has made it to market was produced in Hawaii. The genetically engineered Rainbow and SunUp Papaya, which are resistant to the papaya ringspot virus (PRSV) were successfully commercialized and adopted by farmers in the Puna area of Hawaii [35]. In addition, after 20 years of testing and risk assessment in the laboratory, in greenhouse and field, plum 'HoneySweet' is now used as a GM crop resistant to Sharka disease caused by the plum pox virus (PPV), which has been validated for US cultivation [36]. PPV protection is based on RNAi, and resistance has been shown to be highly effective, stable, durable, and heritable as a dominant trait. Extensive testing has also demonstrated on the safety and the ability of the RNAi technology for fruit production [36]. Chandrasekaran et al. [37] developed a virus-resistant variety of cucumber (*Cucumis sativus L.*) by CRISPR/Cas9 technology. The Ipomovirus infects cucumber and produces a vein yellowing of the leaves. During the infection, the virus requires the plant eIF4E gene (eukaryotic initiation factor of translation 4E) to carry out the recognition and transcription of their genes [38]. With the CRISPR/Cas9 tool, mutations in the eIF4E gene were introduced in cucumber, and the transgenic plants present small deletions or single-nucleotide polymorphisms (SNPs) in the mutated region of eIF4E. By the culture of the plant to next generations, homozygous mutant progeny was selected. The obtained one showed immunity to the Ipomovirus Cucumber vein yellowing infection and increased resistance to Zucchini yellow mosaic virus and papaya ring spot mosaic virus-W. A system for cucumber virus resistance was generated for the first time, without the need to produce a transgenic organism [37].

Regarding the generation of plants resistant to fungal and bacterial diseases, the main strategies are the expression of resistant genes coding for resistance receptors (R) and those involved in the defense mechanisms such as pathogenic-related proteins (PR proteins), peptides, and antimicrobial metabolites, and genes involved in detoxification mechanisms [39]. The plant defense response is triggered by the recognition of pathogen avirulence factors (avr) by the resistance receptors (R) equipped in the host plant [40]. The avr and R interaction activates one or more signal transduction pathways and eventually triggers a local response termed hypersensitive response (HR) and a systemic acquired resistance (SAR). Both of them induced by the signal-molecule salicylic acid (SA) which permits the accumulation of PR proteins through the activation of a signal transduction machinery [41]. Arabidopsis NPR1 gene (PR gene nonexpresser) is well recognized as a key regulator of signal transduction by SA leading to SAR. The NPR1 overexpression in citrus showed a positive response to the increased citrus canker resistance, caused by the bacterial pathogen *Xanthomonas citri* subsp. *Citri* [42]. In apple (*Malus domestica*), transgenic plants overexpressing the MdNPR1 gene exhibited increased resistance to two important fungal pathogens of apple, *Venturia inaequalis* and *Gymnosporangium juniperi-virginianae* [43]. The expression of the R gene was achieved in apples to also overcome the infection by the fungus *V. inaequalis,* which causes Scabies disease which is one of the most serious diseases that hinder the apple crop production. The gene that confers resistance to this disease is termed Rvi6/Vf scab and is present originally in the *Malus floribunda* 821 wild species. This gene has been incorporated in different commercial apple cultivars by classical breeding. However, as *M. floribunda* 821 has not an edible and attractive fruit, the new breeded species that have the resistance gene Rvi6/Vf produce low-quality fruits that do not reach the market [44]. In order to obtain high-resistant species, the Rvi6 gene (formerly HcrVf2) was inserted in susceptible apple cultivar 'Gala', thereby obtaining a commercially attractive variety that is resistant to Scab disease [45]. Unlike other crops, this variety is considered a cisgenic line, as the Rvi6 gene was taken from another variety of apple.

The hydrolytic enzymes chitinase and glucanase, the best characterized class of PR proteins, are able to degrade the cell wall of pathogenic fungi invaders. PR proteins are important components of the response of plant defense against fungal and bacterial pathogens [46, 47]. Transgenic plants expressing genes encoding for chitinase and glucanase showed increased resistance to fungal diseases in many fruit plants [34]. Furthermore, the use of antimicrobial peptides expressed constitutively in plant tissues has been recommended for genetic engineering of plants to increase disease resistance against fungal and bacterial pathogens [39].

Defensins, one of the classic examples of small antimicrobial peptides, play an important role in the response of plant defense against fungi. Defensins produce the permeabilization of the membrane inhibiting the growth of the fungi through the interaction with membrane components of the fungus [48]. Two defensins genes derived from petunia (PhDef1 and PhDef2) were expressed in banana. *In vitro* and *ex vivo* assays clearly suggested that transgenic banana plants were resistant against the pathogenic fungus *Fusarium oxysporum* sp. cubense [49]. Moreover, antimicrobial proteins from other organisms, such as insects or animals, have been used to increase the resistance to pathogens. Some of these nonplant antimicrobial proteins, such as attacin or cecropin from *Hyalophora cecropia* and magainin from *Xenopus laevis*, showed antimicrobial activity and increased resistance to pathogens in transgenic fruit, such as apple, papaya, pear, potato, sugarcane, and grape [39, 50]. Another alternative is to enhance the production of phytoalexins, antimicrobial metabolites that contribute significantly to the resistance against pathogens [51]. Even so, the production of antimicrobial metabolites generally requires coordinated action of a number of biosynthetic enzymes, which means that many genes are required. This feature makes very difficult to increment antimicrobial metabolites to generate plant resistance varieties [39].

Usually, necrotrophic pathogens produce toxins and enzymes that degrade the cell wall to invade the plant cell for a successful infection. Detoxification and degradation of these phytotoxins by the generation of transgenic plants could provide an opportunity to improve resistance to several diseases [52]. However, the strategy to develop disease-resistant plants is not accessible because some phytotoxins are harmful to mammals, and the product of a detoxification reaction could remain toxic in the plant [39].

3.2. Abiotic stress tolerance in fruit crops

Resistance to abiotic stresses is a challenging goal to develop biotechnological fruit cultivars and varieties because many plants face rough conditions of drought, salinity, cold, and heat, among others. These environmental factors are significant plant stressors, and their effects on plant development and productivity are reflected in serious agricultural yield losses [11].

Survival of plants under adverse environmental conditions is realized by structural and metabolic changes into endogenous developmental programs. Therefore, methods for agronomic processes and crop improvement are required to enhance these adaptive responses. Efforts have been made to introduce traits with improved drought tolerance, but in many cases, the strategies involved the insertion of a wide range of genes into plants [53].

Drought, salinity, extreme temperatures, and oxidative stress are interconnected environmental stresses [54] that often activate similar cell signaling pathways [55, 56] and cellular responses, such as the accumulation of compatible solutes, production of stress proteins, and the up-regulation of anti-oxidants [57, 58]. Therefore, plant modification for abiotic-enhanced tolerance is mostly based on the manipulation of one or several genes that are either involved in signaling and regulatory pathways [59, 60] or that encode enzymes present in pathways leading to the synthesis of functional and structural protectants, such as osmolytes and antioxidants [61, 62] or that encode stress-tolerance-conferring proteins [63]. For example, the overexpression of SK3-type DHN gene (ShDHN) in transgenic tomato, which codes for a type of dehydrin (DHN), increased tolerance to drought and cold stresses and improved seedling growth under salt and osmotic stresses [63]. DHN is also known as Group 2 LEA (late embryogenesis abundant) proteins [64], and the overexpression of ShDHN in tomato accumulated more proline, maintained higher enzymatic activities of superoxide dismutase and catalase, and suffered less membrane damage under cold and drought stresses.

Another interesting example in abiotic tolerance is the transformation effect of an important rootstock for lemon, *Citrus macrophylla* W. that constitutively expresses the CBF3/DREB1A transcription factor from *Arabidopsis*. CBF3/DREB1A is a member of transcription factors induced on abiotic stress conditions [65]. Transgenic lemon lines showed normal development and, under salt stress, showed greater growth, better stomatal conductance, and similar accumulation of chloride and sodium in the leaves, in comparison with wild-type plants [66].

The adaptation to stress often affects metabolic and energy requirements that sometimes result in deleterious collateral effects such as yield penalty, which mask and limit its benefit to agriculture. In consequence, some authors have succeeded to enhance the abiotic stress tolerance of agricultural species by combining traditional and molecular breeding [67, 68] with the transformation of specific genes [54] such as those reported in this section. Therefore, these strategies have been applied to other species such as soybean, corn, cotton, and canola, which are currently on the market, although additional research is still necessary to evaluate stress resistance of fruit trees varieties in field trials under real stress conditions [68]. Even more, the problems of high costs on the development and releasing processes and safety requirements in regulatory demands must be solved in this kind of fruit crops.

4. Metabolic engineering for functional fruits development

4.1. Nutritional improvement in fruits

Fruits and their processed derivatives are important nutritional sources, not only for carbohydrates, but also for a wide range of secondary metabolites. That are beneficial to human health. Most important metabolites are carotenoids, flavonoids, and anthocyanins, which are widely known to be powerful antioxidants and anticancer agents. Therefore, since many years ago, a great effort has been done to increase the content of these metabolites in plants to improve the nutritional value of fruits.

4.2. Functional fruits with improved carotenoid content

Carotenoids are the second most abundant pigments found in nature, with more than 750 structurally different compounds responsible for yellow, orange, and red colors [69]. Carotenoids are metabolites synthesized in plants, algae, fungi, and yeasts, and some bacteria where they have photosynthetic, antioxidants, and/or photoprotectant functions [70]. In vertebrates, carotenoids are precursors of vitamin A, and they are also involved in the formation and maintenance of bones and retina. Mammals such as human are not able to synthesize vitamin A, and therefore, they have to include carotenoids in the diet [71]. From a pharmaceutical point of view, these pigments are used as nutritional supplements and antioxidants, highlighting by their protection against UV damage, anticarcinogenic properties, prevention of cardiovascular diseases, cataracts, and macular degeneration [70–73]. Owing to all these important features for human health, the improvement of enhancing carotenoids content in plants and fruits has been carried out since many years, and the production of new varieties of fruits with enhanced amount of carotenoids has been succeeded a few years ago.

The most recognized fruit produced by metabolic engineering is the "super banana," which is part of an Australian project [74]. The aim of this project is to increase the levels of pro-vitamin A in the pulp of commercial banana using metabolic engineering. The overexpression of Psy gene (Apsy2a) from wild banana Asupina, which accumulates carotenoid in the pulp, showed the highest increase in β-carotene levels in the transgenic super banana variety (**Figure 1A**) [75]. Psy gene encodes for the phytoene synthase (PSY), which is the first and the key step in the biosynthetic pathway of carotenoids. Currently, the super banana is in the human-feeding trial stage in the United States. The aim is to transfer the technology to East Africa, where bananas are one of the major basic foods and the levels of vitamin A deficiency are high [76]. Another fruit, which has been modified to improve carotenoid concentration by metabolic engineering, is the kiwifruit (*Actinidia deliciosa*). Transgenic kiwi lines with improved carotenoid content were generated by overexpression of geranylgeranyl diphosphate synthase gene (GGPS), which is part of the metabolic precursors route of carotenoids and Psy of *Citrus unshiu* [77]. Pons et al. showed that it is possible to increase the content of β-carotene in orange fruit (*Citrus sinensis*) through gene silencing of β-carotene hydroxylase gene (CsB-CHX) [78]. This gene is involved in the conversion of β-carotene into xanthophylls. Transgenic fruits showed a dark yellow (golden) phenotype (**Figure 1C**). The levels of β-carotene in transgenic fruit pulp were 36 times higher. Besides, *in vivo* studies performed in *Caenorhabditis elegans* suggested that antioxidant effect in golden fruit was 20% higher than that in conventional fruits [78]. Overexpression of key genes for enhancement of β-carotene synthesis has been extended to other fruits such as tomato (*Solanum lycopersicum*) [79, 80] or the Hong Kong kumquat (*Fortunella hindsii*) [81].

4.3. Functional fruits with improved antocyanin content

Anthocyanins are one of the most important water-soluble plant pigments. They are synthesized by the flavonoid branch in the phenylpropanoid pathway. In plants, anthocyanins are secondary metabolites involved in multiple processes, such as attracting pollinators, protection against damage from UV light, seed dispersal, and pathogen attack [84]. For humans, anthocyanins have taken great importance due to its antioxidant and anti-inflammatory effects [85]. Tomato, which is a red fruit, owes its color to the accumulation of carotenoids in the pulp and peel (mainly lycopene) [86]. However, the content of anthocyanins in tomato is very low. Metabolic engineering has been used to enrich the anthocyanin content in tomatoes. For instance, the expression of Delila (DEL) and Rosea1 (ROS1) transcription factors, from

Antirrhinum majus, in transgenic tomato allowed the increase of anthocyanin content in the fruit [83]. It was observed that DEL and ROS1 transcription factors were able to activate multiple genes related to anthocyanin biosynthesis, such as phenylalanine ammonia lyase (PAL) and flavonoid 3′5′ hydroxylase (F3′5′H). Thus, transgenic tomatoes exhibit intense purple coloration in the pulp and fruit peel as shown in **Figure 1E**, because they contain high concentration of anthocyanin [86]. Several studies on the regulation of the biosynthetic pathway of anthocyanins showed that the family of MYB transcription factors regulates the expression of genes involved in this pathway [87]. Espley et al. reported that greater accumulation of anthocyanins in apple fruits (**Figure 1B**) was obtained by transforming with the transcription factor *MdMYB10* [82]. Moreover, the expression of *FaMYB10* in strawberry generated an increase in anthocyanins in the root, leaf, and strawberry fruit [88].

Figure 1. Examples of transgenic fruits. (A) Fruits of wild banana (*left*) and transgenic golden banana fruits with higher levels of β-carotene (*right*) [75]. (B) Wild apple fruit (*left*) and transgenic fruit that accumulate higher levels of anthocyanins in the pulp (*right*) [82]. (C) Wild apple fruit which presents pulp oxidation (*left*) and arctic apple which has a higher resistance to oxidation of the pulp (*right*). (D) Wild orange fruit (*bottom*) and golden orange fruits with higher levels of β-carotene [78]. (E) Wild tomato fruits (*top*) and transgenic tomato fruits that accumulate higher levels of anthocyanins in the pulp [83].

4.4. Functional fruits with improved folic acid content

Folic acid, also known as vitamin B9, is part of the water-soluble vitamin B complex, which is necessary for the formation of structural proteins and hemoglobin [89]. Folate deficiency is considered as a worldwide problem because it is mainly caused by poor nutrition in poor countries. Folate deficiency during pregnancy causes premature births and babies with low weight and possible defects in neural tube development. In adults, the clearest sign of folate deficiency is anemia while children can also slow growth [89]. Folates are synthesized from pteridine, p-aminobenzoate (PABA) and glutamate precursor [90]. Diaz de la Garza et al. [90] developed transgenic tomatoes, which overexpressed the genes involved in the first steps in the biosynthesis of pteridine and PABA proteins: GTP cyclohydrolase I and amynodeoxychorismate synthase, respectively, in a fruit-specific manner [90]. The amount of folic acid contained in ripe tomato fruits was 25 times higher than in nontransgenic one, and the amount responds to the daily requirement for adult consumers with less than a standard serving.

5. Functional fruits as oral vaccines

Metabolic engineering has been used not only to increase its content of healthy secondary metabolites but also for generating edible oral vaccines. The strategy of oral vaccines offers several advantages over the classical system of injection, as oral vaccine is a cheaper strategy to produce immunization and is a more practical system to be implemented in universal vaccination programs [91]. An example is the strategy to immunize against the enterovirus 71 (EV71), which is known to cause seasonal epidemics of hand, foot, and mouth and can reach fatal neurological complications in young children. Tomato plants expressing the VP1 epitope and the coat protein of the enterovirus 71 (EV71) were produced to induce immunization when eaten [92]. Tomato fruit–expressing VP1 protein was firstly tested in mice, which presented an increment of specific IgA and IgG immunoglobulins against VP1. Besides, the serum from mice fed with transgenic tomato was able to neutralize EV71 infection in rhabdomyosarcoma cell culture and the proliferation of spleen cells in orally immunized mice was also enhanced by VP1 tomatoes, which activated both humoral and cellular immunity. The results of this study not only demonstrated the feasibility of using transgenic tomato as an oral vaccine to generate protective immunity against EV71 in mice but also the likelihood of vaccine development against other kinds of enterovirus [92].

Another example of oral vaccine aims to immunize against the enterobacteria *Yersinia pestis*, which is a gram-negative bacilli, anaerobic facultative and pathogenic to humans that causes pneumonic and bubonic pest. This pathogen affects mainly people in Africa, Asia, and Latin America. Owing to the increasing reports of the emergence of antibiotic-resistant *Y. pestis* strains, the need for a safer and cheaper vaccination system increases. Among all *Y. pestis* antigens, only the F1 and V antigens have generated immunogenicity in conventional vaccines. Alvarez et al. reported the expression of a fusion protein F1-V in tomato plants. The immunogenicity of transgenic tomatoes F1-V was tested in mice, and the immune response of mice vaccinated with antigens of bacterial origin (conventional system) and oral transgenic vaccine was compared. The results showed a similar level of immunization with both strategies [93].

6. Enrichment of organoleptic properties in fruits

Despite the best efforts in metabolic engineering of fruits which have been focused on increasing the nutritional value and defense against biotic and abiotic stress, this technology is also applied in handling the organoleptic properties of fruits such as color, texture, flavor, and aroma to get new varieties more attractive and pleasant to the consumer.

6.1. Color improvement in fruits

The color is a key feature of the quality of fruits and flowers and is often associated with carotenoids, flavonoids, and anthocyanins. As we described in the previous section (nutritional improvement), many fruits have been modified in their metabolism to increase these beneficial molecules for health, but most of them are also responsible for giving color to several organs of the plant. Taken the last example of the "super banana" and tomato with increased content in β-carotene, it is important to note that both fruits have an orange color in the pulp. On the other hand, fruits of tomato, apple, and strawberry modified for increasing anthocyanins accumulation have a more bluish (tomato) [83, 86] and red (apple and strawberry) [82, 88] pulp and peel (**Figure 1**). Thus, the organoleptic property of color can be modified by creating more striking and novel varieties.

6.2. Sweetness increment in fruits

Sweetness is one of the major determinants of the quality of fruits and generally depends on two factors, the composition and content of sugars. In plants, sugar also works as substrates in carbon metabolism and energy [94]. ADP-glucose pyrophosphorylase (AGPase), a key enzyme in the metabolic pathway from sucrose to starch, catalyzes the rate-limiting step of the biosynthesis of starch by generating the ADP glucose and inorganic pyrophosphate from glucose 1-phosphate and ATP. Transgenic plants of strawberry were developed by gene silencing using an antisense sequence for *FaGPS* gene, which codes for AGPase. A decrease in starch content and an increase in total soluble sugar content of 16–37% were obtained in transgenic fruits [95].

Currently, there are many alternative sweeteners that have been approved by the European Union regulators. Some of these are aspartame, saccharin, cyclamate, neohesperidin DC, acesulfame-K, and thaumatin. The first five compounds are low-molecular-weight molecules and are obtained by technology of traditional organic synthesis, although it should be noted that aspartame is an unnatural peptide. In contrast, thaumatin, is a natural protein that is produced normally in the *Thaumatococcus benth* plant, which is native to West Africa. In addition to thaumatin, there are several other sweet proteins in nature. Some of them have been isolated, purified, and characterized. Their genes have been cloned, and in some cases, recombinant versions of the natural protein have been obtained. Consequently, many of these sweet proteins can be used for the development of transgenic plants to enhance sweetness and fruit quality [96]. The tridimensional structure of thaumatin was determined at a high resolution and shows a marked homology with the PR-5 proteins (pathogenesis-related protein-5), which are involved in biotic stress defense [97, 98]. Despite their structural similarity, it has not been reported that any PR-5 proteins produce a sweet taste [99]. Thaumatin is about 100,000 times sweeter than sucrose on a molar basis (about 1600 times on a weight basis). Thus, the threshold sweetness value of thaumatin is about 50 nM [100]. The thaumatin II gene was expressed in cucumber plants. The transgenic fruits accumulating thaumatin II showed a sweet phenotype and a positive correlation between the levels of accumulation of thaumatin and the intensity of sweet taste [101]. In addition, the concentration of E,Z-2,6-nonadienal, which is the main molecular odorant in cucumber, was enhanced in the transgenic cucumber fruits. Thus, transgenic expression of the thaumatin II gene resulted not only in a sweeter taste of fruits compared to control but also in a greater aroma intensity [102]. Tomato, pear, and strawberry were also transformed with the thaumatin II gene, and they also showed a direct correlation between amount of protein expression and the increase in sweetness [103–105]. As a particular case, the strawberry expressing thaumatin II showed a significantly higher resistance to gray mold caused by *Botrytis cinerea* [105].

6.3. Aroma as an important feature in fruits

Aroma has an important influence at the moment of choosing foods. The threshold for human perception of a volatile molecule can be low as 0.007 μg/L in water [106]. Thus, both the unique combination of volatile compound and the specific proportions of each of the volatile components, determine the properties of flavor in fruits and other foods. Additionally, plant volatiles greatly influence pollination and fruit defense responses and are therefore critical for breeders. Thus, the aroma is presented as a complex mixture of a large number of volatile compounds, whose composition is species specific and often for the particular variety of a fruit [107]. Although different fruits share many aromatic characteristics, each fruit has a characteristic aroma, which depends on the combination of volatile compounds, concentration, and perception of volatiles. The most important aromatic compounds are derived from amino acids, lipids, phenols, and mono and sesquiterpenes [107]. Lewinsohn et al. showed that transgenic tomatoes that express the gene for S-linalool synthase (LIS), under the control of the fruit-specific promoter E8, are able to synthesize and accumulate linalool S-terpenoid and 8-hydroxylinalool compounds. No other phenotypic alterations were observed, including levels of other terpenoids such as γ- and α-tocopherols, lycopene, β-carotene, and lutein, and the results show that it is possible to improve levels of monoterpenes in fruit ripening by metabolic engineering [108]. Transgenic tomato with higher levels of other terpenoids was developed by overexpression of geraniol synthase (*GES*) gene from *Ocimum basilicum*, which catalyzed the synthesis of geraniol from geranyl diphosphate, under the direction of the polygalacturonase gene (PG) fruit-specific promoter, it was possible to increase the content of geraniol in tomato fruits, although pigments were decreased. This would indicate that geraniol accumulation occurs at the expense of the accumulation of lycopene probably because geranyl diphosphate is a common precursor for the synthesis of both metabolites. The aroma of transgenic tomatoes was stronger compared to nontransgenic ones. Aroma test was performed to clarify the matter by several panelists, showing that they preferred the aroma of transgenic tomatoes [109].

7. Improvement of the characteristics in post-harvest

A critical decision for fruit growers is the time to harvest a crop. The time depends on factors such as the time required to reach the market and the management in route. Therefore, harvest is carried out when it has reached "harvest maturity" in any type of fresh fruits.

All fruits continue their metabolic processes after harvest. Therefore, the maintenance of the postharvest life is important to avoid that the product become inedible. In other words, many efforts are focused on preserve certain fruit traits, including nutritional value, processing qualities, flavor, and shelf life until the development of the ideal condition for consumption. Otherwise, the time lag of the postharvest life has the risk to expose serious losses in an ever-growing market. Indeed, the postharvest losses of fruits and vegetables, including roots and tubers, reach almost 50% of the production in developing countries, and the ratio of wastage is highest among food products [110, 111]. Also more than 40% of the food products losses occur at postharvest and processing levels in developing countries [112].

Within the postharvest life, fruit ripening occurs through physiological and biochemical reactions that alter visual appearance, flavor, aroma, texture, and fruit firmness [113, 114]. Therefore, many breeders and researchers have studied the complexity of fruit ripening and the development of engineered plants with high quality levels of fruit production in terms of flavor, color, and aroma.

7.1. Inhibition of ripening

The inhibition of fruit ripening has been achieved by reducing ethylene production [115]. Inhibition of ethylene production can be carried out by downregulating genes encoding key enzymes in the biosynthetic pathway of ethylene [116, 117] or by diverging the metabolic flux away from ethylene synthesis through the overexpression of enzymes degrading its immediate precursor, the 1-aminocyclopropane-1-carboxylic acid (ACC) [118, 119]. Even though most tomatoes have successfully lower levels of ethylene and an extended shelf life, most of them also compromised fruit-quality traits. An interesting work employed RNAi technology in which three homologs of 1-aminocyclopropane-1-carboxylate (ACC) synthase (ACS) gene were silenced during the course of ripening. Engineered fruits exhibited delayed ripening, prolonged shelf life for ~45 days, and improved juice quality. Indeed, total soluble solids (TSS) recorded in RNAi-ACS tomatoes increased up to ~40–45% compared to control [120]. In melon and papaya (*Carica papaya*), inhibition of fruit ripening and shelf life extension have also been achieved by silencing of genes coding for the enzymes or regulators of the ethylene biosynthesis pathway [121, 122].

Through Targeting Induced Local Lesions in Genomes (TILLING) approach [123–126], a melon mutant was isolated, which showed delayed fruit ripening and rind yellowing and an increase of fruit firmness and shelf life. The missense mutation G194D occurred in a highly conserved amino acid position of the ethylene biosynthetic enzyme, ACC oxidase 1 (CmACO1), and was predicted to affect the enzymatic activity of CmACO1.

By using RNAseq analysis, it was recently reported that the Polycomb-group (PcG) protein multicopy suppressor of IRA1 (MSI1) negatively regulates a large set of fruit-ripening genes along with the MADS-box protein RIN (ripening inhibitor) and its regulons. In fact, the genetic manipulation of SlMSI1 and RIN transcription factor successfully prolonged the fruit shelf life in tomato [127]. This may be an optimal approach to improve the post-harvest life of functional fruits by employing high-throughput techniques and addressing multiple metabolic pathways such as light-signaling pathways, which have modulatory components to adjust pigmentation during ripening and could be a selective advantage for primeval fleshy fruited plants [128].

7.2. Reduction of softening in ripe fruits

Reduction in fruit firmness due to softening that accompanies ripening plays a major role in determining the cost factor because exacerbates damage during handling and shipping processes having also a pivotal effect on shelf life, palatability, consumer acceptance, and postharvest resistance to pathogens [129–131]. The excessive softening causes losses around 35–40% of fruits and vegetables produced by India known as the second largest producer of these crops in the world [110]. It is suggested that the cell wall modifications are the major determinant of fruit softening induced as a consequence of the increased levels of cell wall–degrading enzymes [132]. Some strategies to control fruit softening include the manipulation of genes coding cell wall–degrading enzymes such as polygalacturonase or β-galactosidase [133–136]. This is the case of the first genetically modified tomato with reduced levels of expression of polygalacturonase gene through PTGS that was marketed as Flavr Savr™ to remark the potentially positive effect on the flavor [137, 138]. Although this strategy was developed to improve the flavor traits, delayed ripening with an expanded shelf life also occurred in the transgenic tomatoes by delaying cell wall softening.

In nonclimateric fruits such as strawberry and capsicum, efforts to control fruit ripening based on slowing down the rate of fruit softening has been successfully achieved by targeting genes involved in cell wall modification. In strawberry (*Fragaria* × *ananassa* Duch.), the down expression of pectate lyase or the fruit-specific polygalacturonase (FaPG1) genes by antisense technology resulted in extended fruit firmness and postharvest shelf life [139–143].

However, the other studies on the suppression of the expression of cell wall–degrading enzymes have not enough impact in prevent softening of genetically engineer fruits [133, 144]. This may be due to the redundant functionality of components taking part of a complex metabolic process [114, 145]. Therefore, the improvement of fruit shelf life constitutes a strong challenge for the identification of new targets to achieve this goal.

7.3. Browning reduction in fruits

The postharvest storage and quality of fresh fruits are also affected by the enzymatic browning having negative effects on color, flavor, taste, and nutritional value. This reaction may be responsible for up to 50% of total losses of fruits and vegetables production [146]. Browning is triggered by the oxidation of phenolic compounds to quinones catalyzed by the polyphenol oxidase (PPO) enzyme [147, 148]. The subsequent nonenzymatic polymerization of the quinones results in the brown pigments formation that induces the postharvest deterioration [149]. This is particularly easy to appear in apples, which are highly susceptible to enzymatic browning and contain high levels of polyphenols [150, 151]. In order to reduce postharvest browning, the silencing of PPO gene in an apple was accomplished. The apples produced less PPO activity, and 50% of browning was inhibited compared to wild type control in golden delicious (GD) and granny Smith (GS) [152, 153]. Transgenic apples can keep the original color of the apple flesh when they are subjected to mechanical damage, such as bruising or slicing. This "nonbrowning" phenotype minimizes shrinkage caused by harvest and postharvest damage and also decreases the need for antibrowning compounds on cut fruit (**Figure 1C**). Similar approaches have been performed to reduce grape berry darkening [149], blackheart in pineapple [154], and the browning process in fruits of Yali pear [61].

8. Conclusion

Metabolic engineering generally involves the redirection of cellular metabolism by modifying the expression of genes and enzymes affecting to the regulatory functions within the cell. For a successful metabolic engineering, rate-limiting step is the target for the increase of specific molecules or newly introduced molecules.

Traditional strategies for modifying gene expression such as using *A. tumefaciens* for overexpressing and silencing genes will be continue to be used but the strategies to manipulate the gene expression have been gradually refined. For instance, the selection markers have been removed, and only the transgene or mutagenic effect remains in the plant. In case of using CRISPR/Cas vectors, this methodology induces site-specific mutations and can produce thereby specific knock-out plants with the absence of external DNAs requirements. Since the modified plants do not contain a transgene, they are not included in the category of GMO.

Even more sophisticated metabolomic tools based on biochemical, genetic, environmental, and developmental parameters will offer the possibility to study the production of metabolites through the improvement of primary and secondary metabolic pathways in fruits. The increasing number of plant genomes sequenced, and the availability of many molecular markers can be used to track candidate genes that are associated with the desired trait and feature. However, traditional plant breeding together with genetic engineering provides greater opportunities to develop fruit crops with the desired amount and/or composition of specific metabolites. The most relevant characteristics for genetic improvement of plants that bear economic interest were discussed in this chapter. These characteristics included higher resistance to biotic and abiotic stress and improvement of pre-harvest and post-harvest features to face those problems that affect the vast cultivars and cause a high degree of economic losses. On the other hand, nutritional and organoleptic improvement of functional fruits are in direct benefit for end consumers.

http://dx.doi.org/10.5772/67219

Despite the benefits of metabolic engineering, the development and marketing of genetically modified fruit plants are hampered by many regulatory and social barriers. From the biosafety and consumers point of view, the presence of selectable marker genes, which are essential for the initial selection of transgenic plants, is undesirable. Therefore, the production of transgenic fruit plants without markers is now an essential requirement for commercial exploitation. The techniques such as RNAi in rootstocks for virus silencing, cisgenesis, or intragenesis show great potential and greater acceptance when generating genetically modified organisms. Additionally, selection marker free plants may improve the confidence and bring the benefits of genetically modified products to consumers.

Acknowledgements

The Chilean Regular Fondecyt 1130245 and Fondef VIU 140049.

Author details

Luis Quiroz-Iturra, Carolina Rosas-Saavedra and Claudia Stange Klein*

*Address all correspondence to: cstange@uchile.cl

Department of Biology, Faculty of Science, Centre of Plant Molecular Biology, University of Chile, Santiago, Chile

References

[1] Diplock AT, Aggett PJ, Ashwell M, Bornet F, Fern EB, Roberfroid MB. Scientific concepts of functional foods in Europe. Consensus document. Br J Nutr. 1999;**81**:S1-27.

[2] Roberfroid MB. What is beneficial for health? The concept of functional food. Food Chem Toxicol. 1999;**37**(9–10):1039-41. DOI: 10.1016/s0278-6915(99)00080-0.

[3] Miranda JM, Anton X, Redondo-Valbuena C, Roca-Saavedra P, Rodriguez JA, Lamas A, et al. Egg and egg-derived foods: effects on human health and use as functional foods. Nutrients. 2015;**7**(1):706-29. DOI: 10.3390/nu7010706.

[4] Bigliardi B, Galati F. Innovation trends in the food industry: the case of functional foods. Trends Food Sci Technol. 2013;**31**(2):118-29. DOI: 10.1016/j.tifs.2013.03.006.

[5] Agarwal S, Rao AV. Tomato lycopene and its role in human health and chronic diseases. CMAJ. 2000;**163**(6):739-44.

[6] Johanningsmeier SD, Harris GK. Pomegranate as a functional food and nutraceutical source. Annu Rev Food Sci Technol. 2011;**2**:181-201. DOI: 10.1146/annurev-food-030810-153709.

[7] Cervantes-Paz B, Victoria-Campos CI, Ornelas-Paz JDJ. Absorption of carotenoids and mechanisms involved in their health-related properties. In: Stange C, editor. Carotenoids in Nature; Biosynthesis, Regulation and Function. Vol. 79: Springer International; 2016. pp. 415-54. DOI: 10.1007/978-3-319-39126-7_16.

[8] Bravo K, Sepulveda-Ortega S, Lara-Guzman O, Navas-Arboleda AA, Osorio E. Influence of cultivar and ripening time on bioactive compounds and antioxidant properties in Cape gooseberry (*Physalis peruviana L.*). J Sci Food Agric. 2015;**95**(7):1562-9. DOI: 10.1002/jsfa.6866.

[9] Flores-Felix JD, Silva LR, Rivera LP, Marcos-Garcia M, Garcia-Fraile P, Martinez-Molina E, et al. Plants probiotics as a tool to produce highly functional fruits: the case of phyllobacterium and vitamin C in strawberries. PLoS One. 2015;**10**(4):e0122281. DOI: 10.1371/journal.pone.0122281.

[10] Cardenosa V, Girones-Vilaplana A, Muriel JL, Moreno DA, Moreno-Rojas JM. Influence of genotype, cultivation system and irrigation regime on antioxidant capacity and selected phenolics of blueberries (*Vaccinium corymbosum L.*). Food Chem. 2016;**202**:276-83. DOI: 10.1016/j.foodchem.2016.01.118.

[11] Tester M, Langridge P. Breeding technologies to increase crop production in a changing world. Science. 2010;**327**(5967):818-22. DOI: 10.1126/science.1183700.

[12] Ricroch AE, Henard-Damave MC. Next biotech plants: new traits, crops, developers and technologies for addressing global challenges. Crit Rev Biotechnol. 2016;**36**(4):675-90. DOI: 10.3109/07388551.2015.1004521.

[13] Bailey JE. Toward a science of metabolic engineering. Science. 1991;**252**(5013):1668-75.

[14] Dudareva N, DellaPenna D. Plant metabolic engineering: future prospects and challenges. Curr Opin Biotechnol. 2013;**24**(2):226-8. DOI: 10.1016/j.copbio.2013.02.002.

[15] Holme IB, Wendt T, Holm PB. Intragenesis and cisgenesis as alternatives to transgenic crop development. Plant Biotechnol J. 2013;**11**(4):395-407. DOI: 10.1111/pbi.12055.

[16] Zupan J, Muth TR, Draper O, Zambryski P. The transfer of DNA from *Agrobacterium tumefaciens* into plants: a feast of fundamental insights. Plant J. 2000;**23**(1):11-28. DOI: 10.1046/j.1365-313x.2000.00808.x.

[17] Gelvin SB. Agrobacterium-mediated plant transformation: the biology behind the "gene-jockeying" tool. Microbiol Mol Biol Rev. 2003;**67**(1):16-37. DOI: 10.1128/mmbr.67.1.16–37.2003.

[18] Mocellin S, Provenzano M. RNA interference: learning gene knock-down from cell physiology. J Transl Med. 2004;**2**(1):39. DOI: 10.1186/1479-5876-2-39.

[19] Mahmood R, Ali I, Husnain T, Riazuddin S. RNA interference: the story of gene silencing in plants and humans. Biotechnol Adv. 2008;**26**(3):202-9. DOI: 10.1016/j.biotechadv.2007.12.002.

[20] Zaratiegui M, Irvine DV, Martienssen RA. Noncoding RNAs and gene silencing. Cell. 2007;**128**(4):763-76. DOI: 10.1016/j.cell.2007.02.016.

[21] Sikora P, Chawade A, Larsson M, Olsson J, Olsson O. Mutagenesis as a tool in plant genetics, functional genomics, and breeding. Int J Plant Genom. 2011;**2011**:314829. DOI: 10.1155/2011/314829.

[22] Wang TL, Uauy C, Robson F, Till B. TILLING in extremis. Plant Biotechnol J. 2012;**10**(7):761-72. DOI: 10.1111/j.1467-7652.2012.00708.x.

[23] Shu QY. Induced Plant Mutations in the Genomics Era. Rome: Food and Agricultural Organization of the United Nations; 2009. 458 p.

[24] Shu QY, Forster BP, Nakagawa H. Plant Mutation Breeding and Biotechnology. Cambridge, Mass: CABI International; 2012. 608 p.

[25] Mali P, Yang L, Esvelt KM, Aach J, Guell M, DiCarlo JE, et al. RNA-guided human genome engineering via Cas9. Science. 2013;**339**(6121):823-6. DOI: 10.1126/science.1232033.

[26] Cong L, Ran FA, Cox D, Lin S, Barretto R, Habib N, et al. Multiplex genome engineering using CRISPR/Cas systems. Science. 2013;**339**(6121):819-23. DOI: 10.1126/science.1231143.

[27] Sorek R, Lawrence CM, Wiedenheft B. CRISPR-mediated adaptive immune systems in bacteria and archaea. Annu Rev Biochem. 2013;**82**:237-66. DOI: 10.1146/annurev-biochem-072911-172315.

[28] Gasiunas G, Barrangou R, Horvath P, Siksnys V. Cas9-crRNA ribonucleoprotein complex mediates specific DNA cleavage for adaptive immunity in bacteria. Proc Natl Acad Sci U S A. 2012;**109**(39):E2579-86. DOI: 10.1073/pnas.1208507109.

[29] Gao J, Wang G, Ma S, Xie X, Wu X, Zhang X, et al. CRISPR/Cas9-mediated targeted mutagenesis in *Nicotiana tabacum*. Plant Mol Biol. 2015;**87**(1–2):99-110. DOI: 10.1007/s11103-014-0263-0.

[30] Gómez-Lim MA, Litz RE. Genetic transformation of perennial tropical fruits. In Vitro Cell Dev Biol Plant. 2004;**40**(5):442-9. DOI: 10.1079/ivp2004547.

[31] Litz RE, Padilla G. Genetic transformation of fruit trees. In: Schnell RJ, Priyadarshan PM, editors. Genomics of Tree Crops. New York, NY: Springer New York; 2012. pp. 117-53. DOI: 10.1007/978-1-4614-0920-5_5.

[32] Bhojwani SS, Dantu PK. Plant Tissue Culture: An Introductory Text. New Delhi; New York: Springer; 2013. 1 online resource p.

[33] Merkle SA, Dean JF. Forest tree biotechnology. Curr Opin Biotechnol. 2000;**11**(3):298-302. DOI: 10.1016/s0958-1669(00)00099-9.

[34] Rai MK, Shekhawat NS. Recent advances in genetic engineering for improvement of fruit crops. Plant Cell Tissue Organ Cult (PCTOC). 2014;**116**(1):1-15. DOI: 10.1007/s11240-013-0389-9.

[35] Gonsalves C, Lee DR, Gonsalves D. Transgenic Virus-Resistant Papaya: The Hawaiian 'Rainbow' was Rapidly Adopted by Farmers and is of Major Importance in Hawaii Today. APSnet Feature. American Pythopathological Society. 2004. DOI: 10.1094/APSnetFeature-2004-0804.

[36] Scorza R, Callahan A, Dardick C, Ravelonandro M, Polak J, Malinowski T, et al. Genetic engineering of plum pox virus resistance: 'honeysweet' plum—from concept to product. Plant Cell Tissue Organ Cult (PCTOC). 2013;**115**(1):1-12. DOI: 10.1007/s11240-013-0339-6.

[37] Chandrasekaran J, Brumin M, Wolf D, Leibman D, Klap C, Pearlsman M, et al. Development of broad virus resistance in non-transgenic cucumber using CRISPR/Cas9 technology. Mol Plant Pathol. 2016;**17**(7):1140-53. DOI: 10.1111/mpp.12375.

[38] Ling KS, Harris KR, Meyer JD, Levi A, Guner N, Wehner TC, et al. Non-synonymous single nucleotide polymorphisms in the watermelon eIF4E gene are closely associated with resistance to zucchini yellow mosaic virus. Theor Appl Genet. 2009;**120**(1):191-200. DOI: 10.1007/s00122-009-1169-0.

[39] Collinge DB, Jorgensen HJL, Lund OS, Lyngkjaer MF. Engineering pathogen resistance in crop plants: current trends and future prospects. Annu Rev Phytopathol. 2010;**48**:269-91. DOI: 10.1146/annurev-phyto-073009-114430.

[40] Heil M, Bostock RM. Induced systemic resistance (ISR) against pathogens in the context of induced plant defences. Ann Bot. 2002;**89**(5):503-12. DOI: 10.1093/aob/mcf076.

[41] Durrant WE, Dong X. Systemic acquired resistance. Annu Rev Phytopathol. 2004;**42**:185-209. DOI: 10.1146/annurev.phyto.42.040803.140421.

[42] Zhang X, Francis MI, Dawson WO, Graham JH, Orbović V, Triplett EW, et al. Over-expression of the *Arabidopsis* NPR1 gene in citrus increases resistance to citrus canker. Eur J Plant Pathol. 2010;**128**(1):91-100. DOI: 10.1007/s10658-010-9633-x.

[43] Malnoy M, Boresjza-Wysocka EE, Norelli JL, Flaishman MA, Gidoni D, Aldwinckle HS. Genetic transformation of apple (*Malus × domestica*) without use of a selectable marker gene. Tree Genet Genomes. 2010;**6**(3):423-33. DOI: 10.1007/s11295-009-0260-7.

[44] Dayton DF, Williams EB. Additional allelic genes in Malus for scab resistance of two reaction types. J Am Soc Horticult Sci. 1970;**95**:735-6.

[45] Joshi SG, Schaart JG, Groenwold R, Jacobsen E, Schouten HJ, Krens FA. Functional analysis and expression profiling of HcrVf1 and HcrVf2 for development of scab resistant cisgenic and intragenic apples. Plant Mol Biol. 2011;**75**(6):579-91. DOI: 10.1007/s11103-011-9749-1.

[46] Neeraja C, Anil K, Purushotham P, Suma K, Sarma P, Moerschbacher BM, et al. Biotechnological approaches to develop bacterial chitinases as a bioshield against fungal diseases of plants. Crit Rev Biotechnol. 2010;**30**(3):231-41. DOI: 10.3109/07388551.2010.487258.

[47] Ceasar SA, Ignacimuthu S. Genetic engineering of crop plants for fungal resistance: role of antifungal genes. Biotechnol Lett. 2012;**34**(6):995-1002. DOI: 10.1007/s10529-012-0871-1.

[48] De Coninck B, Cammue BPA, Thevissen K. Modes of antifungal action and in planta functions of plant defensins and defensin-like peptides. Fungal Biol Rev. 2013;**26**(4):109-20. DOI: 10.1016/j.fbr.2012.10.002.

[49] Ghag SB, Shekhawat UK, Ganapathi TR. Petunia floral defensins with unique prodomains as novel candidates for development of fusarium wilt resistance in transgenic banana plants. PLoS One. 2012;**7**(6):e39557. DOI: 10.1371/journal.pone.0039557.

[50] Cardoso SC, Barbosa-Mendes JM, Boscariol-Camargo RL, Christiano RSC, Filho AB, Vieira MLC, et al. Transgenic sweet orange (*Citrus sinensis L.* Osbeck) expressing the attacin A gene for resistance to *Xanthomonas citri* subsp. *Citri*. Plant Mol Biol Report. 2010;**28**(2):185-92. DOI: 10.1007/s11105-009-0141-0.

[51] Ahuja I, Kissen R, Bones AM. Phytoalexins in defense against pathogens. Trends Plant Sci. 2012;**17**(2):73-90. DOI: 10.1016/j.tplants.2011.11.002.

[52] Punja ZK. Genetic engineering of plants to enhance resistance to fungal pathogens—a review of progress and future prospects. Can J Plant Pathol. 2001;**23**(3):216-35. DOI: 10.1080/07060660109506935.

[53] Golldack D, Li C, Mohan H, Probst N. Tolerance to drought and salt stress in plants: unraveling the signaling networks. Front Plant Sci. 2014;**5**:151. DOI: 10.3389/fpls.2014.00151.

[54] Wang W, Vinocur B, Altman A. Plant responses to drought, salinity and extreme temperatures: towards genetic engineering for stress tolerance. Planta. 2003;**218**(1):1-14. DOI: 10.1007/s00425-003-1105-5.

[55] Knight H, Knight MR. Abiotic stress signalling pathways: specificity and cross-talk. Trends Plant Sci. 2001;**6**(6):262-7.

[56] Zhu JK. Salt and drought stress signal transduction in plants. Annu Rev Plant Biol. 2002;**53**:247-73. DOI: 10.1146/annurev.arplant.53.091401.143329.

[57] Cushman JC, Bohnert HJ. Genomic approaches to plant stress tolerance. Curr Opin Plant Biol. 2000;**3**(2):117-24.

[58] Vierling E, Kimpel JA. Plant responses to environmental stress. Curr Opin Biotechnol. 1992;**3**(2):164-70.

[59] Zhao TT, Zhang J, Liang LS, Ma QH, Chen X, Zong JW, et al. Expression and functional analysis of WRKY transcription factors in Chinese Wild Hazel, *Corylus heterophylla* fisch. PLoS One. 2015;**10**(8):e0135315. DOI: 10.1371/journal.pone.0135315.

[60] Wang RK, Li LL, Cao ZH, Zhao Q, Li M, Zhang LY, et al. Molecular cloning and functional characterization of a novel apple MdCIPK6L gene reveals its involvement in multiple abiotic stress tolerance in transgenic plants. Plant Mol Biol. 2012;**79**(1–2):123-35. DOI: 10.1007/s11103-012-9899-9.

[61] Li G, Qi J, Zhang Y, Gao Z, Xu D, Li H, et al. Construction and transformation for the antisense expression vector of the polyphenol oxidase gene in Yali pear. Front Agric China. 2011;**5**(1):40-4. DOI: 10.1007/s11703-010-1058-y.

[62] Chen X, Han H, Jiang P, Nie L, Bao H, Fan P, et al. Transformation of beta-lycopene cyclase genes from *Salicornia europaea* and Arabidopsis conferred salt tolerance in Arabidopsis and tobacco. Plant Cell Physiol. 2011;**52**(5):909-21. DOI: 10.1093/pcp/pcr043.

[63] Liu H, Yu C, Li H, Ouyang B, Wang T, Zhang J, et al. Overexpression of ShDHN, a dehydrin gene from *Solanum habrochaites* enhances tolerance to multiple abiotic stresses in tomato. Plant Sci. 2015;**231**:198-211. DOI: 10.1016/j.plantsci.2014.12.006.

[64] Battaglia M, Olvera-Carrillo Y, Garciarrubio A, Campos F, Covarrubias AA. The enigmatic LEA proteins and other hydrophilins. Plant Physiol. 2008;**148**(1):6-24. DOI: 10.1104/pp.108.120725.

[65] Oh SJ, Song SI, Kim YS, Jang HJ, Kim SY, Kim M, et al. Arabidopsis CBF3/DREB1A and ABF3 in transgenic rice increased tolerance to abiotic stress without stunting growth. Plant Physiol. 2005;**138**(1):341-51. DOI: 10.1104/pp.104.059147.

[66] Alvarez-Gerding X, Espinoza C, Inostroza-Blancheteau C, Arce-Johnson P. Molecular and physiological changes in response to salt stress in *Citrus macrophylla* W plants overexpressing *Arabidopsis* CBF3/DREB1A. Plant Physiol Biochem. 2015;**92**:71-80. DOI: 10.1016/j.plaphy.2015.04.005.

[67] Kasuga M, Liu Q, Miura S, Yamaguchi-Shinozaki K, Shinozaki K. Improving plant drought, salt, and freezing tolerance by gene transfer of a single stress-inducible transcription factor. Nat Biotechnol. 1999;**17**(3):287-91. DOI: 10.1038/7036.

[68] Dunwell JM. Transgenic approaches to crop improvement. J Exp Bot. 2000;**51**:487-96.

[69] Nisar N, Li L, Lu S, Khin NC, Pogson BJ. Carotenoid metabolism in plants. Mol Plant. 2015;**8**(1):68-82. DOI: 10.1016/j.molp.2014.12.007.

[70] Bartley GE, Scolnik PA. Plant carotenoids: pigments for photoprotection, visual attraction, and human health. Plant Cell. 1995;**7**(7):1027-38. DOI: 10.1105/tpc.7.7.1027.

[71] Farre G, Bai C, Twyman RM, Capell T, Christou P, Zhu C. Nutritious crops producing multiple carotenoids—a metabolic balancing act. Trends Plant Sci. 2011;**16**(10):532-40. DOI: 10.1016/j.tplants.2011.08.001.

[72] Landrum JT, Bone RA. Lutein, zeaxanthin, and the macular pigment. Arch Biochem Biophys. 2001;**385**(1):28-40. DOI: 10.1006/abbi.2000.2171.

[73] Michaud DS, Feskanich D, Rimm EB, Colditz GA, Speizer FE, Willett WC, et al. Intake of specific carotenoids and risk of lung cancer in 2 prospective US cohorts. Am J Clin Nutr. 2000;**72**(4):990-7.

[74] Technology QUo. Super Bananas—World First Human Trial. 2014. Available from: https://www.qut.edu.au/news/news?news-id=74075 [Accessed: 12 Oct 2016].

[75] Dale J. Golden bananas for Africa and Asia. 2013. Available from: http://www.birac.nic.in/webcontent/dr_james.pdf [Accessed: 13 Oct 2016].

[76] Ekesa B, Poulaert M, Davey MW, Kimiywe J, Van den Bergh I, Blomme G, et al. Bioaccessibility of provitamin A carotenoids in bananas (*Musa* spp.) and derived dishes in African countries. Food Chem. 2012;**133**(4):1471-7. DOI: 10.1016/j.foodchem.2012.02.036.

[77] Kim M, Kim SC, Song KJ, Kim HB, Kim IJ, Song EY, et al. Transformation of carotenoid biosynthetic genes using a micro-cross section method in kiwifruit (*Actinidia deliciosa* cv. Hayward). Plant Cell Rep. 2010;**29**(12):1339-49. DOI: 10.1007/s00299-010-0920-y.

[78] Pons E, Alquezar B, Rodriguez A, Martorell P, Genoves S, Ramon D, et al. Metabolic engineering of beta-carotene in orange fruit increases its *in vivo* antioxidant properties. Plant Biotechnol J. 2014;**12**(1):17-27. DOI: 10.1111/pbi.12112.

[79] D'Ambrosio C, Stigliani AL, Giorio G. Overexpression of CrtR-b2 (carotene beta hydroxylase 2) from *S. lycopersicum L.* differentially affects xanthophyll synthesis and accumulation in transgenic tomato plants. Transgenic Res. 2011;**20**(1):47-60. DOI: 10.1007/s11248-010-9387-4.

[80] Ralley L, Schuch W, Fraser PD, Bramley PM. Genetic modification of tomato with the tobacco lycopene beta-cyclase gene produces high beta-carotene and lycopene fruit. Z Naturforsch C. 2016;**71**(9–10):295-301. DOI: 10.1515/znc-2016-0102.

[81] Zhang J, Tao N, Xu Q, Zhou W, Cao H, Xu J, et al. Functional characterization of citrus PSY gene in Hongkong kumquat (*Fortunella hindsii* Swingle). Plant Cell Rep. 2009;**28**(11):1737-46. DOI: 10.1007/s00299-009-0774-3.

[82] Espley RV, Hellens RP, Putterill J, Stevenson DE, Kutty-Amma S, Allan AC. Red colouration in apple fruit is due to the activity of the MYB transcription factor, MdMYB10. Plant J. 2007;**49**(3):414-27. DOI: 10.1111/j.1365-313X.2006.02964.x.

[83] Butelli E, Titta L, Giorgio M, Mock HP, Matros A, Peterek S, et al. Enrichment of tomato fruit with health-promoting anthocyanins by expression of select transcription factors. Nat Biotechnol. 2008;**26**(11):1301-8. DOI: 10.1038/nbt.1506.

[84] Stintzing FC, Carle R. Functional properties of anthocyanins and betalains in plants, food, and in human nutrition. Trends Food Sci Technol. 2004;**15**(1):19-38. DOI: 10.1016/j.tifs.2003.07.004.

[85] Bowen-Forbes CS, Zhang Y, Nair MG. Anthocyanin content, antioxidant, anti-inflammatory and anticancer properties of blackberry and raspberry fruits. J Food Compos Anal. 2010;**23**(6):554-60. DOI: 10.1016/j.jfca.2009.08.012.

[86] Su X, Xu J, Rhodes D, Shen Y, Song W, Katz B, et al. Identification and quantification of anthocyanins in transgenic purple tomato. Food Chem. 2016;**202**:184-8. DOI: 10.1016/j.foodchem.2016.01.128.

[87] Gambino G, Gribaudo I. Genetic transformation of fruit trees: current status and remaining challenges. Transgenic Res. 2012;**21**(6):1163-81. DOI: 10.1007/s11248-012-9602-6.

[88] Lin-Wang K, Bolitho K, Grafton K, Kortstee A, Karunairetnam S, McGhie TK, et al. An R2R3 MYB transcription factor associated with regulation of the anthocyanin biosynthetic pathway in Rosaceae. BMC Plant Biol. 2010;**10**:50. DOI: 10.1186/1471-2229-10-50.

[89] Bailey LB. Folate in Health and Disease. 2nd ed. Boca Raton: Taylor & Francis; 2010. xvii, 583 p.

[90] Diaz de la Garza RI, Gregory JF, 3rd, Hanson AD. Folate biofortification of tomato fruit. Proc Natl Acad Sci U S A. 2007;**104**(10):4218-22. DOI: 10.1073/pnas.0700409104.

[91] Lal P, Ramachandran VG, Goyal R, Sharma R. Edible vaccines: current status and future. Indian J Med Microbiol. 2007;**25**(2):93-102.

[92] Chen HF, Chang MH, Chiang BL, Jeng ST. Oral immunization of mice using transgenic tomato fruit expressing VP1 protein from enterovirus 71. Vaccine. 2006;**24**(15):2944-51. DOI: 10.1016/j.vaccine.2005.12.047.

[93] Alvarez ML, Pinyerd HL, Crisantes JD, Rigano MM, Pinkhasov J, Walmsley AM, et al. Plant-made subunit vaccine against pneumonic and bubonic plague is orally immunogenic in mice. Vaccine. 2006;**24**(14):2477-90. DOI: 10.1016/j.vaccine.2005.12.057.

[94] Nookaraju A, Upadhyaya CP, Pandey SK, Young KE, Hong SJ, Park SK, et al. Molecular approaches for enhancing sweetness in fruits and vegetables. Sci Hortic. 2010;**127**(1):1-15. DOI: 10.1016/j.scienta.2010.09.014.

[95] Park JI, Lee YK, Chung WI, Lee IH, Choi JH, Lee WM, et al. Modification of sugar composition in strawberry fruit by antisense suppression of an ADP-glucose pyrophosphorylase. Mol Breed. 2006;**17**(3):269-79. DOI: 10.1007/s11032-005-5682-9.

[96] Masuda T, Kitabatake N. Developments in biotechnological production of sweet proteins. J Biosci Bioeng. 2006;**102**(5):375-89. DOI: 10.1263/jbb.102.375.

[97] Ko TP, Day J, Greenwood A, McPherson A. Structures of three crystal forms of the sweet protein thaumatin. Acta Crystallogr D Biol Crystallogr. 1994;**50**(Pt 6):813-25. DOI: 10.1107/S0907444994005512.

[98] Koiwa H, Kato H, Nakatsu T, Oda J, Yamada Y, Sato F. Crystal structure of tobacco PR-5d protein at 1.8 A resolution reveals a conserved acidic cleft structure in antifungal thaumatin-like proteins. J Mol Biol. 1999;**286**(4):1137-45. DOI: 10.1006/jmbi.1998.2540.

[99] Liu JJ, Sturrock R, Ekramoddoullah AK. The superfamily of thaumatin-like proteins: its origin, evolution, and expression towards biological function. Plant Cell Rep. 2010;**29**(5):419-36. DOI: 10.1007/s00299-010-0826-8.

[100] Van der Wel H, Loeve K. Isolation and characterization of thaumatin I and II, the sweet-tasting proteins from *Thaumatococcus daniellii* benth. Eur J Biochem. 1972;**31**(2):221-5. DOI: 10.1111/j.1432-1033.1972.tb02522.x.

[101] Szwacka M, Krzymowska M, Osuch A, Kowalczyk ME, Malepszy S. Variable properties of transgenic cucumber plants containing the thaumatin II gene from *Thaumatococcus daniellii*. Acta Physiologiae Plant. 2002;**24**(2):173-85. DOI: 10.1007/s11738-002-0009-5.

[102] Zawirska-Wojtasiak R, Goslinski M, Szwacka M, Gajc-Wolska J, Mildner-Szkudlarz S. Aroma evaluation of transgenic, thaumatin II-producing cucumber fruits. J Food Sci. 2009;**74**(3):C204-10. DOI: 10.1111/j.1750-3841.2009.01082.x.

[103] Bartoszewski G, Niedziela A, Szwacka M, Niemirowicz-Szczytt K. Modification of tomato taste in transgenic plants carrying a thaumatin gene from *Thaumatococcus daniellii* benth. Plant Breed. 2003;**122**(4):347-51. DOI: 10.1046/j.1439-0523.2003.00864.x.

[104] Lebedev VG, Taran SA, Shmatchenko VV, Dolgov SV. Pear transformation with the gene for supersweet protein Thaumatin Ii. Acta Hortic. 2002; 596: 199-202. DOI: 10.17660/ActaHortic.2002.596.27.

[105] Schestibratov KA, Dolgov SV. Transgenic strawberry plants expressing a thaumatin II gene demonstrate enhanced resistance to *Botrytis cinerea*. Sci Hortic. 2005;**106**(2):177-89. DOI: 10.1016/j.scienta.2005.03.016.

[106] Buttery RG, Seifert RM, Guadagni DG, Ling LC. Characterization of additional volatile components of tomato. J Agric Food Chem. 1971;**19**(3):524-9. DOI: 10.1021/jf60175a011.

[107] Schwab W, Davidovich-Rikanati R, Lewinsohn E. Biosynthesis of plant-derived flavor compounds. Plant J. 2008;**54**(4):712-32. DOI: 10.1111/j.1365-313X.2008.03446.x.

[108] Lewinsohn E, Schalechet F, Wilkinson J, Matsui K, Tadmor Y, Nam KH, et al. Enhanced levels of the aroma and flavor compound S-linalool by metabolic engineering of the terpenoid pathway in tomato fruits. Plant Physiol. 2001;**127**(3):1256-65. DOI: 10.1104/pp.010293.

[109] Davidovich-Rikanati R, Sitrit Y, Tadmor Y, Iijima Y, Bilenko N, Bar E, et al. Enrichment of tomato flavor by diversion of the early plastidial terpenoid pathway. Nat Biotechnol. 2007;**25**(8):899-901. DOI: 10.1038/nbt1312.

[110] Meli VS, Ghosh S, Prabha TN, Chakraborty N, Chakraborty S, Datta A. Enhancement of fruit shelf life by suppressing N-glycan processing enzymes. Proc Natl Acad Sci U S A. 2010;**107**(6):2413-8. DOI: 10.1073/pnas.0909329107.

[111] FAO. Save Food: Global Initiative on Food Loss and Waste Reduction. Rome, Italy. 2012. Available from: http://www.fao.org/save-food/resources/keyfindings/en [Accessed: 12 Oct 2016].

[112] FAO. Global Food Losses and Food Waste—Extent, Causes and Prevention. Rome, Italy. 2011. Available from: http://www.fao.org/docrep/014/mb060e/mb060e.pdf [Accessed: 12 Oct 2016].

[113] Brady CJ. Fruit ripening. Annu Rev Plant Physiol. 1987;**38**(1):155-78. DOI: 10.1146/annurev.pp.38.060187.001103.

[114] Brummell DA, Harpster MH. Cell wall metabolism in fruit softening and quality and its manipulation in transgenic plants. Plant Mol Biol. 2001;**47**(1–2):311-40. DOI: 10.1023/a:1010656104304.

[115] Stearns JC, Glick BR. Transgenic plants with altered ethylene biosynthesis or perception. Biotechnol Adv. 2003;**21**(3):193-210. DOI: 10.1016/s0734-9750(03)00024-7.

[116] Oeller PW, Lu MW, Taylor LP, Pike DA, Theologis A. Reversible inhibition of tomato fruit senescence by antisense RNA. Science. 1991;**254**(5030):437-9. DOI: 10.1126/science.1925603.

[117] Hamilton AJ, Lycett GW, Grierson D. Antisense gene that inhibits synthesis of the hormone ethylene in transgenic plants. Nature. 1990;**346**:284-7. DOI: 10.1038/346284a0.

[118] Klee HJ, Hayford MB, Kretzmer KA, Barry GF, Kishore GM. Control of ethylene synthesis by expression of a bacterial enzyme in transgenic tomato plants. Plant Cell. 1991;**3**(11):1187-93. DOI: 10.1105/tpc.3.11.1187.

[119] Good X, Kellogg JA, Wagoner W, Langhoff D, Matsumura W, Bestwick RK. Reduced ethylene synthesis by transgenic tomatoes expressing S-adenosylmethionine hydrolase. Plant Mol Biol. 1994;**26**(3):781-90. DOI: 10.1007/bf00028848.

[120] Gupta A, Pal RK, Rajam MV. Delayed ripening and improved fruit processing quality in tomato by RNAi-mediated silencing of three homologs of 1-aminopropane-1-carboxylate synthase gene. J Plant Physiol. 2013;**170**(11):987-95. DOI: 10.1016/j.jplph.2013.02.003.

[121] Ayub R, Guis M, Ben Amor M, Gillot L, Roustan JP, Latche A, et al. Expression of ACC oxidase antisense gene inhibits ripening of cantaloupe melon fruits. Nat Biotechnol. 1996;**14**(7):862-6. DOI: 10.1038/nbt0796-862.

[122] Nunez-Palenius HG, Cantliffe DJ, Huber DJ, Ciardi J, Klee HJ. Transformation of a muskmelon 'Galia' hybrid parental line (*Cucumis melo L.* var. reticulatus Ser.) with an antisense ACC oxidase gene. Plant Cell Rep. 2006;**25**(3):198-205. DOI: 10.1007/s00299-005-0042-0.

[123] Dahmani-Mardas F, Troadec C, Boualem A, Leveque S, Alsadon AA, Aldoss AA, et al. Engineering melon plants with improved fruit shelf life using the TILLING approach. PLoS One. 2010;**5**(12):e15776. DOI: 10.1371/journal.pone.0015776.

[124] McCallum CM, Comai L, Greene EA, Henikoff S. Targeted screening for induced mutations. Nat Biotechnol. 2000;**18**(4):455-7. DOI: 10.1038/74542.

[125] McCallum CM, Comai L, Greene EA, Henikoff S. Targeting induced local lesions IN genomes (TILLING) for plant functional genomics. Plant Physiol. 2000;**123**(2):439-42. DOI: http://dx.doi.org/10.1104/pp.123.2.439.

[126] Till BJ, Reynolds SH, Greene EA, Codomo CA, Enns LC, Johnson JE, et al. Large-scale discovery of induced point mutations with high-throughput TILLING. Genome Res. 2003;**13**(3):524-30. DOI: 10.1101/gr.977903.

[127] Liu DD, Zhou LJ, Fang MJ, Dong QL, An XH, You CX, et al. Polycomb-group protein SlMSI1 represses the expression of fruit-ripening genes to prolong shelf life in tomato. Sci Rep. 2016;**6**:31806. DOI: 10.1038/srep31806.

[128] Llorente B, D'Andrea L, Rodriguez-Concepcion M. Evolutionary recycling of light signaling components in fleshy fruits: new insights on the role of pigments to monitor ripening. Front Plant Sci. 2016;**7**:263. DOI: 10.3389/fpls.2016.00263.

[129] Brummell DA. Cell wall disassembly in ripening fruit. Funct Plant Biol. 2006;**33**(2):103-19. DOI: 10.1071/fp05234.

[130] Causier B, Kieffer M, Davies B. Plant biology. MADS-box genes reach maturity. Science. 2002;**296**(5566):275-6. DOI: 10.1126/science.1071401.

[131] Vicente AR, Saladié M, Rose JKC, Labavitch JM. The linkage between cell wall metabolism and fruit softening: looking to the future. J Sci Food Agric. 2007;**87**(8):1435-48. DOI: 10.1002/jsfa.2837.

[132] Fry SC. Primary cell wall metabolism: tracking the careers of wall polymers in living plant cells. New Phytologist. 2004;**161**(3):641-75. DOI: 10.1111/j.1469-8137.2004.00980.x.

[133] Brummell DA, Hall BD, Bennett AB. Antisense suppression of tomato endo-1,4-beta-glucanase Cel2 mRNA accumulation increases the force required to break fruit abscission zones but does not affect fruit softening. Plant Mol Biol. 1999;**40**(4):615-22. DOI: 10.1023/a:1006269031452.

[134] Minoia S, Boualem A, Marcel F, Troadec C, Quemener B, Cellini F, et al. Induced mutations in tomato SlExp1 alter cell wall metabolism and delay fruit softening. Plant Sci. 2016;**242**:195-202. DOI: 10.1016/j.plantsci.2015.07.001.

[135] Powell AL, Kalamaki MS, Kurien PA, Gurrieri S, Bennett AB. Simultaneous transgenic suppression of LePG and LeExp1 influences fruit texture and juice viscosity in a fresh market tomato variety. J Agric Food Chem. 2003;**51**(25):7450-5. DOI: 10.1021/jf034165d.

[136] Smith DL, Abbott JA, Gross KC. Down-regulation of tomato beta-galactosidase 4 results in decreased fruit softening. Plant Physiol. 2002;**129**(4):1755-62. DOI: 10.1104/pp.011025.

[137] Sheehy RE, Kramer M, Hiatt WR. Reduction of polygalacturonase activity in tomato fruit by antisense RNA. Proc Natl Acad Sci U S A. 1988;**85**(23):8805-9. DOI: 10.1073/pnas.85.23.8805.

[138] Krieger EK, Allen E, Gilbertson LA, Roberts JK, Hiatt W, Sanders RA. The Flavr Savr tomato, an early example of RNAi technology. HortScience. 2008;**43**(3):962-4.

[139] Jimenez-Bermudez S, Redondo-Nevado J, Munoz-Blanco J, Caballero JL, Lopez-Aranda JM, Valpuesta V, et al. Manipulation of strawberry fruit softening by antisense expression of a pectate lyase gene. Plant Physiol. 2002;**128**(2):751-9. DOI: 10.1104/pp.010671.

[140] Pose S, Paniagua C, Cifuentes M, Blanco-Portales R, Quesada MA, Mercado JA. Insights into the effects of polygalacturonase FaPG1 gene silencing on pectin matrix disassembly, enhanced tissue integrity, and firmness in ripe strawberry fruits. J Exp Bot. 2013;**64**(12):3803-15. DOI: 10.1093/jxb/ert210.

[141] Quesada MA, Blanco-Portales R, Pose S, Garcia-Gago JA, Jimenez-Bermudez S, Munoz-Serrano A, et al. Antisense down-regulation of the FaPG1 gene reveals an unexpected central role for polygalacturonase in strawberry fruit softening. Plant Physiol. 2009;**150**(2):1022-32. DOI: 10.1104/pp.109.138297.

[142] Quesada MA, Posé S, Santiago-Doménech N, Sesmero R, Molina MC, Mousa S, et al. Effect of silencing of cell wall degrading enzymes on strawberry fruit softening. Acta Hortic. 2009;**842**(842):931-4. DOI: 10.17660/ActaHortic.2009.842.206.

[143] Santiago-Domenech N, Jimenez-Bemudez S, Matas AJ, Rose JK, Munoz-Blanco J, Mercado JA, et al. Antisense inhibition of a pectate lyase gene supports a role for pectin depolymerization in strawberry fruit softening. J Exp Bot. 2008;**59**(10):2769-79. DOI: 10.1093/jxb/ern142.

[144] Giovannoni JJ, DellaPenna D, Bennett AB, Fischer RL. Expression of a chimeric polygalacturonase gene in transgenic rin (ripening inhibitor) tomato fruit results in polyuronide degradation but not fruit softening. Plant Cell. 1989;**1**(1):53-63. DOI: 10.1105/tpc.1.1.53.

[145] Giovannoni JJ. Genetic regulation of fruit development and ripening. Plant Cell. 2004;**16** S170-80. DOI: 10.1105/tpc.019158.

[146] Ferreira D, Tomoyuki K, Guerra MP. Enzymatic browning, polyphenol oxidase activity, and polyphenols in four apple cultivars: dynamics during fruit development. HortScience. 2010;**45**(8):1150-4.

[147] Murata M, Noda I, Homma S. Enzymatic browning of apples on the market: relationship between browning, polyphenol content, and polyphenol oxidase. Nippon Shokuhin Kagaku Kogaku Kaishi. 1995;**42**(10):820-6. DOI: 10.3136/nskkk.42.820.

[148] Mayer AM. Polyphenol oxidases in plants and fungi: going places? A review. Phytochemistry. 2006;**67**(21):2318-31. DOI: 10.1016/j.phytochem.2006.08.006.

[149] Thipyapong P, Stout MJ, Attajarusit J. Functional analysis of polyphenol oxidases by antisense/sense technology. Molecules. 2007;**12**(8):1569-95. DOI: 10.3390/12081569.

[150] Coseteng MY, Lee CY. Changes in apple polyphenoloxidase and polyphenol concentrations in relation to degree of browning. J Food Sci. 1987;**52**(4):985-9. DOI: 10.1111/j.1365-2621.1987.tb14257.x.

[151] Amiot MJ, Tacchini M, Aubert S, Nicolas J. Phenolic composition and browning susceptibility of various apple cultivars at maturity. J Food Sci. 1992;**57**(4):958-62. DOI: 10.1111/j.1365-2621.1992.tb14333.x.

[152] Murata M, Nishimura M, Murai N, Haruta M, Homma S, Itoh Y. A transgenic apple callus showing reduced polyphenol oxidase activity and lower browning potential. Biosci Biotechnol Biochem. 2001;**65**(2):383-8. DOI: 10.1271/bbb.65.383.

[153] Carter N. Petition for Determination of Nonregulated Status ArcticTM Apple (*Malus × domestica*) Events GD743 and GS784. United States Department of Agriculture (USDA) Animal and Plant Health Inspection Service (APHIS) Available at: http://wwwaphisusdagov/brs/aphisdocs/10_16101ppdf. 2012.

[154] Ko HL, Campbell PR, Jobin-Décor MP, Eccleston KL, Graham MW, Smith MK. The introduction of transgenes to control blackheart in pineapple (*Ananas Comosus L.*) cv. Smooth Cayenne by microprojectile bombardment. Euphytica. 2006;**150**(3):387-95. DOI: 10.1007/s10681-006-9124-5.

Chapter 4

Food Wastes as Valuable Sources of Bioactive Molecules

Sonia A. Socaci, Anca C. Fărcaş, Dan C. Vodnar and Maria Tofană

Additional information is available at the end of the chapter

http://dx.doi.org/10.5772/66115

Abstract

Food industry produces worldwide millions of tons of plant-derived wastes which can be exploited as sources of high-value components: proteins, fibres, polysaccharides, flavour compounds or different phytochemicals. These bioactive compounds can be valorised as functional ingredients in food, pharmaceutical, health care, cosmetic and other products. Using the recovered bioactive molecules as functional ingredients represents a sustainable alternative of food wastes exploitation as inexpensive source of valuable compounds, while developing innovative food and non-food products with health-promoting benefits and at the same time contributing to an efficient waste reduction management. This chapter gives an overview of the main classes of bioactive compounds recovered from food wastes and their potential applications as functional chemicals, without being exhaustive.

Keywords: bioactive compounds, functional ingredients, food waste exploitation, renewable resources, recovered biomolecules

1. Introduction

Large amounts of wastes are generated annually by the food industry, their efficient management and valorisation representing one of the main objectives of European Union (EU) actions against food waste and towards sustainable development [1, 2]. The Waste Framework Directive [3] emphasised the importance of prevention of waste generation and the exploitation of wastes by reuse and recycling. Thus, in the 'bioeconomy' concept, the possibilities of conversion of renewable biological resources into economically viable products are addressed. In 2014, the European Commission provided the definition for the term 'food waste' as '*food (including inedible parts) lost from the food supply chain, not including food diverted to material uses such as bio-based products, animal feed, or sent for re-distribution'* [4]. The processing by-products are also included among food waste, if these are not used for other high-value functions (e.g. animal feed and industrial uses). In this chapter, we address only the exploitation of plant-derived by-products as sources of bioactive compounds.

Until few decades ago, food wastes, if not discarded into environment, were mainly used as animal feed. Nowadays, this attitude towards wastes changed, especially due to the growing interest in protecting the environment but also due to the increasing awareness of the benefits deriving from their exploitation. The by-products resulted from the processing of raw vegetables contain sometimes appreciable amounts of bioactive compounds such as proteins, dietary fibres, polysaccharides, fatty acids, flavour compounds and phytochemicals (e.g. polyphenols) that can be extracted, purified, concentrated and reused as functional ingredients in food industry or other related sectors (e.g. pharmaceuticals, cosmetics and health-care products) [5, 6].

2. Bioactive compounds recovered from plant-derived wastes and their potential applications

2.1. General overview

The wastes generated from the food industry can be separated into two main categories: plant-derived wastes and animal-derived wastes. The animal-derived wastes can be divided in three subcategories: (i) meat products, (ii) fish and seafood and (iii) dairy products, whereas the plant-derived wastes can be classified into four subcategories: (i) cereals (e.g. rice bran, wheat bran and brewers' spent grain), (ii) root and tubers (e.g. potato peel, sugar beet and molasses), (iii) oil crops and pulses (e.g. sunflower seeds, soybean seed and olive pomace) and (iv) fruit and vegetables (e.g. orange peel, grape pomace, apple pomace, tomato skin and pomace) [5, 7]. We further focus only on the plant-derived wastes chemical characterisation in terms of composition and content in functional compounds. The plant-derived by-products and especially those from fruits, vegetables and oil crops processing are generated in large amounts, some of them being produced in millions of tons annually worldwide [5, 8–10]. Disposal of such quantities of waste represents a challenge and an environmental problem. Apart from being used as animal feeds or fertilisers, the research conducted in the last decades clearly showed that the by-products resulted from processing of plant materials contained valuable nutrients which could be exploited in the development and production of new functional ingredients [11–15].

There is a wide range of extraction techniques used for the isolation and purification of the bioactive compounds from plant-derived wastes, some of them being based on new emerging techniques. The development of new extraction methods as well as the optimisation of existing ones, in order to increase, for example, the extraction yield or the selectivity for a certain compound, or to improve the production of a natural bioactive compound from a waste, has seen a real upsurge in the last decade [16]. Nevertheless, there is no universal extraction

Compound class	Waste origin	By-product source	Extraction techniques	References
Proteins	Cereals	Brewers' spent grain	Ultrasonic-assisted extraction	[17]
			Sequential extraction of proteins and arabinoxylans	[18]
			Enzymatic-assisted extraction	[19]
	Oil crops	Rapeseed meal	Ultrasound-assisted aqueous extraction	[20]
		Sunflower meals	Alkaline solubilization and acid precipitation	[21]
		Hazelnuts meal	Solvent extraction (water, acetone)	[22]
		Canola meals	Alkaline solubilization and acid precipitation (Isoelectric precipitation)	[23, 24]
			Electro-activated solutions (non-invasive extraction method)	[25]
			Salt precipitation	[24]
		Palm kernel cake	Enzymatic hydrolysis	[26]
	Fruits and vegetable	Apricot kernel cake	Alkaline solubilization and acid precipitation	[27]
Polysaccharides	Cereals	Brewers' spent grain	Enzymatic hydrolysis	[28]
			Sequential extraction of proteins and arabinoxylans	[18]
			Acid hydrolysis	[29]
	Fruits and vegetables	Citrus peel and apple pomace	Subcritical water extraction	[30]
		Orange peel	Microwave extraction	[31]
Lipids	Cereals	Brewers' spent grain	Soxhlet extraction	[32]
	Fruit and vegetables	Grape seeds	Pressurized carbon dioxide extraction with compressed carbon dioxide as solvent and ethanol as co-solvent	[33]
			Supercritical fluid extraction	[34]
Polyphenols	Cereals	Brewers' spent grain	Alkaline hydrolysis	[35]
	Oil crops	Rapeseed	Ultrasound-assisted aqueous extraction	[20]
		Olive by-products	Continuous counter-current liquid-liquid extraction	[36]
			Chemical (acid) hydrolysis	[37]
		Sunflower meals	Mild-acidic protein extraction with adsorptive removal of phenolic compounds	[38]
	Fruits and vegetables	Tomato pomace and skin	Enzymatic-assisted extraction/solvent extraction	[16]
		Potato peels and tubers	Pressurized liquid extractor	[39]
			Solvent extraction (stirring)	[40]
			Ultrasound extraction	[41]
		Orange peels	Nanofiltration	[42]
		Forest fruits pomaces	Supercritical fluid extraction	[43]
		Apple pomace	Ultrasound extraction	[44]
		Grape seeds	Supercritical fluid extraction	[45]
Carotenoids	Fruits and vegetables	Tomato pomace and skin	Enzymatic-assisted extraction	[16]
		Citrus peel	Ultrasound extraction	[46]
		Sea buckthorn seeds	Supercritical carbon dioxide fluid extraction	[47]
Essential oils	Fruit and vegetables	Citrus peel	Solvent extraction, distillation, hydrodistillation	[48]

Table 1. Examples of bioactive compounds from plant-derived wastes and the employed extraction techniques.

technique for the bioactive compounds. When an extraction technique is chosen, several criteria have to be considered, such as waste composition, aggregation state, homogeneity, and so on. Also, plant-derived waste is prone to microbial degradation, so an appropriate way of preservation is necessary for its storage and further exploitation. One of the most common and economically feasible methods used for preservation is the drying of the waste and thus reducing the water content and lowering the microbiological activity [11].

In **Table 1**, examples of some of the most common extraction technique for the main classes of high-value compounds and their sources are given.

2.2. Proteins

Proteins are macronutrients with an important role in human nutrition, having high nutritional value. Nowadays, the consumers are more concerned about their health and are starting to realise the tight correlation between health and diet. The trend is towards vegetarianism, and thus finding new plant sources of protein is crucial for the food industry. For a by-product to be considered as a source of protein, it has to fulfil major requirements: to have high protein content and this protein to be quality protein (well-balanced essential amino acid composition) [12]. Also, the allergic or toxic substances that may be present in the by-product must be removed prior to its utilisation as source of protein.

The main wastes with a relatively high content of protein are the defatted meals obtained from oil industry, including sunflower, canola, rapeseed, but also palm and peanuts. The defatted by-products generated from oil refineries (oil cake, stem and grain husk) are not only good sources of proteins but are also available in large quantities and at a low cost.

Sunflower proteins have been extensively evaluated as food ingredients. Sunflower seeds content in proteins ranges between 10% and 27.1% (dry weight (DW) basis), thus making the sunflower oil cake a good source of quality protein. The sunflower protein isolate's or concentrate's characteristic is the relatively high content in phenolics, compounds that may alter the proteins' functional properties and their shelf life [49]. However, the current tendency is not to obtain protein isolates free of phenolics, but to keep these compounds into the isolates due to the antioxidant activity they exert. The protein concentrates containing different concentrations of phenolics were studied and the results showed that they have high water solubility, moderate water-holding capacity, emulsifying, foaming and gelation capacity similar to commercial isolates [21].

Another source of plant protein is the canola seeds. These seeds contain two main types of storage proteins: salt-soluble (cruciferin) and water-soluble (napin), the total protein content in the defatted canola meal being around 32% [24]. The concentration of proteins in canola protein isolates, when conventional direct alkaline extraction is used, ranged between 66% and 76% [23, 24], while using salt precipitation method may increase the concentration of proteins in isolates up to 93% [24]. There are new emerging non-invasive methods, such as electro-activated solutions, that can be used for the extraction of proteins from canola meals with better extraction yields by solubilising the proteins without damaging their native conformations and maintaining their functional properties [25].

Rapeseed stem, the residual biomass remaining after the extraction of oil, represents roughly 30% of the plant and may also be considered to be used for proteins' recovery. The protein concentration in the rapeseed stem extract, using a green solvent (water) in an enhanced ultrasound extraction, was up to 0.03 g BSA/100 g DW. The ultrasound-assisted extraction showed an increase in extractability and at the same offering the possibility of scaling up [20].

Functional proteins can also be extracted from hazelnut cake (contains up to 54.4% proteins). The isolated hazelnut meal protein was found to exert good antioxidant activity (158–461 mmol Trolox/kg), iron chelation (60.7–126.7 mmol EDTA/kg), antiproliferative activity on colon cancer cells (IC_{50}: 3.0–4.6 mg/ml) and good oil absorption (7.4–9.4 g/g) [22].

In the palm oil-producing countries (e.g. Indonesia and Malaysia), the palm kernel cake is one of the main by-products generated by food industry [26]. Palm kernel cake contains in average 15–21% crude protein, but it is deficient in lysine, methionine and tryptophan, and thus has a poor utility being usually used as feed for ruminants [50, 51]. Nevertheless, palm kernel cake is still a potential source of plant protein. The extracted protein isolates have a 68.50% protein concentration when alkaline extraction was used. Attempts in optimisation of extraction technology were carried out in order to transform the extracted protein into a bioactive plant protein (e.g. by enzymatic hydrolysis) by adding functional properties such as antioxidant function [26, 52].

Cereal origin wastes represent another potential source of bioactive molecules, including plant proteins. Brewers' spent grain is the main insoluble residue generated by the brewing industry. This by-product results after the production of wort and it mainly consists in barley grain husks with minor fractions of pericarp and endosperm [53]. Its chemical composition is dependent on several intrinsic and extrinsic factors (barley cultivar, harvest time, type of malt used in the brewing process, mashing conditions, etc.) [54], but regardless of these factors it contains appreciable amounts of valuable compounds (proteins, lipids, carbohydrates, polyphenols and minerals) that remain unexploited in the brewing process. Brewers' spent grain has a high content (18–35.4%, w/w) [18, 55, 56] of quality protein, with lysine accounting for 14.3% of total protein content [55]. The extraction of protein from brewers' spent grain may be performed by classical alkaline extraction, but recently new integrated processes are developed for a more efficient exploitation of this by-product. For example, simultaneous extraction of proteins and arabinoxylans by use of alkaline reagents directly from brewers' spent grain without any pre-treatment [18] has a great potential to be scaled up being an innovative environmental friendly process that allows the recycling of the reagents and at the same time saving 93% in costs [57]. The incorporation of chitosan into the brewers' spent grain protein had as result a composite film with antimicrobial and antioxidant activities which can be used in packaging materials for foods [58].

The apricot kernel press cake, the waste remaining after the oil extraction, contains 34.5% crude protein which may be valorised by as protein isolates. In this case, before the alkaline extraction of proteins, a pre-step of detoxification is required in order to remove the HCN present in the kernel cake. The obtained isolates had a protein concentration of 68.8% and fairly good functional properties, especially water and oil absorption capacity and foaming properties [27].

The proteins recovered from plant-derived wastes have several functional properties when incorporated in food products: emulsifying agents, film-forming properties, flavour binding, viscosity increase by binding the water and gelation properties. The recovered proteins are successfully used for food fortification, especially in meat and milk products, infant formulae, bakery products and pasta products [20, 22, 27, 59].

2.3. Polysaccharides

Polysaccharides are widely distributed in nature, with about 99% being located in plants and vegetables, the representative ones including starch, cellulose, hemicelluloses, pectin and inulin [60]. These compounds are also referred to as dietary fibre and can be divided into two categories based on their water solubility [61]:

1. insoluble dietary fibre—are insoluble in water and resistant to hydrolysis by digestive tract enzymes (cellulose, hemicelluloses, lignin—non-carbohydrate compounds);
2. soluble dietary fibres are soluble in water and well fermented by digestive tract enzymes (pectin, inulin, gums and mucilages).

In plants, polysaccharides have important functional roles: maintaining the living cell structure, and water binding or energy suppliers. These properties are exploited by the food industry and other related fields in the development of new food additives, functional ingredients or materials for bioactive molecules delivery and controlled release. Their suitability for pharmaceutical or medicinal uses is due to their innocuousness, biocompatibility, biodegradability and water solubility. Thus, there is an increasing and constant interest in finding new sources of plant-derived polysaccharides—the bioagro-waste streams being very promising in this sense [60, 62].

The fruit- and vegetable-processing sector produces wastes (peels, pulp and seeds) that are rich, low cost and sustainable sources of polysaccharides. After isolation and purification, the recovered polysaccharides may have manifold applications.

Pectin is a polysaccharide with a heterogeneous structure that depends on the plant origin, the part of the plant where it is located (peels, pulp, seed, etc.) and how it is extracted. The 'building block' is the uronic acid residue link through α-1-4-glycosidic bonds, forming a galacturonyl polymer backbone. The structural diversity of pectin provides a wide range of physico-chemical and functional properties (gelling, emulsifier, thickening agents, film-forming, water-holding, prebiotic activities, etc.) essential for food industry. According to the Joint Food and Agriculture Organization/World Health Organization (FAO/WHO) Expert Committee on Food Additives and the European Commission, a pectic polysaccharide must have a content of minimum 65% in galacturonic acid [60–63]. Wastes such as orange peels or apple pomace are well-known sources of pectins, but there are also other waste streams that can be exploited in this sense. The pectins from 26 vegetable wastes were characterised in a very complex study, in the framework of EU project NOSHAN, including orange peel, onion hulls, parsley, endive roots and leaves, leek leaves, fresh cabbage, pea pod, sugar beet flakes, berries, apple pomace, sea buckthorn pulp, hop, olive pomace, tomato skin, grape pomace, whole pear and shabal. The results showed that the structure of the pectin extracted from wastes is similar to that from the raw matrices, although the methylation and acetylation degrees are lower due to the processing and/or enzymatic actions. The collected data also emphasise the potential of the recovered pectin to be used either as food additives or other applications (if the minimum concentration in galacturonic acid is not reached) [63].

The most important sources of soluble dietary fibres are the wastes derived from citrus fruits processing. The pectin content differs considerably among citrus varieties, but it generally ranges between 20% and 30% of citrus peel dry weight. Cellulose and hemicellulose can also be recovered from citrus waste as it comprises approximately 50–60% of citrus peel weight. The dietary fibres are not only present in high amount in citrus peels but also have important features due to the presence of associated bioactive constituents (flavonoids and vitamin C) with antioxidant properties, which may provide additional health-promoting effects [64, 65]. For example, the pectin extracted from citrus peel and apple pomace by subcritical water extraction (with maximum yields of 22 and 17%, respectively) showed a high antioxidative and anti-tumour activity [30]. Soluble dietary fibres also reduce the intestinal absorption of blood cholesterol, whereas insoluble dietary fibre associates to water absorption and intestinal regulation apart from the well-known probiotic and health benefits [66].

As previously mentioned, brewers' spent grain besides being a source of quality plant protein is also a good source of carbohydrates, their level being up to 50% of the by-product weight [28]. The main carbohydrates in brewers' spent grain are cellulose (~17% dw) [13, 18, 32] and hemicelluloses, mainly arabinoxylan (25–28% dw) [13, 18]. The vegetable matrices being rich in hemicelluloses can be hydrolysed (e.g. with diluted acid) in order to release the monosaccharides (xylose and arabinose) which can be further subjected to a fermentation process to generate valuable products (e.g. xylitol, a sweetener used in food industry) [29]. Arabinoxylans are considered dietary fibres with a broad range of potential uses as functional ingredients in food products. Their extraction from brewers' spent grain may be performed under strong alkali conditions and also by using an innovative fully integrated process that sequentially extracts the proteins and arabinoxylans [18].

2.4. Phenolics

Phenolics are among the most studied phytochemicals in the last decades. The interest showed by the scientific community in finding new and unconventional sources of phenolic compounds is due to the many studies that suggested that there is an association between the consumption of diets rich in phenolic compounds and a reduced risk of cardiovascular and neurodegenerative diseases [37, 66, 67]. Also, the recovery of phenolic compounds from food processing by-products and their use as functional ingredients sustain the increasing efforts for a sustainable food production.

During fruit processing, the beverage industry leaves between 25 and 35% mass of the raw material called fruit pomace. Unfortunately, some part of pomace in the fruit industry still goes to landfill, and causes environmental pollution and huge losses of valuable materials which could be exploited as a great variety of natural additives and many health-promoting ingredients (phenolic compounds, vitamins, carotenoids and dietary fibre) [68, 69]. Phenolic compounds of different plant sources such as grape and apple pomace are known as potent antioxidants and radical scavengers. The wine-making industries produce millions of tons of residues (grape pomace), which represents a management issue from both ecological and economical point of view [70]. Grape pomace is a phenolic-rich dietary fibre matrix that combines the benefits of both fibre and antioxidants in the prevention of cancer and cardiovascular diseases [66]. Moreover, the grape seeds are considered to be a disposable waste material by the majority of wineries. They are usually discarded, burned or used as animal feed [45]. The oil extracted from the grape seed offers a wide range of benefits for human health, due to its high content of unsaturated fatty acids and antioxidant compounds such as monomeric flavan-3-ols, phenolic acids and oligomeric proanthocyanidins, which is the reason why the valorisation of this by-product is of great interest. Crude grape seed oil consists mainly of linoleic and oleic unsaturated fatty acids and also of palmitic and stearic saturated fatty acids [33, 34]. A study regarding the chemical characterisation of the grape seed extracts obtained by supercritical CO_2 extraction showed that their content in trans-resveratrol was similar to the contents reported in the literature for red wines. This demonstrates that a considerable amount of trans-resveratrol remains unexploited in grape seeds after the fermentation process [33]. An alternative of reuse of grape seeds is as flour incorporated in food products. For example, formulations of frankfurters with grape seed flour showed a decrease in oxidation processes (due to the strong antioxidant activity of the flour), increased total dietary fibre content and water-holding capacity of the final product [59], while the addition of apple pomace extract in meat products reduces the number of synthetic antioxidants needed to be added, and increases the health-promoting properties of the finished product [68].

Besides being a serious environmental problem, olive by-products can also represent a precious resource of potentially valuable molecules. It is worth mentioning that 98% of olive fruit phenols are lost during oil extraction. These compounds are distributed between the olive mill wastewaters (OMWs) phase (approximately 53%) and the solid phase—the 'pomace' (approximately 45%). Consequently, only a 2% fraction of the phenolic classes remains the oil phase depending on the extraction system and olive variety [71]. The evidence relating to decreased prevalence of chronic heart diseases, atherosclerosis or other diseases caused by oxidative stress, through a Mediterranean diet, has oriented scientific research towards the best use of olive-processing by-products (olive leaves and olive mill wastewaters) in order to produce purified natural antioxidants or high antioxidant-rich preparations that could be incorporated in foods, cosmetics and pharmaceuticals [37, 67]. The studies on chemical constituents of olive leaves revealed that phenolic compounds stand out as predominant micronutrients, hydroxytyrosol and oleuropein considered as majority [72]. For example, the hydroxytyrosol-rich olive leaf extract had an inhibitory activity against breast cancer cell proliferation [37]. Also, phenolic-rich extract from OMW and hydroxytyrosol and oleuropein extracts from olive leaves had very pronounced hypocholesterolaemic effects, hypoglycaemic effect, protective action against lipid peroxidation and enhanced antioxidant defence system [73, 74].

Sunflower seeds contain high amounts of polyphenols such as caffeoylquinic and caffeic acids, accounting up to 4% dw. Among all, 5-*O*-caffeoylquinic acid (chlorogenic acid) is the predominant compound. To achieve sustainability of sunflower processing and complete utilisation of by-products arising from sunflower oil production, polyphenols co-extracted during sunflower protein recovery from the expeller were recovered by adsorption technology. In addition, an integrated process was optimised in order to enhance the recovery of polyphenolics as by-products of protein production from sunflower press cake [38, 75].

Other unconventional source of phenolic compounds is the potato peels. Phenolic acids are the most abundant phenolic compounds in potatoes peels, the main representative being the chlorogenic acid (up to 95–98% of phenolic compounds) [39, 76]. It is present in the form of three main isomers: chlorogenic acid (5-*O*-caffeoylquinic acid), neochlorogenic acid (3-*O*-caffeoylquinic acid) and cryptochlorogenic acid (4-*O*-caffeoylquinic acid) [76]. Its extraction from potato peels may be performed by conventional solvent extraction [40], ultrasound-assisted extraction [41] or using an optimised solvent extraction using pressurised liquid extractor [39]. The optimisation of an extraction method is a crucial step for researchers to accurately quantify the content in phenolic compounds and also to be able to estimate their potential health benefits when incorporated in food as functional ingredients. The extracted quantity of phenolic acids from potato peels depends not only on the method parameters but also on genetic factors. While the total phenolics content varies between cultivars and geographical regions, the most abounding isomer of chlorogenic acid was in all cases the 5-*O*-caffeoylquinic acid [39, 40].

2.5. Carotenoids

Carotenoid compounds are known for their health-promoting effects, especially due to their high free radical-scavenging activity. Being powerful antioxidants, when ingested they protect the human body from the damaging actions of the reactive oxygen species and thus lowering the risk of several chronic diseases (cardiovascular diseases, diabetes and cancer). They are fat-soluble pigments which are responsible for the bright-yellow colour of many fruits and vegetables [77].

Lycopene is the main carotenoid found in tomatoes. Some studies suggested that a direct correlation may be established between the consumption of foods rich in lycopene and a low risk of prostate cancer [78].

Tomato (*Solanum lycopersicum* L.) is the second-most consumed vegetable in the world [79]. The solid by-products resulted from its processing into food products such as tomato juice, paste, puree, ketchup and sauce reaching up to 50,000 tons per year [16]. Their exploitation as a source of carotenoids (mainly lycopene) may provide economic benefits. Several techniques are used for the extraction of lycopene from tomato by-products of which enzymatic-assisted process is a promising one. When enzymatic method is used, the tomato by-products are pre-treated by crude enzyme extracts with pectinolytic, cellulolytic and cutinolytic activities prior to their conventional solvent extraction. The results showed an enhancement in the extraction of lycopene from tomato by-products (2.7 mg/100 g) and also a higher overall antioxidant activity for the enzymatic extract (even higher than that of BHA) compared to the one obtained by conventional ethanol extraction [16].

In general, bioaccessibility of carotenoids is low. However, in some fruits, such as mango and papaya, they are present in oil droplet in an esterified form with fatty acids. This kind of structure enhances their extraction and bioavailability during digestion [80]. Poor postharvest technology is one of the major inconveniences in mango annual production, accounting for nearly 60–80% of losses. Therefore, processing mango into flour represents a viable alternative for its use as a functional ingredient and to reduce wastage. The carotenoid content of mango flours ranged from 56.46 to 160.64 μg/g and was found to be higher in ripe mango flours than in green mango flours. In addition, the flour processed from the mango peel has been found to contain significant superior qualities than that from mango pulp in terms of total phenolic, anthocyanins, flavonoids and vitamin C contents and antioxidant activities [81].

Citrus waste is voluminous, heterogeneous, chemically complex and highly biodegradable; therefore, it cannot be disposed of in a landfill without a previous valorisation, in order to avoid both economic loss and environmental pollution issues. About 40–50% of the quantity of this fruit is processed for juice and marmalade production and approximately 50–60% w/w of the processed fruit becomes waste. This by-product contains a wide range of bioactive compounds, such as essential oils, carotenoids, fibre, hesperidin and limonin, which have many applications in food, cosmetic and pharmaceutical industry. After the production of orange juice, the remaining outer layer called flavedo contains considerable amounts of the natural carotenoids. These bioactive compounds comprise approximately 0.1–0.5% of citrus peel dry weight. The major carotenoids available in citrus are α- and β-carotene, lutein, zeaxanthin and β-cryptoxanthin, which are known to be responsible for a wide range of functional properties, mainly offering protection against the reactive oxygen species damaging actions at the cellular level [64, 82–84].

2.6. Other compounds

The wastes from fruits and vegetables can be exploited by microbial processing in order to obtain valuable enzymes such as amylolytic enzymes from banana waste, mango kernels; pectinolytic enzymes from orange peel, lemon peel; tannase from grape seeds; protease from mango peel, potato peel; lipase from coconut cake, lemon peel; and invertase from orange peel, banana peel. The microbial treatment can also be used for the production of organic acids, including lactic acid, citric acid, succinic acid and acetic acid from wastes of potatoes, banana, mango, apple, pineapple and many others [85]. These valuable chemicals can be further exploited as raw materials for other processes or as functional ingredients for newly developed food products and so on [86]. Another example of valuable products recovered from fruit wastes, more exactly, from citrus fruits peels (orange, mandarin, lime, lemons, etc.), is the essential oils. Citrus essential oils extracted from the peels discarded after the fruits processing can be valorised: as flavouring agents in different food products (e.g. soft drinks and confectioneries), perfumes, personal care products, household products; in food preservation enhancing the product's shelf life due to their antioxidant and antimicrobial properties, and thus representing an attractive alternative to synthetic antioxidants and preservatives; and as functional chemicals in agriculture as insects repellent and other more uses [48, 87–89].

3. Conclusion

Food wastes are renewable resources of high-value extractable or convertible chemicals which can be exploited for the development of new functional ingredients, respectively, for the generation of bio-fuels. The scientific research is focused on finding new ways of valorisation of food industry by-products by identifying or optimising the most appropriate extraction methods for the recovery of the biomolecules, as well as by strengthening the cooperation with food industry partners in implementing adequate solutions for a sustainable development and increased competitiveness.

The 'zero-waste' desiderate can be reached by reusing the high-value compounds from by-products in innovative and unconventional ways which may generate profits in a sustainable food production system. The recovered biomolecules are also of great interest for pharmaceutical industry (e.g. carrier agents and controlled release), cosmetics, agriculture, chemical industry and so on.

Acknowledgements

This work was supported by a grant of the Romanian National Authority for Scientific Research, CNCS–UEFISCDI, project number PN-II-RU-TE-2014-4-0842. Authors S.A. Socaci and A.C. Fărcaş contributed equally to this work.

Author details

Sonia A. Socaci*, Anca C. Fărcaş, Dan C. Vodnar and Maria Tofană

*Address all correspondence to: sonia.socaci@usamvcluj.ro

University of Agricultural Sciences and Veterinary Medicine, Cluj-Napoca, Romania

References

[1] EU actions against food waste. Available from: http://ec.europa.eu/food/safety/food_waste/eu_actions/index_en.htm [Accessed: 2016-08-12]

[2] HLPE. Food losses and waste in the context of sustainable food systems. A report by the High Level Panel of Experts on Food Security and Nutrition of the Committee on World Food Security, Rome 2014. Available from: http://www.fao.org/3/a-i3901e.pdf [Accessed:2016.08.12]

[3] Directive of the European Parliament and of the Council 2008/98/EC. Available from:http://eur-lex.europa.eu/legalcontent/EN/TXT/PDF/?uri=CELEX:32008L0098&from=EN [Accessed: 2006-08-12]

[4] Amendment of Directive 2008/98/EC. Available from: http://eur-lex.europa.eu/legal-content/EN/TXT/?uri=CELEX%3A52014PC0397[Accessed: 2006-08-12]

[5] Ravindran R, Jaiswal AK. Exploitation of food industry waste for high-value products. Trends in Biotechnology. 2016;34(1):58–69. DOI: 10.1016/j.tibtech.2015.10.008

[6] Gil-Chavez GJ, Villa JA, Ayala-Zavala JF, Heredia JB, Sepulveda D, Yahia EM, Gonzalez-Aguilar GA. Technologies for extraction and production of bioactive compounds to be used as nutraceuticals and food ingredients: An overview. Comprehensive Reviews in Food Science and Food Safety. 2013;12:5–23. DOI: 10.1111/1541-4337.12005

[7] Galanakis CM. Recovery of high added-value components from food wastes: Conventional, emerging technologies and commercialized applications. Trends in Food Science and Technology. 2012;26:68–87. DOI: 10.1016/j.tifs.2012.03.003

[8] Baiano A. Recovery of biomolecules from food wastes—A review. Molecules. 2014;19:14821–14842. DOI: 10.3390/molecules190914821

[9] Djilas S, Čanadanović-Brunet J, Ćetković G. By-products of fruits processing as a source of phytochemicals. Chemical Industry #x0026; Chemical Engineering Quarterly. 2009;15(4):191-202. DOI: 10.2298/CICEQ0904191D

[10] Pfaltzgraff LA, De Bruyn M, Cooper EC, Budarin V, Clark JH. Food waste biomass: A resource for high-value chemicals. The Royal Society of Chemistry. 2013;15:307–314. DOI: 10.1039/c2gc36978h

[11] Farcas AC, Socaci SA, Tofana M, Dulf FV, Mudura E, Diaconeasa Z. Volatile profile, fatty acids composition and total phenolic content of brewers' spent grain by-product with potential use in the development of new functional foods. Journal of Cereal Science. 2015;64:34–42. DOI: 10.1016/j.jcs.2015.04.003

[12] Oreopoulou V, Tzia C. Utilization of plant by-products for the recovery of proteins, dietary fibers, antioxidants, and colorants. In: Oreopoulou V, Russ W, editors, volume 3. Utilization of By-Products and Treatment of Waste in the Food Industry. Springer; New York, USA. 2007, p. 209–232. DOI: 10.1007/978-0-387-35766-9_11

[13] Mussatto SI. Generating biomedical polyphenolic compounds from spent coffee or silverskin. In: Preedy V, editors. 1st ed. Coffee in Health and Disease Prevention. Elsevier; San Diego, USA. 2015, p. 93–106. DOI: 10.1016/B978-0-12-409517-5.00011-5

[14] Sormoli ME, Langrish T. Spray drying bioactive orange-peel extracts produced by Soxhlet extraction: Use of WPI, antioxidant activity and moisture sorption isotherms. LWT - Food Science and Technology. 2016;72:1–8. DOI: 10.1016/j.lwt.2016.04.033

[15] Teixeira A, Baenas N, Dominguez-Perles R, Barros A, Rosa E, Moreno DA, Garcia-Viguera C. Natural bioactive compounds from winery by-products as health promoters: A review. International Journal of Molecular Sciences. 2014;15:15638–15678. DOI: 10.3390/ijms150915638

[16] Azabou S, Abid Y, Sebii H, Felfoul I, Gargouri A, Attia H. Potential of the solid-state fermentation of tomato by products by Fusarium solani pisi for enzymatic extraction of lycopene. LWT- Food Science and Technology. 2016;68:280–287. DOI: 10.1016/j.lwt.2015.11.064

[17] Tang DS, Tian YJ, He YZ, Li L, Hu SQ, Li B. Optimisation of ultrasonic-assisted protein extraction from brewer's spent grain. Czech Journal of Food Sciences. 2010;28:9–17.

[18] Vieira E, Rocha MAM, Coelho E, Pinho O, Saraiva JA, Ferreira I, Coimbra MA. Valuation of brewer's spent grain using a fully recyclable integrated process for extraction of proteins and arabinoxylans. Industrial Crops and Products. 2014;52:136–143. DOI: 10.1016/j.indcrop.2013.10.012

[19] Niemi P, Martins D, Buchert J, Faulds CB. Pre-hydrolysis with carbohydrases facilitates the release of protein from brewer's spent grain. Bioresource Technology. 2013;136:529–534. DOI: 10.1016/j.biortech.2013.03.076

[20] Yu X, Gouyo T, Grimi N, Bals O, Vorobiev E. Ultrasound enhanced aqueous extraction from rapeseed green biomass for polyphenol and protein valorization. Comptes Rendus Chimie. 2016;19:766–777. DOI: 10.1016/j.crci.2016.03.007

[21] Salgado PR, Molina Ortiz SE, Petruccelli S, Mauri AN. Functional food ingredients based on sunflower protein concentrates naturally enriched with antioxidant phenolic compounds. Journal of the American Oil Chemists' Society. 2012;89:825–836. DOI: 10.1007/s11746-011-1982-x

[22] Aydemir LY, Gökbulut AA, Baran Y, Yemenicioglu A. Bioactive, functional and edible film-forming properties of isolated hazelnut (*Corylus avellana* L.) meal proteins. Food Hydrocolloids. 2014;36:130–142. DOI: 10.1016/j.foodhyd.2013.09.014

[23] Manamperi WAR, Wiesenborn DP, Chang SKC, Pryor SW. Effects of protein separation conditions on the functional and thermal properties of canola protein isolates. Journal of Food Science. 2011;76(3):266–273. DOI: 10.1111/j.1750-3841.2011.02087.x

[24] Karaca AC, Low N, Nickerson M. Emulsifying properties of canola and flaxseed protein isolates produced by isoelectric precipitation and salt extraction. Food Research International. 2011;44:2991–2998. DOI: 10.1016/j.foodres.2011.07.009

[25] Gerzhova A, Mondor M, Benali M, Aider M. Study of the functional properties of canola protein concentrates and isolates extracted by electro-activated solutions as non-invasive extraction method. Food Bioscience. 2015;12:128–138. DOI: 10.1016/j.fbio.2015.10.002

[26] Ng KL, Ayob MK, Said M, Osman MdA, Ismail A. Optimization of enzymatic hydrolysis of palm kernel cake protein (PKCP) for producing hydrolysates with antiradical

capacity. Industrial Crops and Products. 2013;43:725– 731. DOI: 10.1016/j.indcrop.2012.08.017

[27] Sharma PC, Tilakratne BMKS, Gupta A. Utilization of wild apricot kernel press cake for extraction of protein isolate. Journal of Food Science and Technology. 2010;47(6): 682–685. DOI: 10.1007/s13197-010-0096-z

[28] Niemi P, Faulds CB, Sibakov J, Holopainen U, Poutanen K, Buchert J. Effect of a milling pre-treatment on the enzymatic hydrolysis of carbohydrates in brewer's spent grain. Bioresource Technology. 2012;116:155–160. DOI: 10.1016/j.biortech.2012.04.043

[29] Mussatto SI, Roberto IC. Acid hydrolysis and fermentation of brewer's spent grain to produce xylitol. Journal of the Science of Food and Agriculture. 2005;85(14):2453–2460. DOI: 10.1002/jsfa.2276

[30] Wang X, Chen Q, Lü X. Pectin extracted from apple pomace and citrus peel by subcritical water. Food Hydrocolloids. 2014;38:129–137. DOI: 10.1016/j.foodhyd.2013.12.003

[31] Maran JP, Sivakumar V, Thirugnanasambandham K, Sridhar R. Optimization of microwave assisted extraction of pectin from orange peel. Carbohydrate Polymers. 2013; 97:703–709. DOI: 10.1016/j.carbpol.2013.05.052

[32] Niemi P, Tamminen T, Smeds A, Viljanen K, Ohra-Aho T, Holopainen-Mantila U, Faulds CB, Poutanen K, Buchert J. Characterization of lipids and lignans in brewer's spent grain and its enzymatically extracted fraction. Journal of Agricultural and Food Chemistry. 2012b;60:9910-9917. DOI: 10.1021/jf302684x

[33] Dalmolin I, Mazutti MA, Batista EAC, Meireles MAA, Oliveira JV. Chemical characterization and phase behavior of grape seed oil in compressed carbon dioxide and ethanol as co-solvent. Journal of Chemical Thermodynamics. 2010;42(6):797–801. DOI: 10.1016/j.jct.2010.02.003

[34] Prado MJ, Dalmolin I, Carareto ND, Basso RC, Meirelles AJ, Oliveira JV, Batista EA, Meireles MA. Supercritical fluid extraction of grape seed: Process scale-up, extract chemical composition and economic evaluation. Journal of Food Engineering. 2012;109:249–257. DOI: 10.1016/j.jfoodeng.2011.10.007

[35] Mussatto SI, Dragone G, Roberto IC. Ferulic and p-coumaric acids extraction by alkaline hydrolysis of brewer's spent grain. Industrial Crops and Products. 2007;25:231–237. DOI: 10.1016/j.indcrop.2006.11.001

[36] Allouche N, Fki I, Sayadi S. Toward a high yield recovery of antioxidants and purified hydroxytyrosol from olive mill wastewaters. Journal of Agricultural and Food Chemistry. 2004;52(2):267–273. DOI: 10.1021/jf034944u

[37] Bouallagui Z, Han J, Isoda H, Sayadi S. Hydroxytyrosol rich extract from olive leaves modulates cell cycle progression in MCF-7 human breast cancer cells. Food and Chemical Toxicology. 2011;49:179–184. DOI: 10.1016/j.fct.2010.10.014

[38] Weisz GM, Carle R, Kammerer DR. Sustainable sunflower processing—II. Recovery of phenolic compounds as a by-product of sunflower protein extraction. Innovative Food Science and Emerging Technologies. 2013;17:169–179. DOI: 10.1016/j.ifset.2012.09.009

[39] Luthria DL. Optimization of extraction of phenolic acids from a vegetable waste product using a pressurized liquid extractor. Journal of Functional Foods. 2012;4:842–850. DOI: 10.1016/j.jff.2012.06.001

[40] Ieri F, Innocenti M, Andrenelli L, Vecchio V, Mulinacci N. Rapid HPLC/DAD/MS method to determine phenolic acids, glycoalkaloids and anthocyanins in pigmented potatoes (*Solanum tuberosum* L.) and correlations with variety and geographical origin. Food Chemistry. 2011;125:750–759. DOI: 10.1016/j.foodchem.2010.09.009

[41] Singhai PK, Sarma BK, Srivastava JS. Phenolic acid content in potato peel determines natural infection of common scab caused by *Streptomyces* spp. World Journal of Microbiology and Biotechnology. 2011;27:1559–1567. DOI: 10.1007/s11274-010-0608-z

[42] Conidi C, Cassano A, Drioli E. Recovery of phenolic compounds from orange press liquor by nanofiltration. Food and Bioproducts Processing. 2012;90:867–874. DOI: 10.1016/j.fbp.2012.07.005

[43] Laroze LE, Díaz-Reinoso B, Moure A, Zúñiga ME, Domínguez H. Extraction of antioxidants from several berries pressing wastes using conventional and supercritical solvents. European Food Research and Technology. 2010;231:669–677. DOI: 10.1007/s00217-010-1320-9

[44] Pingret D, Fabiano-Tixier AS, Bourvellec CL, Renard CMGC, Chemat F. Lab and pilot-scale ultrasound-assisted water extraction of polyphenols from apple pomace. Journal of Food Engineering. 2012;111:73–81. DOI: 10.1016/j.jfoodeng.2012.01.026

[45] Agostini F, Bertussi RA, Agostini G, Atti dos Santos AC, Rossato M, Vanderlinde R. Supercritical extraction from vinification residues: Fatty acids, α-tocopherol, and phenolic compounds in the oil seeds from different varieties of grape. Scientific World Journal. 2012;790–486. DOI: 10.1100/2012/790486

[46] Sun Y, Liu D, Chen J, Ye X, Yu D. Effects of different factors of ultrasound treatment on the extraction yield of the all-trans-β-carotene from citrus peels. Ultrasonics Sonochemistry. 2011;18: 243–249. DOI: 10.1016/j.ultsonch.2010.05.014

[47] Kagliwal LD, Patil SC, Pol AS, Singhal R S, Patravale VB. Separation of bioactives from sea buckthorn seeds by supercritical carbon dioxide extraction methodology through solubility parameter approach. Separation and Purification Technology. 2011;80:533–540. DOI :10.1016/j.seppur.2011.06.008

[48] Thongnuanchan P, Benjakul S. Essential oils: Extraction, bioactivities, and their uses for food preservation. Journal of Food Science. 2014;79(7):1231–1249. DOI: 10.1111/1750-3841.12492

[49] Gonzalez-Perez S, Vereijken JM. Sunflower proteins: overview of their physicochemical, structural and functional properties. Journal of the Science of Food and Agriculture. 2007;87:2173–2191. DOI: 10.1002/jsfa.2971

[50] Chong CH, Zulkifli I, Blair R. Effects of dietary inclusion of palm kernel cake and palm oil, and enzyme supplementation on performance of laying hens. Asian-Australian Journal of Animal Science. 2008;21(7):1053–1058. DOI: 10.5713/ajas.2008.70581

[51] Saenphoom P, Liang JB, Ho YW, Loh TC, Rosfarizan M. Effect of enzyme treatment on chemical composition and production of reducing sugars in palm (*Elaeis guineenis*) kernel expeller. African Journal of Biotechnology. 2011;10(68):15372–15377. DOI: 10.5897/AJB11.1211

[52] Zarei M, Ebrahimpour A, Abdul-Hamid A, Anwar F, Saari N. Production of defatted palm kernel cake protein hydrolysate as a valuable source of natural antioxidants. International Journal of Molecular Sciences. 2012;13:8097–8111. DOI: 10.3390/ijms13078097

[53] Mussatto SI. Brewer's spent grain: a valuable feedstock for industrial applications. Journal of the Science of Food and Agriculture. 2014;94:1264–1275. DOI: 10.1002/jsfa.6486

[54] Santos M, Jiménez JJ, Bartolomé B, Gómez-Cordovés C, del Nozal MJ. Variability of brewer's spent grain within a brewery. Food Chemistry. 2003;80:17–21. DOI: 10.1016/S0308-8146(02)00229-7

[55] Waters DM, Jacob F, Titze J, Arendt KE, Zannini E. Fibre, protein and mineral fortification of wheat bread through milled and fermented brewer' s spent grain enrichment. European Food Research and Technology. 2012;235:767–778. DOI: 10.1007/s00217-012-1805-9

[56] Farcas A, Socaci S, Tofana M, Muresan C, Mudura E, Salanta L, Scrob S. Nutritional properties and volatile profile of brewer's spent grain supplemented bread. Romanian Biotechnological Letters. 2014;19(5):9705–9714.

[57] Coimbra MA, Pinto MI, Martins MA, Ferreira ME, Saraiva JM, Castro Pinho OM. Integrated process for extracting proteins and arabinoxylans from brewer's spent grain. Available from: https://www.google.com/patents/WO2012069889A1?cl=en [Accessed: 2016-07-21]

[58] Lee J, Lee J, Yang HJ, Song KB. Preparation and characterization of brewer's spent grain protein-chitosan composite films. Journal of Food Science and Technology. 2015;52:7549. DOI: 10.1007/s13197-015-1941-x

[59] Ozvural EB, Vural H. Grape seed flour is a viable ingredient to improve the nutritional profile and reduce lipid oxidation of frankfurters. Meat Science. 2011;88(1):179–183. DOI: 10.1016/j.meatsci.2010.12.022

[60] Poli A, Anzelmo G, Fiorentino G, Nicolaus B, Tommonaro G, Di Donato P. Polysaccharides from wastes of vegetable industrial processing: New opportunities for their

eco-friendly re-use. In: Elnashar M, editors. Biotechnology of Biopolymers. InTech; Rijeka, Croatia. 2011, p. 32–56. DOI: 10.5772/16387

[61] Dhingra D, Michael M, Rajput H, Patil RT. Dietary fibre in foods: A review. Journal of Food Science and Technology. 2012;49(3):255–266. DOI: 10.1007/s13197-011-0365-5

[62] Di Donato P, Poli A, Taurisano V, Nicolaus B. Polysaccharides: Applications in biology and biotechnology/Polysaccharides from bioagro-waste new biomolecules-life. Polysaccharides. 2014;1–29. DOI: 10.1007/978-3-319-03751-6_16-1

[63] Müller-Maatsch J, Bencivenni M, Caligiani A, Tedeschi T, Bruggeman G, Bosch M, Petrusan J, Droogenbroeck B, Elst K, Sforza S. Pectin content and composition from different food waste streams. Food Chemistry. 2016;201:37–45. DOI: 10.1016/j.foodchem.2016.01.012

[64] El-Sharnouby GA, Aleid SM,Al-Otaibi MM. Conversion of processed citrus wastes into nutritional components. Journal of Food Processing and Technology. 2013;4:8. DOI: 10.4172/2157-7110.1000259

[65] Ferrentino G, Asaduzzaman Md, Scampicchio MM. Current technologies and new insights for the recovery of high valuable compounds from fruits by-products. Critical Reviews in Food Science and Nutrition. DOI: 10.1080/10408398.2016.1180589

[66] Zhu F, Du B, Zheng L, Li J. Advance on the bioactivity and potential applications of dietary fibre from grape pomace. Food Chemistry. 2015;186:207–212. DOI: 10.1016/j.foodchem.2014.07.057

[67] Bouallagui Z, Bouaziz M, Han J, Boukhris M, Rigane G, Friha I, Jemai H, Fki I, Ghorbel H, Isoda H, Sayadi S. Valorization of olive processing by-products: Characterization, investigation of chemico-biological activities and identification of active compounds. Journal of Arid Land Studies. 2012;22(1):61–64.

[68] Peiretti PG, Gai F. Fruit and pomace extracts: applications to improve the safety and quality of meat products. In: Owen JP, editors. Fruit and Pomace Extracts Biological Activity, Potential Applications and Beneficial Health Effects. Nova Science; New York, USA. 2015, p. 1–28.

[69] Kruczek M, Drygas B, Habryka C. Pomace in fruit industry and their contemporary potential application. World Scientific News. 2016;48:259–265.

[70] Fontana AR, Antoniolli A, Bottini R. Grape pomace as a sustainable source of bioactive compounds: Extraction, characterization, and biotechnological applications of phenolics. Journal of Agricultural and Food Chemistry. 2013;61(38):8987–9003. DOI: 10.1021/jf402586f

[71] Rodis PS, Karathanos VT, Mantzavinou A. Partitioning of olive oil antioxidants between oil and water phases. Journal of Agricultural and Food Chemistry. 2002;50(3):596–601. DOI: DOI: 10.1021/jf010864j

[72] Bouaziz M, Grayer RJ. Simmonds Monique SJ, Damak M, Sayadi S. Identification and antioxidant potential of flavonoids and low molecular weight phenols in olive cultivar chemlali growing in Tunisia. Journal of Agricultural and Food Chemistry. 2005;53:236–241. DOI: 10.1021/jf048859d

[73] Jemai H, Fki I, Bouaziz M, Bouallagui Z, El Feki A, Isoda H, Sayadi S. Lipid-lowering and antioxidant effects of hydroxytyrosol and its triacetylated derivative recovered from olive tree leaves in cholesterol-fed rats. Journal of Agricultural and Food Chemistry. 2008;56(8):2630–2636. DOI: 10.1021/jf072589s

[74] Jemai H, El Feki A, Sayadi S. Antidiabetic and antioxidant effects of hydroxytyrosol and oleuropein from olive leaves in alloxan-diabetic rats. Journal of Agricultural and Food Chemistry. 2009;57(19):8798–8804. DOI: 10.1021/jf901280r

[75] Weisz GM, Schneider L, Schweiggert U, Kammerer DR, Carle R. Sustainable sunflower processing — I Development of a process for the adsorptive decolorization of sunflower (*Helianthus annuus* L.) protein extracts. Innovative Food Science and Emerging Technologies. 2010;11:733–741. DOI: 10.1016/j.ifset.2010.05.005

[76] Akyol H, Riciputi Y, Capanoglu E, Caboni MF, Verardo V. Phenolic compounds in the potato and its byproducts: An overview. International Journal of Molecular Science. 2016;17:835. DOI:10.3390/ijms17060835

[77] Rao AV, Rao LG. Carotenoids and human health. Pharmacological Research. 2007;55:207–216. DOI:10.1016/j.phrs.2007.01.012

[78] Rao LG, Guns E, Rao AV. Lycopene: Its role in human health and disease. AgroFood Industry Hi-Tech. 2003;55:35–30.

[79] Socaci SA, Socaciu C, Muresan C, Fărcas A, Tofană M, Vicas S, Pintea A. Chemometric discrimination of different tomato cultivars based on their volatile fingerprint in relation to lycopene and total phenolics content. Phytochemical Analysis. 2014;25(2): 161–169. DOI: 10.1002/pca.2483

[80] Courraud J, Berger J, Cristol JP, Avallone S. Stability and bioaccessibility of different forms of carotenoids and vitamin A during in vitro digestion. Food Chemistry. 2013;136:871–877. DOI: doi:10.1016/j.foodchem.2012.08.076

[81] Abdul Aziz NA, Wong LM, Bhat R, Cheng LH. Evaluation of processed green and ripe mango peel and pulp flours (*Mangifera indica* var. Chokanan) in terms of chemical composition, antioxidant compounds and functional properties. Journal of the Science of Food and Agriculture. 2012;92:557–563.DOI: 10.1002/jsfa.4606

[82] Wilkins MR, Suryawati L, Maness NO, Chrz D. Ethanol production by *Saccharomyces cerevisiae* and *Kluyveromyces marxianus* in the presence of orange-peel oil. World Journal of Microbiology and Biotechnology. 2007a;23:1161–1168. DOI:10.1007/s11274-007-9346-2

[83] Negro V, Mancini G, Ruggeri B, Fino D. Citrus waste as feedstock for bio-based products recovery: Review on limonene case study and energy valorization. Bioresource Technology. 2016;214:806–815. DOI: 10.1016/j.biortech.2016.05.006

[84] Siles JA, Vargas F, Gutiérrez MC, Chica AF, Martín MA. Integral valorisation of waste orange peel using combustion, biomethanisation and co-composting technologies. Bioresource Technology. 2016;211:173–182. DOI: 10.1016/j.biortech.2016.03.056

[85] Panda SK, Mishra SS, Kayitesi E, Ray RC. Microbial-processing of fruit and vegetable wastes for production of vital enzymes and organic acids: Biotechnology and scopes. Environmental Research. 2016;146:161–172. DOI: 10.1016/j.envres.2015.12.035

[86] Laufenberg G, Kunz B, Nystroem M. Transformation of vegetable waste into value added products: (A) the upgrading concept; (B) practical implementations. Bioresource Technology. 2003;87:167–198. DOI: 10.1016/S0960-8524(02)00167-0

[87] Bakry AM, Abbas S, Ali B, Majeed H, Abouelwafa MY, Mousa A, Liang L. Microencapsulation of oils: A comprehensive review of benefits, techniques, and applications. Comprehensive Reviews in Food Science and Food Safety. 2016;15:143–182. DOI: 10.1111/1541-4337.12179.

[88] Calo JR, Crandall PG, O'Bryan CA, Ricke SC. Essential oils as antimicrobials in food systems—A review. Food Control. 2014;54:111–119. DOI: 10.1016/j.foodcont.2014.12.040

[89] Miguel MG. Antioxidant and anti-inflammatory activities of essential oils: A short review. Molecules. 2010;15:9252–9287. DOI: 10.3390/molecules15129252

Chapter 5

Selected Superfoods and Their Derived Superdiets

Beatrice Nakhauka Ekesa

Additional information is available at the end of the chapter

http://dx.doi.org/10.5772/67239

Abstract

Despite the reported decline in undernourishment in developing regions from 23.3% to 12.9% within 25 years, sub-Saharan Africa is the most malnourished region in the world, and the situation could get even worse depending on how the continent's love affairs with the few popular foods play out. The irony is that there are millions of nutrient-rich edible plants, insects and animals within tropical Africa, but due to modernization, only 3% of these foods are utilized within diets. Through a comprehensive literature review, this chapter will explore eight of the most feasible superfoods with an objective of using a systems approach to further look into their derived superdiets. Superfoods are naturally occurring plant or animal-based foods dense in nutrients, antioxidants and healthy fats, whilst superdiets are defined as feasible dishes prepared based on selected superfoods, incorporating other food ingredients and using appropriate processing and cooking techniques. The selected superfoods will include amaranth, teff, fonio, moringa leaves, baobab fruit, tamarind and hibiscus leaves. With the dense vitamins, minerals, healthy fats and antioxidants, these superfoods and more importantly their derived dishes have great potential in boosting the immune system, reducing risk of chronic diseases and promoting a healthy and productive population.

Keywords: superfoods, superdiets, moringa, hibiscus, tamarind, amaranth, teff, fonio, baobab

1. Introduction

The number of hungry people in the world has dropped to 795 million from 1 billion in 1990/1992 to the latest state of food insecurity in the world as reported in 2015 [1]. In addition, in the developing regions, the prevalence of undernourishment declined to 12.9% of the population, down from 23.3% a quarter century ago [1]. Despite the progress, sub-Saharan Africa is the region with the highest prevalence of undernourishment in the world, and the situation could get better—or worse—depending on how the continent's love affairs with some of its increasingly popular foods play out [2]. The irony is that there are millions of edible plants, insects and animals, and just like the Amazon, tropical Africa is still hiding most of the food items considered to be superfoods. Although superfood is a term originally used just as an advertisement and marketing tool, a superfood can be defined as a nutrient-dense, antioxidant-rich, natural-food product that is minimally processed and bioavailable in numerous, potent nutritive constituents. Consumption of superfoods increases energy and vitality, regulates cholesterol and blood pressure and may help to prevent or fight cancer and other diseases [3]. Superfoods are generally beneficial for health and well-being. Basing on the above definitions, there is a big range of foods considered to be superfoods, but for this chapter, focus will be on those mainly available within Africa and with higher potential of integration within existing food and diet systems. The objective is to downplay on the 'superfood' and emphasize more on 'super diets', where the emphasis is on a healthy balanced diet. The foods that will be explored include moringa leaves, hibiscus, amaranth, baobab fruit, tamarind, teff and fonio.

2. Selected superfoods

2.1. Amaranth

Botanically referred to as *Amaranthus*, this crop was cultivated by the Aztecs 8000 years ago and is still a native crop in Peru. The ancient history of amaranth can be traced to Mexico and the Yucatan Peninsula. The name for amaranth comes from a Greek word *amarantos* meaning 'one that does not wither' or 'never fades'; this is true as amaranth's bushy flowers retain their vibrancy even after harvesting and drying. In addition, some varieties of ornamental amaranth do not produce the fancy flowers but produce flashy foliage, sprouting leaves [4]. *Amaranthus* are now grown in Africa, India, China, Russia, throughout South America and North America. Amaranth is tall about 6 feet, has broad leaves with colours ranging from deep blood red to light green with purple veins and has around 60 different species [4], several of which are cultivated as leaf vegetables, grains or ornamental plants [5]. It is commonly known as pigweed (English), hanekam (Afrikaans), thepe (Sesotho), imbuya (isiZulu), mchicha (Swahili), terere (Gikuyu, Meru and Embu of Kenya), doodo (Luganda), shoko (Yoruba) [5] and lengalenga (Democratic Republic of Congo and Burundi) [6]. Both the amaranth leaves and seeds are useful in terms of human health [5]. Whether you choose to consume amaranth as a leaf vegetable, a cereal grain or grain flour, considering the versatility and high concentration of antioxidants and nutrients, amaranth is one of the most of valuable health foods that you may have never heard of [5].

2.1.1. Vegetable/leafy amaranth

Vegetable amaranths are probably the most widely eaten boiled or steamed greens throughout Africa's humid lowlands. They secure the food supply for millions. The leaves and stems make excellent boiled or steamed vegetables as stew or sauce; they have a soft texture, mild flavour and no trace of bitterness [7]. As already indicated, raw vegetables have higher nutrient levels than cooked vegetables, but it is also obvious that not all vegetables can be consumed raw. Amaranth leaves are one of those vegetables that have to be cooked. It is therefore important that when cooking the amaranth, the cooking time should not exceed 5 min, and in the case water is used, it should be used in minimal quantities and not discarded as most nutrients leach into the water. If the amaranth is being cooked together with other food items that require longer cooking time such as legumes and meats, it should be added to the food just a few minutes before the meal is ready. Therefore, following appropriate cooking methods, amaranth leaves have great nutrition value.

Cooked amaranth leaves are packed with antioxidants and an excellent source of several nutrients especially vitamins and minerals [8]. According to the FAO West African food composition table, 100 g of boiled leafy amaranth contains 4.6 g protein, 380 mg calcium, 4.9 mg iron, 58 mg magnesium, 54 mg potassium, 42 mcg folate, 19 mg vitamin C, 0.25 mg vitamin E and 228 mcg RAE of vitamin A [9]. There are very few leafy vegetables with high levels of calcium, and therefore amaranth is an absolute superfood in terms of boosting bone strength and preventing osteoporosis, thus extending your 'active life' well into old age. The significant level of carotenoids and vitamin A in amaranth leaves is a major boost for eye health, as these antioxidants can prevent macular degeneration and slow or stop the development of cataracts [10]. By lowering oxidative stress in the ocular system, amaranth keeps your vision healthy and strong. In addition, vitamin A plays a major role in boosting the immune system, thus reducing the likelihood of contracting infections and the severity if contracted. The type of vitamin E in this leafy vegetable is tocotrienols, a type which helps in reducing bad cholesterol (low density lipoprotein-LDL) levels in the body and prevents coronary heart diseases [10]. The high levels of potassium and magnesium are crucial for maintaining proper electrolyte balance in the body, and the presence of significant amounts of dietary fibre aids in the management and prevention of hypertension [10]. The antioxidant property of vitamin E, vitamin C and lysine in addition to other essential nutrients makes it possible for these leaves to fight against harmful free radicals and prevent the formation of malignant cells responsible for cancer [10]. Basing on the important role that folate plays especially in preventing neural tube defects in newborns, including amaranth vegetables to your diet would help protect your newest addition to the family.

2.1.1.1. Amaranth-based diets

Amaranth leaves come in different colour shades ranging from dark green and reddish green to deep red and purple, but the most popular variety in Africa especially East Africa is the dark green leafy type. Although there is no standard recipe of cooking amaranth, the basic ingredients include amaranth leaves and small parts of the stem, cooking oil, tomatoes, onions and salt. Depending on the culture and economic ability, other ingredients that could be added include meat, small fish, groundnuts, African eggplant, green pepper, garlic or red beans.

The most popular use of amaranth vegetable is as a vegetable sauce accompanying starchy staples such as steamed/boiled/stewed banana; 'ugali' (African polenta or cornmeal mush) also called sima, sembe, kaunga, dona, banku, kenkey pup, posho, fufu; rice; and potatoes or sweet potatoes. As detailed in **Table 1**, it can also be cooked together with other food items to form a complete meal. In Burundi, amaranth is cooked together with beans and bananas and sometimes small fish locally called 'dagala' added. This provides a very nutritious meal able to meet a good range of both macro- and micronutrient needs.

Dish name	Ingredients	Cooking procedure
Basic amaranth sauce	-Two bunches of amaranth leaves and small parts of the stem washed and drained (three to four cups) -One medium onion (chopped) -One tbsp of cooking oil -One medium tomato (chopped) -Water -Salt to taste -Seasoning (optional)	Add oil to a hot pan and immediately add the onions. Saute until translucent; do not let them turn brown Add the leaves and stir to prevent them from burning Cook covered on low heat for 3 min Add the chopped tomatoes and cook covered for 1 min Add seasoning of your choice Serve hot with starch accompaniment such as banana, 'ugali' potatoes, rice, etc.
Amaranth in groundnut sauce	-One onion, chopped -Two tbsp oil -Three tomatoes, cubed -1 kg amaranth, washed and chopped -Three tbsp groundnut paste -¼ tbsp salt -Water NB: one can either use flour made from grinding uncooked groundnuts or paste made from slightly roasted groundnuts)	-Add oil to a hot pan and immediately add the onions. Saute until translucent; do not let them turn brown -Add the leaves and stir to prevent them from burning -Cook covered on low heat for 2 min -Meanwhile mix the groundnut powder in warm water, and stir into a smooth paste and keep aside -Add the chopped tomatoes to the amaranth and cook covered for 1 min -Stir the tomatoes in the vegetables and add the smooth groundnut paste into the amaranth, add some little water, keep stirring, season with salt, and cook covered on low heat for 5 more minutes -Serve with main dish of choice
Beans, cooking banana and amaranth leaves	-One cup dried red kidney beans -Five green cooking bananas -One bunch of amaranth leaves with small part of stems, washed and drained -One tbsp vegetable oil -Medium onion, thinly sliced -Two tomatoes slices into cubes -One tsp salt to taste -Hot chilli powder (optional)	-Soak the beans for 6–8 hrs or overnight in lots of water -Drain, place in a pan, cover with new water, and boil until tender -Drain the remaining water and reserve -Peel and slice the bananas into desired sizes. Add the oil to a hot pan and immediately add the onion. Fry the onions until translucent -Add the beans and plantains to the oil, add some salt and if desired chilli pepper, and fry for about 2 min, stirring constantly. -Add about four cups of water and let the food boil -Reduce to a simmer and cook until plantains are soft -Add the washed and drained amaranth leaves on the cooking mixture, cook covered for another 2 min, stir in the vegetables to mix with the other ingredients, and cook covered for 2 more minutes NB: The amaranth can be cooked with beans excluding the bananas and sometimes with bananas excluding the beans. All the options can be served alone or with rice, ugali or any other starchy staple

Table 1. Selected amaranth leaf-based dishes.

2.1.2. Grain amaranth

Amaranth grain is reported to have been domesticated between 6000 and 8000 years ago, and it has a long and colourful history in Mexico [11]. When ground, the amaranth flour is generally a pale ivory shade, but the red 'buds' can be ground as well for a red-tinged and very healthful grain. One of the most important aspects of this tiny grain is that it is gluten-free [11, 12], thus providing a viable wheat alternative for millions of people suffering from celiac disease or gluten intolerance.

At about 13–14%, grain amaranth easily trumps the protein content of most other grains, and you may hear the protein in amaranth referred to as 'complete' because it contains lysine, an amino acid missing or negligible in many grains [12]. It also contains other primary proteins called albumin and globulins, which, in comparison with the prolamins in wheat, are more soluble and digestible [12]. A 100 g of raw amaranth contains 14 g of protein, 15 mg of iron, 159 mg of calcium, 4 mg of vitamin C and 18 mg of fibre [9]. The high fibre level results in smooth digestion of food and facilitates an efficient uptake of minerals level. At 105% of the daily value per serving, the manganese in amaranth is off the charts, yet it contains fewer carbohydrates [4]. Amaranth contains 6–10% oil, predominantly unsaturated, or around of which 77% are unsaturated fatty acids, including linoleic acid, acid that is required for optimum nutrition [4]. With all the above nutrients, amaranth grain is a true powerhouse, likely to prevent a number of chronic health conditions such as diabetes, heart disease, cancer, and stroke.

2.1.2.1. Grain amaranth-based dishes

Grain amaranth has been used for food by humans in a number of ways. Being extremely dense, amaranth is too heavy to be used by itself. Although it can be popped and eaten like popcorn or flaked like oatmeal, it is best used with other grains for a lighter texture. The ground grain is used as an enrichment to staple-based diets such as porridge, soups, ugali, etc., thus supplying more nutrients to vulnerable population groups. For instance, amaranth grain porridge (1 cup) combined together with moringa leaf powder (1 tbsp) from moringa leaves not only provides an excellent nutritional food for individuals with compromised immune system (HIV/AIDs), but also those consuming the amaranth/moringa combination are able to take anti-retroviral drugs with no complications [13].

The ground grain is also mixed with wheat flour and used in making more nutritious breads, noodles, pancakes, cereals, cookies and other flour-based products. There are more than 40 products containing amaranth and that can be used by consumers of different social economic/cultural backgrounds. Amaranth grain flour is also used as an exceptional thickener for sauces, soups, stews and even jellies; the four can be made from freshly ground grains of by sprouting/germinating the grain, drying it and milling/grinding it into flour. Eaten as a snack, amaranth can have a light, nutty or peppery-crunchy texture and flavour.

Cooking amaranth grain alone is comparable to cooking pasta or rice: boil plenty of water (six cups of water per one cup of amaranth), measure the grain into it, cook and stir for 15–20 min, drain, rinse, and its ready to eat. **Table 2** gives details of selected dishes based on amaranth grain.

Dish name	Ingredients	Cooking procedure and serving recommendations
Amaranth polenta with mushrooms	-½ cup dried mushrooms -One tbsp cooking oil -One cup amaranth grain -¼ tsp salt -Freshly ground pepper to taste -One medium onion Two cups of hot water	-Soak the dried mushrooms in a covered contained with hot water for 10 min to soften -When soft chop any large pieces into preferred sizes. Meanwhile, heat the oil in a heavy saucepan, add the onions, and cook till soft and translucent -Stir in the amaranth; add the soaked chopped mushrooms with the soaking liquid, taking care to leave any grit on the bottom of the cup -Bring to a boil, reduce the heat, cover, and simmer for 15 min -Stir in the salt and pepper to taste -Continue simmering, covered, until the mixture is like porridgy and the amaranth is tender (about 10–15 min more) -(Tender amaranth should still be crunchy but should not taste hard or gritty) -If mixture is too thick before the amaranth is cooked, stir in more boiling water -Serve in small bowls
Enriched porridge	-One cup maize meal flour/any other common porridge flour -½ cup amaranth flour -2 ½ cups water -2 ½ cups milk (optional) -Two tbsp. margarine (optional)	-Combine amaranth flour and maize meal flour in a medium-sized saucepan -Add in water and stir into a smooth paste; add in all the water; keep stirring -Put the saucepan over high heat and simmer for 15 min, stirring occasionally or until the mixture becomes very thick -Add in the milk; let it simmer for additional 5 min -Add in margarine, cook for 1 min and remove from heat, cool whilst stirring, and serve
Enriched 'ugali'	-One cup of maize meal flour/any other common ugali flour (cassava flour/ cassava and millet flours) -One cup of amaranth flour -Three cups of water	-Mix the flours thoroughly -Add the mixture to boiling water little by little as you stir in the flour with a wooden spoon, keep stirring, and begin to knead with the same spoon as it gets stiffer -Reduce heat slightly and cover to let it cook for about 2 min -Uncover and continue kneading for another 2 min until you begin to smell burning corn -Remove it from heat and turn it over in a serving plate -Can be served with stir-fried amaranth leaves or any other vegetable or/ and any protein dish
Chapati	-Wheat flour (1400 g) -Popped powdered grain amaranth (600 g) -Water (1000 ml) -Grated carrots (156 g) -Cooking oil (35 ml) -Salt to taste (15 g)	-Mix the flour, salt and the popped grain together in a bowl -Add in the grated carrots and onions -Using your finger tips, mix in the oil gently -Add the water and knead into dough -Divide the dough into portions of your preferred size -Roll each portion into flat round shapes; size can be different based on fying pan or personal taste -Put some oil on the frying pan; when oil is hot, reduce heat and put the flat round dough on the pan -Cook one side until cooked (showing spots of light brown) -Flip it over and cook the other size -Serve hot with legume or meat stew

Dish name	Ingredients	Cooking procedure and serving recommendations
Pancakes	-Cassava flour (500 g) -Popped/germinated grain amaranth flour (250 g) -Freshly peeled sweet bananas (560 g) -Cooking oil (1 litre)	-Mix the cassava flour and popped/germinated grain amaranth flour thoroughly -Sift the mixture -Mix in sweet banana and any other spices -Knead the mixture using hands or blender to form a hard dough -Roll the dough flat on a smooth clean surface -Cut into desired shapes -Deep fry until golden brown

Table 2. Examples of amaranth grain-based dishes.

2.2. Teff

Teff is a tiny fine ivory, red/brown or mixed (ivory, red/brown) grain. Red/brown teff has a subtle hazelnut, almost chocolate-like flavour and a moist texture similar to millet (but more exotic). Ivory teff has a milder flavour than the brown. This grain is the national pride of Ethiopia, where it has been consumed for more than 1000 years (way BC). Teff is scientifically known as Eragrostis teff. 'Teffa', the Amharic name for 'lost', is so named because of 'teff's' small size; it is the smallest grain in the world and often is lost in the harvesting and threshing process [14]. It is now starting to get global attention which is good news for all of us especially because it is a durable crop that can grow in almost every climate.

With its subtle nutty flavour, the same flexibility holds also in the kitchen. Teff leads all the known grains by a large margin. Its small size means that the germ and the bran—the most nutrient-dense layers—make up a large proportion of the overall seed as the grain cannot be separated into bran, germ and endosperm. Apart from its gluten-free nature which makes it a delicious wheat alternative, the teff grain is also known for its superior amino acid profile, being high in lysine, a protein essential for muscle repair [15]. Teff is the primary carbohydrate source for most Ethiopians. It has an estimated 20–40% resistant starch and high fibre; these particular components are important in dealing with diabetes and assisting with blood sugar control [15, 16]. 100 g of edible portion of raw teff has 13 g protein, 8 g fibre, 180 mg Calcium and 8 mg iron [9]. Just a cup of cooked teff contains 123 mg of calcium, about the same as half a cup of spinach [2, 15, 16]. It is also high in iron, calcium and vitamin C. It is also packed full of B vitamins, which makes it great for energy. Last but not least, teff packs a little something that the others do not 'vitamin K', a fat soluble vitamin which is required for blood clotting and also bone health [17].

2.2.1. Teff-based dishes

As shown in **Table 3**, Teff has the versatility of corn meal and millet. Delicious in porridge, stews, stuffing and pilaf, teff can be cooked alone or in combination with other grains and vegetables. Alone, teff's cooking time is 20 min, and for each cup of grain, you need three cups of water. All you need to do is combine teff and water in a pot and bring to a boil. Reduce heat, cover and simmer for 20 min, until water is absorbed. You may stir occasionally towards the end of cooking.

Dish name	Ingredients	Cooking procedure and serving recommendations
Injera	-1 1/2 cups of teff flour -Two cups pure water -1/2 tsp baking powder -Cooking oil -1/4 tsp salt, to taste	-Putt teff flour in a large bowl, add water, and stir well till smooth -Cover with a cheesecloth or towel and place on the counter and let it sit for 1 day/24 hrs without any stirring or agitation -After 24 hours, the batter will appear alive and fermenting, stir in the salt, and season with any other preferred spices until you can barely detect the saltiness -Stir in the baking powder -Bring a pan to medium heat; lightly coat the pan with cooking oil -Pour enough batter into the pan to fill the whole surface; cover with a lid or a cookie sheet as it is important to keep a lot of moisture in the pan or the injera will crack -Do not flip or brown its underside -When you see the top bubble like pancakes and start to dry out and the edges begin to curl/dry, remove the injera from the pan -Place on a plate and repeat layering cooked injera with parchment paper until you use up all the batter -Serve with chickpea and sweet potato wat or Ethiopian lentils with berbere spice
Teff banana pancakes	-Two eggs -Two ripe bananas peeled -1 1/2 cups of milk -One tbsp honey -1 1/2 tsp oil -1 1/2 cups of teff flour -One tbsp baking powder -1/2 tsp cinnamon (optional) -1/4 tsp salt	-Mix eggs, banana, milk, vanilla, honey and 1/2 tsp oil, and beat or blend well -In a large mixing bowl, put teff flour, baking powder, salt and cinnamon, and stir in banana the milk mixture -Place the frying pan over medium heat; when the pan is hot, brush on one tsp of oil -Using a tablespoon, scoop up the batter and pour it on the hot pan, one heaping tablespoon for each pancake -Cook pancakes for 3–4 min on the first side or until the top forms bubbles and begins to dry -Flip them over and cook for another minute or two -Feel free to substitute maple syrup for honey, and use juice for milk. Ground-up flax seeds easily take the place of egg -Serve plain or dipped into yoghurt
Teff grain/ flour porridge	-½ cup teff grains/flour -1 ½ cup water -One tbsp butter/margarine (optional) -¼ cup milk (optional)	-If using grains pop the grains first -Add water into the flour or popped grains, and stir well -Put the mixture in a saucepan on medium heat; stir well for about 3 min -Bring to a gentle boil, cover, and cook for about 15–20 min -If the porridge is too thick, add some water or milk, and give it a few more minutes -Stir in the cream or milk; sweeten with sugar, honey or maple syrup to taste

Table 3. Examples of teff-based dishes.

As the preferred staple in the Ethiopian and Eritrean dishes, teff flour is used in making engera/injera (pronounced en-jer-a and sometimes spelled injera), a flat sour-like fermented pancake that is used with 'wot', a stew made with spices, meats and pulses, such as lentils, beans and split peas.

In combination with other ingredients which is a better option as enhances nutrient-nutrient interaction, teff grain and teff flour are wonderful alternatives to wheat, barley and rye for those on a gluten-free diet. Teff flour will expand food choices beyond potato, corn and rice flour!.

2.3. Fonio

There are two types, white fonio (*Digitaria exilis*) and black fonio (*Digitaria iburua*), but both are actually a type of millet grain. White fonio is grown in the Sahel area that borders the Sahara Desert, and it grows well in dry and grassy savannah as well as in richer climates. Black fonio is found in Benin, Niger, Nigeria and Togo and is generally less common (and even more nutritious). Although fonio is found all over West Africa, it is especially prized in the Fouta Djallon region of Guinea and Senegal and the Akposso region of Togo and Central Nigeria [18].

Like teff, fonio matures quickly, producing grain in just 6–8 weeks, which makes it the world's fasted maturing cereal. It can therefore be relied upon in semiarid areas with poor soil and unreliable rainfall [2, 18]. After they are mature, fonio's tiny grains must be dried and removed from their husk before they are ready to cook. Before machines did the dehusking, the fonio was dehusked in a mortar and pestle, where the grains were pounded with sand. Fonio could also be slowly toasted in a large pan until it popped out of its husk and then pounded to separate the grain from its covering [18].

A staple in African cuisine and diets, it is prepared steamed as the anchor in many meals and is also milled into flour to be used in baking. It is called the 'seed of the universe' in Malian mythology [19]. It provides 3.6 calories per gramme of grain that is similar to other cereals [20]. Just as teff and amaranth grain, it is also gluten-free making it another great wheat replacement. Fonio is simpler to digest making it suitable for children and older people [20]. Fonio is consumed as a whole grain; the barn and germ of fonio are full of nutrients. A 100 g of boiled whole-grain black fonio has 3.7 g protein, 21 mg calcium, 4.1 mg iron, 9 mg folate, 181 mg magnesium and 3.1 g of fibre [9]. In addition, fonio has essential amino acids methionine and cysteine which jointly help liver function and help in detoxing process. The high fibre content makes it necessary to keep the digestive system smooth. It helps in bowel motions and helps prevent constipation; in certain parts of Africa, fonio is provided as food to individuals struggling with stomach problems [20]. Fonio has got lower glycemic index. It really is absorbed in body gradually and therefore effect on blood sugar increment is gradual. The presence of essential amino acids helps in preventing liver damage and colon cancer and is also useful in drug removal symptoms, whilst the high levels of folic acid as well as iron play an important role in iron metabolism [20]. It is good in avoiding anaemia. This particular nourishing food is typically suitable for pregnant as well as lactating women in Africa [20]. Moreover, because of its insulin-secreting properties, fonio products have found that diabetics are their key customers.

2.3.1. Fonio-based dishes

The small grains are beloved in Burkina Faso, Guinea, Mali and Nigeria, Senegal and Togo, where fonio is a staple part of most people's diets. Fonio is a favourite in salads, stews and porridges [18]. In Togo, fonio is cooked with black-eyed peas and, in other places, mixed with nutrient-dense sesame seeds to add even more vegetarian nutrition [18].

Fonio grains can be cooked whole, or it can be ground into a gluten-free flour and used as a substitute for wheat flour [18]. **Table 4** provides details of selected fonio-based dishes.

Dish name	Ingredients	Cooking procedure and serving recommendations
Simple fonio	-Two cups of fonio grains -Two cups of water -Salt to taste -(Serves four)	-Bring the water to a boil in a pot -Add the salt and the fonio, and cover -Remove from the heat and let stand for 10 min or until the fonio has absorbed all its water -If the fonio is still too chewy for your taste, add a little more boiling water to the pot, cover, and leave for a few more minutes as it absorbs. If there is still water in the pot, put it back on high heat and cook, stirring, until the water boils off -This will substitute rice for a protein-rich, delicious side dish, serve with vegetables, stew or sauce
Paleo pancakes using fonio flour	-Three tsp baking powder -One tsp salt -1 ¼ cup fonio flour -¼ cup cooking oil/margarine -One tsp vanilla extract -One egg -One cup milk/water -One tbsp of maple syrup -(Serves four)	-Mix the fonio flour baking powder and salt -Add ½ of the cooking oil/margarine into the flours and shorted/rub in using finger tips -Add water/milk gradually as you keep mixing to a consistency of your liking -Beat in the egg -Heat a pan; if the pan is non-stick, no need to use oil; if using ordinary pan, put some little oil on the pan; let it heat -Pour a scoop of the mixture in the pan and let it spread or slightly tilt the pan to get it to your preferred size -Cook one side till bubbles appear and begin to dry -Flip it over and cook for a short whilst -Can be served with maple/pancake syrup and/or with ripe bananas and a glass of milk or juice
Fonio Pilaf	-Two cups of fonio rinsed and drained -Two cups of meat cubed -One carrot diced -One cup cabbage sliced -One cup green peas -Two tbsp cooking oil -One sliced onion -Two peeled sliced tomatoes -Pepper/turmeric/garlic (optional) -Ten cups of water -(Serves six to eight)	-Heat oil in the pan and add some sliced onion; cook till the onions are translucent -Add tomato, pepper, turmeric and meat pieces -Add salt to taste, mix everything, and cook for about 2 min -Add to this about ten cups of water and let the ingredients simmer for some time (10–15 min) -Add recently cut cabbage, peas and carrot to mixture and let simmering for additional time -Put the cooked vegetables aside -Add to the mixture two cups of fonio and cook for an additional 5 min -Serve the complete dish
Fonio cake	-1/2 kg fonio flour -1/2 kg wheat flour -650 g sugar -Three eggs -1/2 Cup dried milk -1/2 margarine -10 g yeast -½ litre water -(Serves six to eight)	-Add sugar into the ½ litre water and allow it to dissolve -Add three eggs, ½ cup of dried milk and ½ cup of margarine and mix properly -In a separate bowl, mix the fonio flour and the wheat flour -Gradually add the flour into the mixture and keep kneading with a spatula, up until paste is smooth -Allow this particular paste to stay for half an hour -Preheat oven to 220°C -Grease the cake tins with butter and keep aside -Add the yeast to the dough knead lightly to make sure the yeast is well incorporated -Fill the cake tins with the dough -Put the cake tin in oven for about 10–15 min at 180°C -Let the cakes cool and serve with a few hot chocolate sauces

Dish name	Ingredients	Cooking procedure and serving recommendations
Fonio pudding	-One cup cleaned fonio grains -A pinch of salt -Six cups of water (or more) -Four tbsp. granulated sugar (or as desired) -Milk (as desired) -Raisins/cashew nuts/groundnuts (as desired)—optional -(Serves three to four)	-Place the washed fonio grains in a medium-sized sauce pan, and add about six cups of water to it -Leave to cook on medium heat whilst stirring continuously at the early stage of cooking to prevent lumps from forming. The grains will then lose their grittiness and become fluffy and soft -When the pudding appears thick, remove from heat; add the sugar and milk -You can also add in any other toppings like raisins, cashew nuts or groundnuts -Serve the pudding whilst still hot as it tends to solidify after sitting at room temperature for a whilst -Can be eaten alone or served with akara (bean balls), plantain or even bread

Table 4. Examples of fonio-based dishes.

2.4. Moringa

The *Moringa oleifera* tree is a small tree that is native to the Himalayas of India and was being used in Indian medicine around 5000 years ago. There are also accounts of it being utilized by the ancient Greeks, Romans and Egyptians [21]. Although there are technically 13 different species of moringa tree [21], for simplicity, this chapter is in reference to the *Moringa oleifera* tree and using the common name 'moringa'. This tree was, and still is, considered a panacea and is referred to as the 'The Wonder Tree', 'The Divine Tree' and 'The Miracle Tree' amongst many others. It is priced as a multipurpose tree with all parts usable either as a raw or cooked nutriment, medicine or as a water purification additive. It is also known for its long twisted pods, from which it derives its name. 'Murungai' means 'twisted pod' in the Tamil language [21].

Moringa is beneficial for both food and medicine because of its ability to grow in virtually all countries. Currently, its growth is most prevalent in Africa, Central and South America and Asia. But its effects are being felt around the world [21].

The leaves typically the most common part of the plant are especially high in protein and considered a multivitamin-mineral complex. A 100 g of mature moringa leaves contain 5.7 g protein, 15 mg β-carotene, 459 mg vitamin C, 25 mg vitamin E, 9.2 mg iron and 638 mg calcium [22]. The moringa leaves also contain 18 amino acids, including the 9 essentials: histidine, isoleucine, leucine, lysine, methionine, phenylalanine, threonine, tryptophan and valine [23].

The leaves are harvested and steamed as a green vegetable but are also dried and ground into a powder used in sauces, soups and cooked grains. In these different forms, moringa is used as a vitamin-mineral supplement that is extremely effective at balancing nutritional uptake needed for greater dietary balance. With the high nutrition content, moringa leaves are gaining popularity as a remedy for malnutrition in Africa, especially amongst infants, children and nursing mothers [2]. Moringa could also be a great nutrient safety net because the tree is in full leaf at the end of the dry season when other foods are typically scarce [2]. The seeds and pods can also be eaten just as you would green beans, and the flowers and buds make a nice tea although they contain a laxative effect [24].

2.4.1. *Moringa-based foods*

Whilst the leaves can be steamed or boiled and eaten as a green leafy vegetable with a slight 'bite' taste to them, the moringa leave powder has a mild, somewhat spinach-like taste and works well in smoothies, green juices and soups or sprinkled over most anything. **Table 5** provides details on how moringa can be used as leaves and as powder in selected recipes.

Dish name	Ingredients	Cooking procedure and serving recommendations
Moringa sprinkles	-Moringa powder -Any prepared food	-Sprinkle one tsp of powder on a plate of food, a cup of soup or on a sandwich -Do this a few times throughout the day, and you are set -You can also add the moringa powder to juice or milkshake -Do not cook moringa powder; it diminishes the nutritional value -Honey works well to mask the moringa taste
Moringa, lemon, honey, mint leaves, ginger and ice blend	-Two tsp moringa powder -1 litre chilled water -Two limes, sliced thinly -Two lemons, sliced thinly -Large handful of fresh mint leaves -Three tbsp honey	-Mix as many of the indicated ingredients as you have in a blender -Add plenty of water -How much powder you can handle versus how much honey you need to sweeten the drink depends on your preference -Experiment with the ratios, but know these ingredients all play nice together -To add some more flavour, swap out the water for coconut milk -Leave the mixture to sit in refrigerator for about 2 hours before serving with plenty of ice
Stir fried moringa leaves	-Five bunches washed and drained moringa leaves -One tsp cooking oil -Mustard seeds -Red chillies -A medium-size sliced onion -Salt to taste	-In a hot pan, add the oil -When oil is hot, add mustard seeds; when the mustard seeds begin to pop, add slices onions cook till the onions are slightly browned -Add the moringa leaves bit by bit, constantly stirring to ensure they do not stick at the bottom or clump -When the leaves begin to welt and get soft, add salt -Keep sauntering the leaves, until all the water that tend to come from the leaves during cooking is evaporated -Serve with a starch diet of your choice, i.e. rice and some curry, ugali, matooke, etc.
Scrambled eggs with moringa leaves	-Two eggs -Freshly picked moringa leaves -Cooking oil -Salt to taste -Minced fresh garlic cloves or garlic salt (optional)	-Beat the eggs in a bowl add salt and any other spices -Put oil in a pan; heat slightly -Pour the egg on the hot pan with oil; keep stirring with a cooking spoon -As the eggs solidify and are almost ready, add the moringa leaves -As soon as the moringa leaves turn bright green and wilt a bit, your dish is ready
Moringa leaves with maize/corn and beans	-Two cups of fresh corn off the cob Two cups of fresh beans from the pods -One onion sliced -Two tomatoes sliced -Cooking oil	-Boil the beans for about 30 min separately until they are soft and cooked -Sauté the onions in the oil until they are translucent -Add the corn, the beans and sauté all together, until the corn just starts to brown on the edges a bit -Stir in the moringa leaves; cook for about 3 min covered -The meal is ready and can be consumed alone

Table 5. Examples of moringa-based dishes.

2.5. Baobab fruit

The iconic baobab is a common tree in eastern and southern Africa's savannahs [2]. It is one of the most nutrient-dense wholefoods on the planet. It is a 100% organic, raw superfruit that dries naturally on the branch [25]. In Africa, the baobab fruit has been used medicinally for centuries to treat everything from fevers, malaria and gastrointestinal problems to vitamin C deficiency [26].

The unusually high levels of vitamin C in moringa fruit are what contribute to the great potential health benefits of moringa fruit powder and its fresh counterpart [26]. A 100 g of baobab fruit pulp contains 247 mg vitamin C (nearly four times of the daily requirements); therefore, a single serving of baobab powder (10 g or two to three tsp) will have about 24.7 mg of vitamin C providing about 40% of your daily Vitamin C requirement making baobab fruit one of the best sources of vitamin C in the world [25].

Vitamin C plays a crucial role in our bodies. It contributes to normal collagen formation supporting healthy gums, teeth, skin, bones, cartilage and blood vessels; energy release, energy-yielding metabolism and reduction of fatigue; immune function; functioning of the nervous system and psychological function; and protection of cells from oxidative stress [25].

There are two types of fibre that your body needs: soluble and insoluble and baobab being 50% fibre contains equal quantities of both. The soluble fibre in baobab helps to slow down the release of sugar into the bloodstream, thus reducing energy spikes. Fibre also helps maintain a healthy digestive system including bowel regularity, and the fact that it helps you feel fuller for longer, it can be helpful for weight management [25].

Baobab powder has twice the antioxidants gramme per gramme of goji berries and more than blueberries and pomegranates combined, thus having the highest antioxidant content amongst all fruits [25]. Antioxidants are essential for protecting, repairing and preventing cell damage; supporting the ageing process of the skin particularly over the long term; and neutralizing free radicals, unstable atoms and molecules that can cause damage to your body at the cellular level, increasing the risk of degenerative diseases and other signs of ageing, including wrinkles and fine lines on the skin. Antioxidants counteract oxidative stress and the effects of free radicals (unstable molecules that damage collagen causing skin dryness, fine lines, wrinkles and premature ageing). When fresh baobab pulp is used in cooking or concentrated baobab fruit powder added to dishes, it boosts the supply of beneficial minerals including calcium, copper, iron, magnesium, potassium and zinc. These minerals act both individually and synergistically to perform hundreds of tasks in the human body. A 100 g of fresh baobab pulp contains 295 mg calcium, 1.6 mg copper, 9.3 mg iron, 90 mg magnesium, 1240 mg potassium, 27.9 mg sodium and 1.8 mg zinc [26].

2.5.1. Baobab-based dishes

Baobab fruit is very dry so it keeps almost indefinitely, and it is used to make juice by soaking the fruit and straining out the pulp and seeds [2]. The fruit powder (or fresh baobab fruit if you can get) can be added to your diet (liquid or solid) to enhance your body's fat-burning capacities, especially if you are working on losing weight and your current diet is not rich in vitamin C. Absorption of iron and vitamin C actually increases your body's absorption of iron, which is why vitamin C-rich baobab and iron-rich moringa work so well together. See **Table 6** for selected Baobab-based recipes.

Dish name	Ingredients	Cooking procedure and serving recommendations
Baobab nutri-shake	-One glass of water -Two scoops (6 g) of baobab fruit powder	-Add the baobab powder to the water and shake or stir -It is ready to drink -NB: Use it as a sports drink when working out, or take it in the morning in place of your coffee; sip it mid-afternoon for a pick me up
Tropical baobab-papaya smoothie	-One tsp baobab powder -One cup papaya (pawpaw) -1/2 cup pineapple -Seven cashew nuts/groundnuts (optional)	-Add all the ingredients to a blender and blend until smooth -Add a little water or coconut water if needed to help everything blend or for a thinner texture

Table 6. Examples of baobab-based dishes.

2.6. Hibiscus

The hibiscus plant (*Hibiscus sabdariffa*) is thought to originate from the areas surrounding central Africa presumably Angola. It is also widely cultivated throughout many tropical and subtropical regions, particularly Mexico, India, Thailand and China [27, 28]. Hibiscus is recognized for its large, colourful flowers that are often used as decorative pieces in gardens and homes. When the petals of the hibiscus flower begin to detach from the main plant, underneath they reveal flower bud-like structures known as calyces. These deep red buds are subsequently used to produce hibiscus tea and hibiscus extract. Whilst the tea is popular with health-conscious consumers, the extract is more versatile. It can be used in a number of different food and beverages and still maintaining its health-promoting properties [27].

Hibiscus is a rich source of vitamins and microelements including 13 organic acids. A 100 g of hibiscus tea will provide approximately 6, 31 and 48% of the daily values of vitamin A, vitamin C and iron [28]. These nutrients and microelements boost your immune system. It makes your blood vessels stronger, lowers the blood pressure and the level of 'bad' cholesterol and even has an antibacterial effect; this makes it prevent and reduce symptoms of metabolic syndrome (a combination of diabetes, obesity and high blood pressure), thus reducing the risk of developing heart disease and stroke [29]. It is also good for the pancreas and liver [28]. Hibiscus erases post-effects of alcohol intoxication and contains antioxidant elements similar to those of red wine; this helps the extract act as an anti-solar agent, by absorbing skin-damaging ultraviolet radiation from the sun. The antioxidants also minimize cell damage from free radicals, which may help to slow down the natural ageing process and reduce the risk of developing a number of age-related diseases [28, 29].

Hibiscus extract is also thought to promote a healthy digestive system. It has antibacterial properties, which may help to maintain a favourable gut flora. Hibiscus can also act as a mild laxative, which may help with symptoms of constipation and indigestion. One study also found that the extract demonstrates anti-urolithiatic activity, meaning that it can help reduce the formation of kidney stones. All these properties explain why the Arabs call carcade 'the remedy from all illnesses'. It is said that three cups of red tea per day is enough to get the most of it. On the other hand, like any remedy, it is not good for everyone. People with low blood pressure, ulcer or gastritis should be very careful with it.

2.6.1. *Hibiscus-based dishes*

When dried hibiscus flowers are steeped in hot water, the dark red hibiscus tea is called karkadeh/karkady in Arabic and is popular in North Africa, particularly Egypt and Sudan where it is used to not only maintain normal body temperature, support heart health and encourage fluid balance [2, 29] but also at wedding celebrations as a toasting drink [2]. In West Africa, it is known as bissap, tsoborodo or wonjo; bissap is called the 'national drink of Senegal'. It is either served hot (it loses a bit of its characteristic sour) or can also be served chilled with ice [2].

The recipes in **Table 7** show that hibiscus powder is added to hot or cold water to make a simple, slightly tart-tasting tea, but it is also increasingly used as a functional ingredient in many applications, from sorbets to confectionary. It can also be used as a colouring and flavouring agent in jams, relishes, sauces and baked goods. In addition, hibiscus extract has been applied as a colourant and antioxidant in the skin and hair care applications.

Dish name	Ingredients	Cooking procedure and serving recommendations
Hibiscus water	-Six cups of water -One cup of dried hibiscus petals -Simple syrup to taste -(Syrup—bring one cup of water and one cup of sugar to a boil until sugar dissolves)	-Boil six cups of water -Remove from heat/turn off the heat and add the hibiscus -Let it sit in the hot water for 3 min -Drain the water and add the simple syrup to taste -Let it cool down and serve cold with a lot of ice
Hibiscus iced tea	-1/2 cup dried hibiscus flowers (about 1/2 ounce or 15 g) -One cinnamon stick -Four cups of cold water -Two tbsp to 1/4 cup simple syrup -Lime wedges (optional, for serving)	-Put the hibiscus and cinnamon stick in a large container or bowl and add water -Cover and refrigerate overnight (8–12 hours). Add simple syrup to taste -Strain out the solids and serve with ice and a squeeze of lime, if desired -The tea can be taken immediately or brewed covered in the refrigerator for up to one week -You can use four hibiscus tea bags instead of the loose flowers -You can use honey or a sweetener in place of the syrup
Hibiscus smoothie	-One tbsp dried hibiscus flowers -1/2 cup boiling water for brewing -Five strawberries/raspberries -One ripe banana	-Brew the hibiscus tea in the boiling water for about 5 min. Allow tea to cool -Once tea is cooled, add the brewed tea, 1/2 of the soaked flowers and the rest of the ingredients to a blender, and process till smooth. Add water to thin out if desired

Table 7. Examples of hibiscus-based foods.

2.7. Tamarind

Tamarind trees are native to tropical Africa but found in tropical regions throughout the world [2]. The tree produces an abundance of long, curved, brown pods filled with small brown seeds, surrounded by a sticky pulp that dehydrates naturally to a sticky paste. The pods look a bit like huge, brown, overly mature green beans [30].

Just as the other ancient foods do, tamarind has a long history of being used as a medical remedy. It has been known to ease stomach discomfort, aid digestion and act as a laxative [30]. Tamarind preparations are used to relieve fevers, sore throat, rheumatism, inflammation and sunstroke. Dried or boiled tamarind leaves and flowers are made into compresses used for swollen joints, sprains, boils, haemorrhoids and conjunctivitis. Similar to the natural gums and pectins found in other foods, the tamarind sticky pulp contains non-starch polysaccharides, which contribute to its high dietary fibre content (5.1 g/100 g fruit pulp). They bind with bile to help flush waste through the colon, decreasing the chances of it sticking around, thus reducing chances of colon cancer. Prized for its sweet-and-sour flavour, tamarind (also known as ukwaju in Swahili) is used to make juice and is rich in vitamins, minerals and antioxidants [2]. 100 g of tamarind contain 36% of the thiamin, 35% of the iron, 23% of magnesium and 16% of the phosphorus recommended for a day's worth of nutrition [30]. Other prominent nutrients include niacin, calcium, vitamin C, copper and pyridoxine. Tamarinds also contain high levels of tartaric acid, just as citrus fruits contain citric acid, providing not just a zing to the taste buds but evidence of powerful antioxidant action against harmful free radicals floating through your system [30].

Other phytochemicals found in tamarinds include limonene, safrole (a natural oil also found in sassafras), geraniol (a natural antioxidant with a rose-like scent), methyl salicylate (a plant essence with counterirritant properties), cinnamic acid, pyrazine and alkyl-thiazoles (natural flavours and fragrances derived from plants and vegetables). Each of these phytochemicals brings their own healing property and flavour to the fruit's overall make-up [30]. In addition, due to its ability to restore electrolyte imbalance during dehydration, many East African coastal communities will serve a glass of ukwaju (tamarind juice) to a guest coming in from a hot day or as a hangover remedy [2].

2.7.1. Tamarind-based dishes

In addition to being used alone as a drink/juice/tea, **Table 8** shows that the pulp from tamarind fruit is also used as a spice and souring agent in sauces, marinades, salads, stir fries, even sorbets and cool refreshing summer drinks. The English word 'tamarind' is taken from the Arabic tamar-hindi, meaning 'Indian date', and it is popular in equatorial cuisines, such as Indian, Mexican and Thai. Also known as imli, tamarind is used as a souring agent in many cuisines, especially those of South and Southeast Asia. There, you will find it in curries, stirred into drinks, made into relishes and sauces and even cooked down into a sweet-and-spicy dessert paste. The pulp can be pressed to form a 'cake' or processed to make a paste. When used in marinades as indicated earlier, besides adding flavour, the fruit's natural acidity helps to tenderize tougher cuts of beef by breaking down the fibres in the meat. Marinated overnight in a tamarind-tinged liquid, beef becomes succulent and tender. But it is important to note that when marinating fish or chicken, if left in the marinade too long, the tamarind will begin to chemically 'cook' it.

Dish name	Ingredients	Cooking procedure and serving recommendations
Tamarind juice	•80 g tamarind •2 l water •Sugar or honey to taste	-Bring water to a boil -Remove the brown shell from the tamarind to expose the pulp, and place the tamarind pulp in a heatproof container/mixing bowl -Pour boiling water over the tamarind, cover, and allow to sit for about 15 min -Leave to cool completely; then using your hands, separate seeds from the pulp -Once done, strain with a strainer -Dilute the juice with more water and add sugar or honey to taste -Refrigerate and serve chilled -Cinnamon, vanilla or grated ginger can be added to enhance the taste
Easy tamarind rice	-One tamarind -Small lemon ball size (without seeds) -One cup boiled rice, one tsp salt, ¼ tsp hing, ½ tbsp cooking oil -A handful of roasted peanuts -¼ tsp turmeric powder—two or three red chillies	-Take 150 ml of water in a container, add turmeric powder and tamarind -Pressure cook this for 2–3 whistles or boil covered for 30 min -Once cooled mash the contents nicely and keep aside -Heat oil in a non-stick saucepan -Add peanuts and any other spices and fry till golden brown -Add the mashed contents and salt -Let the mixture boil for 5–7 min and tamarind paste is ready -Add the required quantity of tamarind paste to cooked rice and mix well -Serve with fried potato chips/ugali/cooked banana -NB: Tastes best if eaten 3–4 hours after it is made

Table 8. Examples of tamarind-based dishes.

3. Conclusions

Just as in the Amazon, there are millions of edible plants, insects and animals within tropical Africa that are not only nutrient rich but also contain essential elements beneficial in preventing and/or managing a range of health conditions that are of great public concern. Amaranth, teff, fonio, moringa leaves, baobab fruit, tamarind and hibiscus leaves just to mention a few are some of the superfoods that can be used alone but more importantly transformed into health superdiets to provide simple, acceptable and sustainable remedies that not only address malnutrition but also play a major role in the battle against non-communicable diseases like cardiovascular diseases (heart attacks and stroke), cancers, chronic respiratory diseases and diabetes.

Acknowledgements

The author first acknowledges the authors of all articles cited in this chapter. Secondly, Bioversity International www.bioversityinternational.org is thanked for the staff time spent during the writing of this chapter. Last but not least, the author acknowledges her family for bearing with her as she spent some family time in putting this chapter together.

Author details

Beatrice Nakhauka Ekesa

Address all correspondence to: b.ekesa@cgiar.org

Bioversity International, Kampala, Uganda

References

[1] FAO, IFAD and WFP. The State of Food Insecurity in the World 2015. Meeting the 2015 International Hunger Targets: Taking Stock of Uneven Progress. 2015. Rome, FAO. Available from http://www.fao.org/3/a-i4646e.pdf [Accessed: 24th May 2016]

[2] Christine M. The Secret is Out: 10 African Super Foods you Need to Eat and Drink Right Now [Internet]. 1st June 2015. Available from: http://mgafrica.com/article/2015-05-30 [Accessed: 24th May 2016]

[3] What is a Superfood. Global Health Centre. 3rd June 2014.[Internet] Available from http://www.globalhealingcenter.com/natural-health/what-is-a-superfood/ [Accessed: 2nd December 2016]

[4] Joseph M, Food Facts, What is Amaranth Good for? 2016. [Internet] Available from http://foodfacts.mercola.com/amaranth.html [Accessed 29th September 2016]

[5] Republic of South Africa (2010). Amaranthus Production Guideline. Department of Agriculture, Forestry and Fisheries, South Africa. 2010. Available from http://www.nda.agric.za/docs/Brochures/Amaranthus.pdf [Accessed: 29th September 2016]

[6] DianaBuja's Blog. Amaranth Greens (Lenga-Lenga)—Politically Correct, Easy to Grow, and Delicious. 16th November 2012 [Internet]. Available from https://dianabuja.wordpress.com [Accessed: 23rd September 2016]

[7] National Academy of Sciences. Amaranth, Lost Crops of Africa, Volume II: Vegetables. 2016. Available from https://www.nap.edu/read/11763/chapter/3#45 [Accessed: 10th August 2016]

[8] Innovateus. The 12 Health Benefits of Eating Amaranth Leaves. Available from http://www.innovateus.net/health/12-health-benefits-eating-amaranth-leaves [Accessed: 10th September 2016]

[9] FAO. West African Food Composition Tables. 2012. Available from http://www.fao.org/ [Accessed: 11th August 2016]

[10] Organic Facts. Health Benefits of Amaranth. [Internet] Available from https://www.organicfacts.net/health-benefits/vegetable/amaranth.html Accessed: 10th August 2016]

[11] Whole Grains Council. Amaranth-May Grain of the Month. May 2016. Available from http://wholegrainscouncil.org/whole-grains-101/easy-ways-enjoy-whole-grains/grain-month-calendar/amaranth-may-grain-month [Accessed: 10th August 2016]

[12] Cholorexa. Amaranth (Kiwicha) [Internet]. Available from http://colorexa.com/products/amaranth-kiwicha/ [Accessed: 13th August 2016]

[13] ECHO. Amaranth Grain and Vegetable Type. ECHO Technical Note. 2008. [Internet] Available from https://c.ymcdn.com/sites/echocommunity.site-ym.com/resource/collection/E66CDFDB-0A0D-4DDE-8AB1-74D9D8C3EDD4/Amaranth_Grain_&_Vegetable_Types_ [Accessed: 13th August 2016]

[14] DorisPiccinin,MSRD,DepartmentofNutritionandFoodServiceContributor(s):Tsegazeab Woldetatios, PhD, Agronomy, Medical Interpreter at Harborview Medical Center. More About Ethiopian Food: Teff. 14th December 2010. [Internet] Available from https://ethnomed.org/clinical/nutrition/more-about-ethiopian-food-teff [Accessed:14th August 2016]

[15] Patel A. Teff Benefits: 8 Healthy Reasons to Add Teff to Your Diet, The Huffingtone Post, Canada. 2nd June 2014. Available from http://www.huffingtonpost.ca/2014/02/06/teff-benefits-_n_4740219.html [Accessed: 16th September 2016]

[16] Whole Grains Council. Teff and Millet-November Grains of the Month. November 2016. Available from http://wholegrainscouncil.org/whole-grains-101/easy-ways-enjoy-whole-grains/grain-month-calendar/teff-and-millet-%E2%80%93-november-grains [Accessed 16th September 2016]

[17] Danielle C. Step Aside Quinoa: Is Teff the New 'Superfood' Grain? Teff Tribe. 23rd March 2015. Available from http://tefftribe.com.au/2016/01/08/good-food-step-aside-quinoa-is-teff-the-new-superfood-grain/ [Accessed: 16th September 2016]

[18] African Epicure. Fonio. Available from http://www.africanepicure.com/about-african-food/starches-and-artisanal-grains/fonio/ [Accessed: 16th September 2016]

[19] FARAFENA. Fonio Flour, Superfoods. Available from https://farafena.com/fonio-flour/ [Accessed: 16th September 2016]

[20] Value Food. Health Benefits of Fonio, Pulses and Cereals. Nutrition and Health Information Portal. Available from http://www.valuefood.info/1765/health-benefits-of-fonio/ [Accessed: 16th September 2016]

[21] Wellness Mama. Is Moringa Really a Superfood? 5th January 2017. [Internet] Available from http://wellnessmama.com/119304/moringa-superfood/ [Accessed: 25th January 2017]

[22] Yang R, Chang L, Hsu J, Weng BBC, Palade MC, Chadha ML, Virginie L. Nutritional and Functional Properties of Moringa Leaves-From Germplasm, to Plant, to Food, to Health. Moringa and Other Highly Nutritious Plant Resources: Strategies, Standards and Markets for a Better Impact on Nutrition in Africa. Accra, Ghana, November 16–18, 2006. Available from http://miracletrees.org/moringa-doc/from-germplasm-to-plant-to-food-to-health.pdf [Accessed: 17th September 2016]

[23] Moringa Leaf Company. Moringa: A Great Source of Protein for Vegetarians and Vegans.19th June 2016. [Internet] Available from http://www.moringaleafcompany.com/blogs/moringa-news/tagged/high-protein-foods [Accessed: 18th September 2016]

[24] One Green planet. All About Moringa: The Uber Nutritious Superfood that could Combat Malnutrition.[Internet] 13th October 2014. Available from http://www.onegreenplanet.org/natural-health/all-aboutmoringa-the-uber-nutritious-superfood-that-could-combat-malnutrition/ [Accessed: 22nd October 2016]

[25] Aduna. Baobab Superfruit: Health and Beauty Benefits. Available from https://aduna.com/pages/baobab-benefits [Accesed: 12th September 2016]

[26] Heal With Food. Health Benefits of Baobab Fruit Powder (A Superfood Packed with Vitamin C). Available from http://www.healwithfood.org/health-benefits/baobab-fruit-powder-superfood.php [Accessed: 12th September 2016]

[27] Superfoods Ingredients (SFI). Wholesale Superfood Ingredients in the Spotlight: High Antioxidant Hibiscus Extract. Available from http://www.sfisuperfoods.com/superfood-ingredients-Hibiscus.html [Accessed; 11th September 2016]

[28] Jon J. 10 Surprising Health Benefits of Hibiscus Tea, Foods and Nutrition. Doctors Health Press. 2nd August 2015. Available from http://www.doctorshealthpress.com/food-and-nutrition-articles/hibiscus-tea-benefits [Accessed: 11th September 2016]

[29] Hudson T The Surprising Health Benefits of Hibiscus, Gaia Herbs. 9th July, 2013. Available from http://www.gaiaherbs.com/articles/detail/42/The-Surprising-Health-Benefits-of-Hibiscus [Accessed: 12th September 2016]

[30] Joseph M, Food Facts, What Is Tamarind Good For? 2016. [Internet] Available from http://foodfacts.mercola.com/tamarind.html [Accessed: 12th September 2016]

Chapter 6

The Mediterranean Diet in the Prevention of Degenerative Chronic Diseases

Elisabetta Della Valle, Francesco Cacciatore, Eduardo Farinaro, Francesco Salvatore, Roberto Marcantonio, Saverio Stranges and Maurizio Trevisan

Additional information is available at the end of the chapter

http://dx.doi.org/10.5772/67119

Abstract

Degenerative chronic diseases are a problem related to the aging phenomenon of industrialized countries due to the increase of risk factors and related comorbidity such as overweight, obesity, metabolic syndrome, diabetes, hypertension and hyperlipidemia with a consequent increased risk of cardiovascular disease (CVD) and cancer. Moreover, the significant reduction of physical activity in daily life and the huge growth in food availability have considerably increased the risk of such diseases. Particular attention should be paid to primary prevention by means of health strategies based on improvement in lifestyle intervention such as implementation of Mediterranean diet and promotion of physical activity programs. In this chapter, the protective effect of Mediterranean diet and the role of certain foods and/or their constituents are analyzed; the possible mechanisms by which Mediterranean diet is effective in the prevention of cardiovascular and other chronic diseases are presented, in particular the effects exerted by antioxidants, polyphenols, fibers, unsaturated fatty acids, and alcohol. The genetic revolution in the past decades has produced new fields of study where the interaction between foods, nutrients, and our genetic makeup is investigated. The relationship between nutrigenetics and nutrigenomics and the Mediterranean diet are the future area that research should discover.

Keywords: functional foods, Mediterranean diet, cardiovascular diseases, chronic diseases, prevention

1. Introduction

The progressive improvement of socioeconomic conditions, which occurred in the second half of the last century in industrialized countries, has produced a major change in lifestyle with a significant reduction of physical activity, due to the mechanization of work activities and transport systems, and thus of total energy expenditure, and a contemporary huge increase in food availability. Eating habits have changed substantially and have acquired two characteristics: excess and inadequacy. The biggest change is the adoption of a high-calorie diet, rich in animal fats, cholesterol, refined sugars, salt, and alcohol, with a low ratio of nutrients to calories. These aspects of modern life have favored an increase in overweight and obesity and consequently also the frequency of diabetes, hypertension, and hyperlipidemia. The genetic constitution, meanwhile, has remained the same as that of primitive man as the human genome has not had the time to adapt to the new environment that instead has changed rapidly. The natural selection has favored the appropriate mechanisms to address the deficiency of food rather than those to limit weight gain; consequently, most of the current diseases are the result of a precarious genetic adaptation to the new environment created by man [1]. In addition, improved sanitary conditions, together with the introduction of antibiotic therapy and vaccination, have led to an increase in lifespan: humanity in general has aged and presents all the issues related to aging. There has been a gradual, steady increase of chronic degenerative diseases and in some countries there has been a real epidemic of these diseases and in particular of cardiovascular diseases (CVDs). The social impact of these diseases is considerable, especially for costs, both direct and indirect. The first type of cost is related to medical interventions necessary to the care (hospitalization, drugs, and rehabilitation), while the indirect costs are related to the loss of productivity and the need to replace the person affected by the disease in a period of his/her life characterized by a high professional qualification in its business work. CVDs are still the leading cause of death and disability in many industrialized countries. Recently, also in developing countries or economies in transition, we are observing a continuous and rapid increase in CVD. Smoking is considerably widespread, especially among women, and among the youngest which has further increased the risk of CVD.

These are already the first cause of death worldwide with the exception of Sub-Saharan Africa [2]. They represent a crucial issue in public health and, consequently, their prevention, especially at primary level, is an essential point in the choice of health strategies developed by the governments of several countries.

The importance of environmental factors in the pathogenesis of CVD is recognized as certain and among the most important environmental factors is lifestyle, defined as eating habits and physical activity.

The genetic revolution in the past decades has produced new fields of study focused on the interaction between foods, nutrients, and human genetic make-up to investigate our predisposition and ability to prevent or treat CVD, cancer, diabetes, obesity, cognitive decline and dementia, and inflammatory bowel disease [3–5]. Experimental data demonstrated that environment and foods could regulate gene expression and structure [6]. Food constituents and nutrients may induce the change of structure and function of genes and may be able to prevent or cause specific diseases; these new areas of study are called nutrigenomics and nutrigenetics [7]. Nutrigenomics aims at relating in the population the effects of certain foods on human health on the basis of genetic predisposition. It will therefore be possible, in the next few years, to identify the best strategy for the prevention of many chronic degenerative diseases, and with specific tests, it will be possible to understand which foods are the most suitable and which ones need to be avoided. Nutrigenetics specifically investigates the modifying effects of inheritance in nutrition-related genes on micronutrient uptake and metabolism as well as dietary effects on health. In this way, it is possible to hypothesize a diet tailored to the patient, based on his/her genotype, the quality, and quantity of the daily required nutrients to his/her body, with determination of minimum and maximum amounts needed to obtain the most benefits. These two branches of science can combine genetics with nutrition trying to play an active preventive role in defense of the organism; this is the new pathway for genetics applied to nutrition. A new frontier has been opened and has created a new scientific approach to prevention based on genetics. Proper and targeted feeding combined with the genotypic diversity of each individual will allow us to clarify the guidelines for the prevention of a large number of diseases and will allow the development of new experimental therapies, aimed at the treatment of complex pathologies such as metabolic diseases. A proper and balanced diet is essential for a long and healthy life, but it is not the same for everyone; modern genetic testing allows us to determine the best suited diet to each individual. This systemic approach, based on genetics, once fully operational, will provide results that can be functional to the improvement of human well-being. The typing of biomolecules with enhanced nutritional properties will be reflected on the dietary recommendation with a more accurate and effective action of prevention and population health protection.

Actually, nutrigenomics and nutrigenetics are still in a beginning phase, without definite scientific evidence that the effects observed in experimental and small clinical studies have real clinical implications. We have to rely on existing scientific evidence, suggesting nutritional models known to be effective on the health of individuals and communities. One of the most widespread diets is the "*Mediterranean diet (MD)*."

2. Mediterranean diet

The term Mediterranean diet has become a synonymous of a healthful and tasteful pattern of eating. The MD is a way to "enjoy food" while ensuring a long and healthy life. The increasing interest is the consequence of numerous studies conducted around the world in the last 50 years, when the famous nutritionist Ancel Keys launched and organized the Seven Countries Study, an epidemiological study that analyzed the role of diet and other cardiovascular risk factors on cardiovascular disease and death [8–10]. The study originated from the observations that in nations such as Greece, Italy, and Japan the cases of myocardial infarction (at least those in the hospitals) were much lower than those he had observed in Minnesota, in Netherlands, and in Finland.

The observation was conducted in 16 cohorts enrolled in Finland, Greece, Italy, Japan, the Netherlands, USA, and former Yugoslavia. The Seven Countries Study was one of the first examples of international collaboration in medical research and has represented, over the years, the groundbreaking evidence on the effect of diet on health and in cardiovascular and chronic disease epidemiology. The main findings of the study were the demonstration of a significant association between coronary heart disease (CHD) and diet, particularly positive correlation between the consumption of saturated fatty acids and CHD and relevant inverse relationship between the consumption of monounsaturated fat and CHD. The 15-year mortality follow-up demonstrated an inverse association between coronary deaths and the ratio of the dietary monounsaturated/saturated fats [8–10]. Olive oil has been considered one of the principal components of the MD. Wine, garlic, fish, vegetables, legumes, almonds, and other nuts, other constituents of this dietary pattern, have also been identified to have beneficial effects on health [11, 12]. The data of 15-year follow-up of the Seven Countries Study have been followed up by numerous evidences showing important inverse relationships between the MD, and/or its elements, and either CVD or its risk factors [13–16].

More recent studies have demonstrated that MD and its components may be a powerful aid against certain conditions, such as diabetes, stroke, dementia, colorectal cancer (CRC), and mortality. The greater part of the findings up to now comes from epidemiological studies, even if the cause-effect relationship is not so clear. A recent clinical trial study, based on randomized population, showed the positive outcomes of the MD in the prevention of CVD in individual at a high risk for this disease [17]. A recent study aimed to evaluate the effects of adherence to MD on survival on a large sample of 71,333 Swedish men and women, followed up for 14 years, demonstrated a linear dose-response association between the MD score average and the length of life with the higher score associated with longer survival. The difference in the average length of life between subjects with extremes scores (0 vs. 8) of MD was up to 2 years [18]. The PREDIMED trial was performed using an energy-unrestricted MD, enriched with nuts or extra-virgin oil; the relative risk of cardiovascular events was the reduction of approximately 30% in people who were free of CVD at the beginning of the study, reinforcing the evidence of the MD in the primary prevention of CVD with relevant risk reduction [17]. MD is also effective in reducing the rate of cardiovascular complications after myocardial infarction in the secondary prevention as demonstrated in the Lyon Diet Heart Study where a large reduction in rates of coronary heart disease events was observed with a modified MD enriched with alpha-linolenic acid (a key constituent of walnuts) [19]. More recently, the ATTICA study carried out between 2001 and 2002 on 3024 prevalently male individuals between 20 and 89 years living in the province of Attica (Greece) demonstrated on individuals free of CVD or chronic viral infections that higher the level of adherence to the traditional MD pattern lower the risk of left ventricular systolic dysfunction in patients affected by acute coronary syndrome [20, 21].

3. Effective components contained in the Mediterranean diet

The apparent ability of the traditional MD to reduce the risk of CVD, cancer, and degenerative diseases development and progression has been attributed, at least in part, to the content of micronutrients and compounds with antithrombotic and antioxidant capacity.

3.1. Antioxidants

It is conceivable that the protective effect of the MD, which guarantees a regular intake of substances with antioxidant activity (ascorbic acid, α-tocopherol, retinol, and β-carotene), estimated that 10–100 mg per day is to be ascribed essentially to its ability to maintain constantly high antioxidant capacity in the blood [22, 23]. The abundance of fruits and vegetables, along with extra-virgin olive oil, red wine, aromatic herbs (oregano, parsley, and rosemary), garlic, onion, and pepper (ingredients generously used in Mediterranean cuisine), offers a number of phenolic compounds with a strong antioxidant action that is hardly possible to achieve with other types of diet. Examples are allyl sulfides, which are present in garlic and raw onions, give cardiovascular benefits, improve cognitive ability, and have chemopreventive activity; it was shown that certain isothiocyanates (degradation products of glucosinolates, compounds present in caper berries) can affect the cell cycle and induce apoptosis in HT-29 human colon cancer cells and other isothiocyanates, present in high concentration in cruciferous vegetables (cabbage and broccoli) [24, 25], have the capacity to modulate the metabolism of carcinogens; kaemferolo and flavonoids quercetin and hydrocinnamic acids from capers have well-known anti-inflammatory and antioxidant effects and chives also rich in phenolic compounds with diuretic, antihypertensive, anti-inflammatory, and antioxidant substances [26–28]; catechins fruit (e.g., apple skin and grape) antioxidant molecules prevent the production of reactive oxygen species generated by oxidative stress; the anthocyanins, plant pigments, give the red or blue color to fruits and vegetables (berries, eggplant, black grapes, and red beet), are antioxidants, photoprotective, and are able to inhibit angiogenesis. One other major constituent of MD is vitamin E, which contains a group of eight isomers: four tocopherols (α, β, γ, δ-tocopherol) and four tocotrienols (α, β, γ, δ-tocotrienol). There are several studies demonstrating healthful effects of α-tocopherol, while little is known on γ-tocopherol, the main form of vitamin E in food. In the last 20 years, much of the supposed beneficial effects of antioxidant vitamins were not confirmed in controlled clinical trials [29–31]. However, it is hard to believe that such vitamins may adverse the development of CVD events when administered in patients with advanced stages of the disease, while a protective effect could be supposed in population in which this nutrient is present throughout the life.

Lycopene, a carotenoid present in tomatoes and tomato products, is a dietary antioxidant that has received great consideration. Epidemiological studies have demonstrated a lower incidence of CVD in those with higher consumption of tomatoes and lycopene, confirmed also by lycopene levels in serum and adipose tissue [32–34]. A protective effect on acute myocardial infarction (AMI) with an odds ratio of 0.75 was found in one of the earlier studies that investigated the serum antioxidant status and lycopene [35]. The most remarkable population-based evidence from a multicenter case-control study (EURAMIC) [36] indicated lycopene levels, and not β-carotene, to be protective against myocardial infarction with an odds ratio of 0.52 comparing the 10th to the 90th percentiles. In the Malaga region, the component of EURAMIC study adipose tissue lycopene levels showed an odds ratio of 0.39 [37]. In Atherosclerosis Risk in Communities (ARIC) case-control study, fasting serum antioxidant levels were inversely related to the intima-media thickness with an odds ratio of 0.81 [38]. Although these epidemiological studies provide convincing evidence for the role of lycopene in CVD prevention, they can only suggest but not prove a causal relationship between lycopene intake and the risk of CVD. Such a proof can be obtained only by performing controlled clinical dietary intervention studies where both the biomarkers of the status of oxidative stress and the disease are measured.

3.2. Polyphenols

Polyphenols are the most abundant antioxidants in the diet, present in fruits and plant-derived beverages such as fruit juices, tea, coffee, red wine, cereals, chocolates, and dry legumes. The total dietary intake could be as high as 1 g/d; this is 10 times higher than the intake of vitamin C and 100 times higher than the intakes of vitamin E and carotenoids [39].

Despite the wide distribution in plants, the effects of polyphenols on health have come to the consideration of nutritionists only in recent times. Polyphenols and other antioxidants were considered to protect cell constituents against oxidative damage, through scavenging of free radicals for many years. Nowadays, this theory is drastically changed; polyphenols give signals principally through the receptors or enzymes related to signal transduction and the signal may lead to modification of the redox status of the cell, and may activate a series of redox-dependent responses [40]. Many evidences on the prevention of diseases exerted by polyphenols derives from in vitro or animal experiments, which are often done with higher doses than those humans exposed with a regular diet [41, 42].

Epidemiological studies are necessary to establish the effects of polyphenol consumption on CVD [43]. Moreover, it was shown that short- and long-term black tea consumption increases plasma flavonoids and reverses endothelial dysfunction in CVD patients [44].

All these observations suggest that polyphenols can protect vascular damage via antioxidant effects and nitric oxide restoration. However, clinical trials using different antioxidants have failed to demonstrate preventive effects on major CVD events. One imaginable explanation for this discrepancy is that experimental studies are not comparable to real life in humans, although very useful to understand pathophysiological mechanisms [45]. Moreover, antioxidants quantity used in studies conducted in humans may not have been appropriate, and/or the state of disease too severe to evaluate the protective effect that could probably be existent in a preclinical state.

3.3. Dietary fiber

Dietary fiber (DF) has been widely studied and numerous evidences support the health benefits of its consumption. Several prospective studies have demonstrated the inverse association between DF intake and cardiovascular risk. An important pooled analysis of 10 cohort studies demonstrated that DF consumption was inversely related to coronary heart disease. Thence, the introduction of functional foods enriched in DF—alone or in combination with other bioactive compounds—in the diet may represent a useful strategy to improve the cardio-metabolic profile in high-risk subjects preventing cardio-metabolic diseases. The promotion in use of both natural and functional foods might facilitate adherence to a healthy diet with a higher fiber intake compared with the common nutritional conducts of western populations.

Cohort studies have found a consistent protective effect of dietary fibers on glucose control and serum lipoproteins in diabetic patients [46] and in turn on CVD [47].

However, the biologic mechanisms of fibers on the cardiovascular system have yet to be fully elucidated. In the Nurses Health Study, women in the highest quintile of fiber intake had an age-adjusted relative risk for major coronary events that was 47% lower than women in the lowest quintile [48].

Practical recommendations for CVD prevention include food-based approach favoring an increased intake of whole-grain and dietary fiber (especially soluble fiber), fruits, and vegetables providing a mixture of different types of fibers [49].

3.4. Unsaturated fatty acids

Dietary sources of n-3 polyunsaturated fatty acids (PUFAs) include fish oils, rich in eicosapentaenoic acid (EPA) and docosahexaenoic acid (DHA), along with plants rich in a-linolenic acid. Regular consumption of fish, characteristic of the MD, allows satisfying the need for omega-3 fatty acids. PUFAs contained in fish regulate effectively hemostatic factors, cancer, and hypertension, and play a crucial role in the maintenance of neural functions in humans and in the prevention of certain psychiatric disorders; evidence from epidemiological and clinical secondary prevention trials suggests a significant role of fish consumption on long-term (20-year) mortality from coronary heart disease and n-3 PUFA in the prevention of CHD and arrhythmias [50, 51]. Prospective randomized trials show a favorable impact on CV health of both fish and plant sources of n-3 PUFAs. The omega-6 is present mainly in vegetable oils (sunflower and corn oils which, however, should not be cooked since these oils are thermolabile). Among them, the linoleic acid content in nuts, grains, legumes, corn and sunflower oil, synthesized, comes from the gammalinoleic acid (or GLA) [52]. Randomized secondary prevention clinical trials with fish oils and a-linolenic acid have demonstrated a reduction in risk that compare favorably to those seen in landmark secondary prevention trials with lipid-lowering drugs. A meta-analysis of randomized trials involving patients with cardiac disease showed that supplementation with the marine n-3 fatty acids EPA and DHA reduced the rate of death from coronary heart disease by 20% [53].

The beneficial effects of olive oil on cardiovascular disease risk factors are now recognized and often attributed only to the high levels of monounsaturated fatty acids (MUFAs). The olive oil is a functional food. Secondary components of olive oil, which constitute only 1–2% of the total virgin olive oil content, are classified into two types: the unsaponifiable fraction, defined as the fraction of the oil extracted after saponification through the use of solvents, and the soluble fraction which includes the phenolic residual. The unsaponifiable fractions of the components are hydrocarbons (squalene), tocopherols, fatty alcohols, triterpene alcohol, 4-methylsterols, sterols, terpene, and other polar compound pigments (chlorophyll and pheophytins). The accumulation of scientific evidence suggests that flavoring and seasoning foods with olive oil bring great health benefits including reducing the risk of coronary heart disease and preventing various cancers (by inhibiting proliferation, inducing apoptosis, and minimizing DNA damage). Also, it appears to have a role in bone mineralization reducing the risk of osteopenia and osteoporosis [54].

3.5. Alcohol

Although the excessive consumption of alcohol must be discouraged due to the significant health damage to individuals and societies [55, 56], increasing evidence shows that moderate consumption of alcoholic beverages may decrease CVD [57]. A dose-response relation between wine intake and vascular risk resulted in a J-shaped curve, with a significant risk reduction at moderate (one to two drinks) consumption (trend analysis $p = 0.032$) [58].

Data derived from PREDIMED demonstrated that moderate red-wine consumption is associated with a lower prevalence of the metabolic syndrome in an elderly Mediterranean population at a high cardiovascular risk [59].

The protective effects of alcohol have been primarily explained by an action on blood lipids (increase in high-density lipoprotein (HDL) levels) and platelets (decreased aggregation) resulting in a reduced rate of coronary artery obstruction [58]. Moderate drinking may improve the early outcomes after AMI and prevent sudden cardiac death, suggesting a direct effect of ethanol on the ischemic myocardium that has been referred to as "ethanol preconditioning" [60].

Moreover, a protective effect of moderate alcohol intake is demonstrated by the Italian Longitudinal Study on Aging (ILSA). In this study, participants with moderate cognitive alterations who consumed approximately 15 g of alcohol a day (moderate drinkers) experienced a decreased rate of progression toward dementia compared to non-drinkers. In the same study, alcohol consumption in older age is associated with healthier hematological values of fibrinogen, HDL cholesterol, Apo A-I lipoprotein, and insulin [61].

4. Effects of the Mediterranean diet on health

4.1. Effects of the Mediterranean diet on cancer and other degenerative diseases

As life expectancy increases, there are an increased number of elderly individuals suffering from cardiovascular disease, dementia, and cancer.

In sedentary people eating Western-type diets, aging is associated with several chronic diseases, including type 2 diabetes mellitus, cancer, and cardiovascular diseases. About 80% of elderly (over 65 years of age) have at least one chronic disease, and 50% have at least two chronic diseases, with an increase in disability related to comorbidity [62]. Data from epidemiological studies and clinical trials indicate that many age-associated chronic diseases can be prevented, and even reversed, by the implementation of healthy lifestyle interventions [63]. Recent data demonstrate that higher Mediterranean-type diet adherence and higher physical activity were independently associated with a reduced risk for Alzheimer disease [64].

Epidemiological burden of cancer in Mediterranean countries is lower when compared to other states, such as the UK and the USA. There is increasing evidence that Mediterranean dietary adherence reduces the risk of several cancer types and cancer mortality. Particularly, high consumption of fruits and vegetables, whole grains, and little assumption of processed meat, characteristic aspects of the Mediterranean diet, is inversely related to the risk of tumor pathogenesis at different cancer sites. Observational studies provide new evidence suggesting that high adherence to a MD is associated with a reduced risk of overall cancer mortality as well as a reduced risk of incidence of cancers of the colorectum, aerodigestive tract, breast, stomach, pancreas, prostate, liver, and head and neck [65]. A recent review and meta-analysis of 23 observational studies with an overall population of 1,784,404 demonstrated that the highest adherence to MD was significantly associated with a 13% lower risk of all-cause cancer mortality, 17% colorectal cancer, 7% breast cancer, 27% gastric cancer, 4% prostate cancer, 42% liver cancer, 60% head and neck cancer, 52% pancreatic cancer, and 90% respiratory cancer. The meta-analyses confirm a prominent and consistent inverse association provided by adherence to MD in relation to cancer mortality and the risk of several cancer types [66].

The Healthy Ageing: a Longitudinal Study in Europe (HALE) Study, which evaluated 3496 participants in 10 European countries, reported that individuals between 70 and 90 years who follow up an MD experienced more than 50% reduction in all-cause mortality [67]. The EPIC study designed to clarify the relationship between diet, environmental factors, lifestyles, and the incidence of cancer and other chronic diseases demonstrated in the Spanish cohort that a lower incidence of cancer (12% reduction) is observed in those with a greater adherence to MD after an 8-year follow-up. The EPIC study has also shown that the contemporary consumption of more components of the diet has a greater effect than the single-component assumption [68].

In an epidemiological study investigating the role of both Dietary Inflammatory Index (DII) and Mediterranean Diet Score (MDS), the DII was positively associated with a risk of lung cancer in current smokers while the MDS was inversely associated with lung cancer risk overall (hazard ratio (HR) = 0.64) and for current smokers (HR = 0.38), demonstrating a protective effect even more evident in high-risk patients [69].

The Women's Health Initiative Observational Study assessed the association between diet quality index scores on Healthy Eating Index 2010 (HEI-2010), Alternative HEI-2010, alternative Mediterranean Diet Index, and the Dietary Approaches to Stop Hypertension (DASH and colorectal cancer (1993–2012)), a US study of postmenopausal women. During an average of 12.4 years of follow-up, there were 938 cases of CRC and 238 CRC-specific deaths. Closer adherence to HEI-2010 and DASH dietary recommendations was inversely associated with a risk of CRC in this large cohort of postmenopausal women [70].

Data from PREDIMED find an effect of a long-term dietary intervention on breast cancer incidence, suggesting a beneficial effect of an MD supplemented with extra-virgin olive oil in the primary prevention of breast cancer. The multivariate-adjusted hazard ratios were 0.32 for the MD with extra-virgin olive oil group and 0.59 for the MD with nuts group. In analyses with yearly cumulative updated dietary exposures, the hazard ratio for each additional 5% of calories from extra-virgin olive oil was 0.72 [71]. Mediterranean diet appeared to exert a protective effect also on hip fracture in two Swedish cohort studies consisting of 37,903 men and 33,403 women (total $n = 71,333$, mean age 60 years) free of previous cardiovascular disease and cancer who answered a medical and a food-frequency questionnaire in 1997. One unit increase in modified Mediterranean diet score (mMED; range 0–8 points) was associated with 6% lower hip fracture. Comparing the highest quintile of adherence to the mMED (6–8 points) with the lowest (0–2 points) conferred an adjusted HR of hip fracture of 0.78 [72].

4.2. Reduction in caloric intake in lowering incidence of degenerative disease

A more drastic nutritional interventions and implementation of physical activity programs may have additional beneficial effects on several metabolic and hormonal factors, implicated in the etiology of degenerative diseases and aging [73–75].

The traditional MD means also a diet with a reduced caloric intake, at least referred to the past century; caloric restriction (CR) can be defined as the reduction of all dietary nutrients, except vitamins and minerals (to avoid malnutrition), and has recently emerged as the most promising pro-longevity/anti-aging candidate measure); in fact, it is a highly robust phenomenon capable of slowing aging [76]. Moderate CR can prevent or reverse the damaging effects of visceral obesity, insulin resistance, type 2 diabetes, high blood pressure, dyslipidemia, and inflammation. Energy deficits induced by CR and physical activity in overweight and obese subjects are accompanied by similar improvements in glucose tolerance and insulin action, and similar reductions in several major CHD risk factors, with a loss after 2 years of intervention of 14 kg [77].

CR improves metabolic status also in normal-weight individuals. Data from a series of studies conducted in a group of self-imposed CR (approximately 30% reduction in daily calories) show that a prolonged CR determines sustained beneficial effects on lipid profile, blood pressure, and carotid artery intima-media thickness [78].

Finally, weight loss obtained with an energy deficit of 500–750 kcal per day from their daily energy requirement and exercise was effective in improving the score of physical functioning in obese elderly with frailty, improving, body composition, bone mineral density, physical functions, and quality of life [79].

4.3. Dietary supplements

Dietary supplementation has increased significantly in the last years because of the perception that antioxidant vitamins and minerals may reduce the risk of CVD, cancer, and other chronic diseases. However, no clear evidence in chronic disease prevention is demonstrated for dietary supplements, at least among healthy individuals in the general population [80]. Nevertheless, from a public health perspective, it is extremely important to understand the effects of nutrients on health. At the moment, the use of selenium as supplement in the diet to prevent cardio-metabolic disease is not justified and thus not to be encouraged.

Longitudinal epidemiological studies have led to the identification of recognized functional foods and dietary patterns as beneficial in the primary prevention of cardio-metabolic diseases. The mechanisms by which these foods exert their protective effects are complex and probably related to the macro- and micronutrient contents of the food [81]. The benefits may depend on the clinical status due to risk factors, and state of diseases, may be dose dependent, and may be affected by the food preparation. The benefits of functional food have been reproduced using isolated components of foods as supplements. At the moment, randomized, double-blind, placebo-controlled trials of clinical end points are necessary to establish the efficacy in modifying cardiovascular risk profile in humans.

5. Future directions of nutrigenetics and nutrigenomics in the Mediterranean diet

Increasing evidence enhances the idea that functional foods may improve health status by means of physiologically active components [81]. This area of research is now developing and additional studies are necessary to demonstrate the potential benefit of those foods for which the diet-health relationships are not yet scientifically validated.

A personalized diet based on specific nutrition strategies exerts a pivotal role in the treatment of phenylketonuria, galactosemia, and fructose intolerance, diseases known as "single-gene autosomal recessive disorders." More than 6000 human monogenic disorders have been identified, including over hundreds of protein-based metabolic disorders. Some are rare and complex dietary diseases, namely fatty-acid oxidation disorders, organic acid metabolism disorders, urea cycle defects, and glycogen storage disease. Patients may reduce their intake of the dietary substrates or metabolites that accumulate in these conditions and nutrigenetics will improve prevention and treatment by identifying specific mutations or haplotype combinations that modulate the dietary response in affected patients [81]. In multifactorial pathologies such as CVD, obesity, type 2 diabetes mellitus, cancer, and so on, nutrigenomic studies have shown that dietary intervention may modulate the onset and progression of the disease.

Recently, there has been notable progress in gene-environment interaction evaluation; this field is now accessible to patients to help them to improve their health. Therefore, the current challenge for nutritional genomics is to clarify the role of food and the human microbiota in human health, to better understand the relationship between them and to use this knowledge to promote and preserve a healthy status [82, 83].

6. Conclusions

Scientific research and the wider dissemination of its results made aware the industrialized countries population of the strong connection between nutrition and health, and the role of certain foods and/or their constituents in maintaining this balance. This helps to clarify the role of diet in the prevention and control of morbidity and premature mortality caused by non-communicable diseases. Adaptations to the diet can not only influence today's health but also act in determining whether a person will develop or not, in the course of his/her life, diseases such as cancer, cardiovascular diseases, or diabetes. A healthy diet based on the balance between nutrients represents the first preventive intervention to protect the health and physical harmony. As a result, today, nutrition has new meanings. The concept of food has undergone a radical modification to the point of attributing to each food, in addition to its intrinsic nutritional and sensory properties, an important role in maintaining health and the psycho-physical well-being. Improving eating habits and increasing physical activity levels will reduce the risk of death and disability related to chronic diseases. The practical implications of these recommendations should lead to increased consumption of fruits, vegetables, and fish, and to change the quality of fats and oils, as well as the amount of sugar and starch, by acting as much as possible to match the Mediterranean diet dictates that is found in the Hippocrates famous quote "Let food be thy medicine and let thy medicine be food," the needed action for the twenty-first century population.

Author details

Elisabetta Della Valle[1], Francesco Cacciatore[2], Eduardo Farinaro[1]*, Francesco Salvatore[4], Roberto Marcantonio[1], Saverio Stranges[3] and Maurizio Trevisan[5]

*Address all correspondence to: elisabetta.dellavalle@unina.it

1 Department of Public Health, Federico II University, Naples, Italy

2 U.O. of Cardiac Rehabilitation, Salvatore Maugeri Foundation, IRCCS, Telese Terme Institute, Benevento, Italy

3 Department of Epidemiology and Biostatistics, Schulich School of Medicine & Dentistry, University of Western Ontario, London, Canada

4 CEINGE-Advanced Biotechnologies, Naples, Italy

5 CUNY School of Medicine, City College of New York, New York, USA

References

[1] Neel JV. Diabetes mellitus: a "thrifty" genotype rendered detrimental by "progress"? Am J Hum Genet 1962;4:353–62.

[2] Mathers CD, Boerma T, Ma Fat D. Global and regional causes of death. Br Med Bull 2009;92:7–32.

[3] Müller M, Kersten S. Nutrigenomics: goals strategies. Nat Rev Genet 2003;4:315–22.

[4] Ordovas JM, Mooser V. Nutrigenomics and nutrigenetics. Curr Opin Lipidol 2004;15:101–8.

[5] Roberfroid MB. Concepts and strategy of functional food science: the European perspective. Am J Clin Nutr 2000;71(Suppl.):1660Se4S.

[6] Saladin R, de Vos P, Guerre Millotá M, Leturqueá A, Girardá J, Staels B, Auwerx J. Transient increase in obese gene expression after food intake or insulin administration. Nature 2002;377:527–8.

[7] Debusk RM, Fogarty CP, Ordovas JM. Nutritional genomics in practice: where do we begin. J Am Med Assoc 2005;105:589–98.

[8] Keys A. Seven countries. A multivariate analysis of death and coronary heart disease. Cambridge, MA: Harvard University Press; 1980.

[9] Keys A, Aravanis C, Van Buchem F, Blackburn H, Buzina R, Djordjevic BS, et al. The diet and all causes death rate in the Seven Countries Study. Lancet 1981;2:58e61.

[10] Keys A, Menotti A, Karvonen MJ, Aravanis C, Blackburn H, Buzina R, et al. The diet and 15year death rate in the Seven Countries Study. Am J Epidemiol 1986;124:903e15.

[11] Matalas AL, Zampelas A, Stavrinos V, Wolinsky I. The Mediterranean diet: constituents and health promotion. Boca Raton, FL: CRC Press; 2001.

[12] Hoffman R, Gerber M. The Mediterranean diet: health and science. Hoboken, NJ, Wiley Blackwell; November 2011.

[13] Trevisan M, Krogh V, Freudenheim J, Blake A Muti P, Panico S, et al. Consumption of olive oil, butter, and vegetable oils and coronary heart disease risk factors. JAMA 1990;263:688–92.

[14] Mente A, de Koning L, Shannon HS, Anand SS. A systematic review of the evidence supporting a causal link between dietary factors and coronary heart disease. Arch Intern Med 2009;169(7):659–69.

[15] de Lorgeril M, Renaud S, Salen P, Monjaud I, Mamelle N, Martin JL, et al. Mediterranean alph-alinolenic acid rich diet in secondary prevention of coronary heart disease. Lancet 1994;343:1454–9.

[16] Trichopoulou A, Bamia C, Trichopoulos D. Mediterranean diet and survival among patients with coronary heart disease in Greece. Arch Intern Med 2005 Apr 25;165:929–35.

[17] Estruch R, Ros E, Salas Salvadó J, Covas MI, Corella D, Arós F, et al for the PREDIMED Study Investigators. Primary prevention of cardiovascular disease with a Mediterranean diet. N Engl J Med 2013;368:1279–90.

[18] Bellavia A, Tektonidis TG, Orsini N, Wolk A, Larsson SC. Quantifying the benefits of Mediterranean diet in terms of survival. Am J Epidemiol 2016 Jul 1;184:23–32.

[19] de Lorgeril M, Salen P, Martin JL, Monjaud I, Delaye J, Mamelle N. Mediterranean diet, traditional risk factors, and the rate of cardiovascular complications after myocardial infarction: final report of the Lyon Diet Heart Study. Circulation 1999;99:779–85.

[20] Chrysohoou C, Panagiotakos DB, Pitsavos C, Das UN, Stefanadis C. Adherence to the Mediterranean diet attenuates inflammation and coagulation process in healthy adults: The ATTICA Study. J Am Coll Cardiol 2004; 44:152–8

[21] Rees K, Hartley L, Flowers N, Clarke A, Hooper L, Thorogood M, Stranges S. 'Mediterranean' dietary pattern for the primary prevention of cardiovascular disease. Cochrane Database Syst Rev 201312;(8):CD009825.

[22] Trichopoulou A, Costacou T, Bamia C, Trichopoulos D. Adherence to a Mediterranean diet and survival in a Greek population. N Engl J Med. 2003;348:2599–608.

[23] Trichopoulou A, Bamia C, Trichopoulos D. Anatomy of health effects of Mediterranean diet: Greek EPIC prospective cohort study. BMJ 2009;338:2337.

[24] Gamet-Payrastre L, Li P, Lumeau S, Cassar G, Dupont MA, Chevolleau S, Gasc N, Tulliez J, Tercé F. Sulforaphane, a naturally occurring isothiocyanate, induces cell cycle arrest and apoptosis in HT29 human colon cancer cells. Cancer Res 2000;60:1426–33.

[25] Parnaud G, Li P, Cassar G, Rouimi P, Tulliez J, Combaret L, Gamet-Payrastre L. Mechanism of sulforaphane-induced cell cycle arrest and apoptosis in human colon cancer cells. Nutr Cancer 2004;48:198–206.

[26] Kulisic-Bilusic T, Schmöller I, Schnäbele K, Siracusa L, Ruberto G. The anticarcinogenic potential of essential oil and aqueous infusion from caper (Capparis spinosa L). Food Chem. 2012 May 1;132:261–7.

[27] Storniolo CE, Casillas R, Bulló M2,, Castañer O4,, Ros E,3, Sáez GT6,3, Toledo E, Estruch R, Ruiz-Gutiérrez V, Fitó M, Martínez-González MA, Salas-Salvadó J, Mitjavila MT, Moreno JJ. A Mediterranean diet supplemented with extra virgin olive oil or nuts improves endothelial markers involved in blood pressure control in hypertensive women. Eur J Nutr 2015;8:1–9.

[28] Urpi-Sarda M, Casas R, Chiva-Blanch G, Romero-Mamani ES, Valderas-Martínez P, Arranz S, Andres-Lacueva C, Llorach R, Medina-Remón A, Lamuela-Raventos RM, Estruch R. Virgin olive oil and nuts as key foods of the Mediterranean diet effects on inflammatory biomarkers related to atherosclerosis. Pharmacol Res 2012;65:577–83.

[29] Napoli C, Cacciatore F. Novel pathogenic insights in the primary prevention of cardiovascular disease. Prog Cardiovasc Dis 2009;51:503–23.

[30] Dai J, Miller AH, Bremner JD, Goldberg J, Jones L, Shallenberger L, Buckham R, Murrah NV, Veledar E, Wilson PW, Vaccarino V. Adherence to Mediterranean diet is inversely associate with circulating Interleukin-6 among middle aged men. A Twin Study. Circulation 2008;117:169–75.

[31] Napoli C, Palinski W. Clinical and pathogenic implications of maternal hypercholesterolemia during pregnancy for the later development of atherosclerosis. Eur Heart J 2001;22:4–9.

[32] Murr C, Winklhofer-Roob BM, Schroecksnadel K, Maritschnegg M, Mangge H, Böhm BO, Winkelmann BR, März W, Fuchs D. Inverse association between serum concentrations of neopterin and antioxidants in patients with and without angiographic coronary artery disease. Atherosclerosis. 2009;202:543–9.

[33] Ohira T, Hozawa A, Iribarren C, Daviglus ML, Matthews KA, Gross MD, Jacobs DR Jr. Longitudinal association of serum carotenoids and tocopherols with hostility: the CARDIA Study. Am J Epidemiol 2008;167:42–50.

[34] Hozawa A, Jacobs DR Jr, Steffes MW, Gross MD, Steffen LM, Lee DH. Relationships of circulating carotenoid concentrations with several markers of inflammation, oxidative stress, and endothelial dysfunction: the Coronary Artery Risk Development in Young

Adults (CARDIA)/Young Adult Longitudinal Trends in Antioxidants (YALTA) study. Clin Chem 2007;53:447–55.

[35] Street DA, Comstock GW, Salkeld RM, Schüep W, Klag MJ. Serum antioxidant and myocardial infarction: are low levels of carotenoids and alpha-tocopherol risk factors for myocardial infarction? Circulation 1994;90:1154–61.

[36] Kohlmeier L, Kark JD, Gomez-Gracia E, Martin BC, Steck SE, Kardinaal AF, Ringstad J, Thamm M, Masaev V, Riemersma R, Martin-Moreno JM, Huttunen JK, Kok FJ. Lycopene and myocardial infarction risk in the EURAMIC Study. Am J Epidemiol 1997;146:618–26.

[37] Gomez-Aracena J, Sloots J, Garcia-Rodriguez A. Antioxidants in adipose tissue and myocardial infarction in Mediterranean area. The EURAMIC study in Malaga. Nutr Metab Cardiovasc Dis 1997;7:376–82.

[38] Iribarren C, Folsom AR, Jacobs DR Jr, Gross MD, Belcher JD, Eckfeldt JH. Association of serum vitamin levels, LDL susceptibility to oxidation, and autoantibodies against MDA-LDL with carotid atherosclerosis. Arterioscler Thromb Vasc Biol 1997;17:1171–77.

[39] Manach C, Scalbert A, Morand C, Rémésy C, Jiménez L. Polyphenols: food sources and bioavailability. Am J Clin Nutr 2004;79:727–47.

[40] Aviram M, Volkova N, Coleman R, Dreher M, Reddy MK, Ferreira D, Rosenblat M. Pomegranate phenolics from the peels, arils, and flowers are antiatherogenic: studies in vivo in atherosclerotic apolipoprotein e-deficient (E 0) mice and in vitro in cultured macrophages and lipoproteins. J Agric Food Chem 2008;56:1148–57.

[41] Vita JA. Polyphenols and cardiovascular disease: effects on endothelial and platelet function. Am J Clin Nutr 2005;81:292S–97S.

[42] Sies H, Schewe T, Heiss C, Kelm M. Cocoa polyphenols and inflammatory mediators. Am J Clin Nutr 2005;81:304S–412S.

[43] Arts ICW, Hollman PCH. Polyphenols and disease risk in epidemiologic studies. Am J Clin Nutr 2005;81:317S–25S.

[44] Duffy SJ, Keaney JF Jr, Holbrook M, Gokce N, Swerdloff PL, Frei B, Vita JA. Short- and long-term black tea consumption reverses endothelial dysfunction in patients with coronary artery disease. Circulation 2001;104:151–6.

[45] Esposito K, Marfella R, Ciotola M, Di Palo C, Giugliano F, Giugliano G, D'Armiento M, D'Andrea F, Giugliano D. Increased alpha-linolenic acid intake lowers C-reactive protein, but has no effect on markers of atherosclerosis. Eur J Clin Nutr 2004;58:1083–9.

[46] Rivellese A, Riccardi G, Giacco A, Pacioni D, Genovese S, Mattioli PL, Mancini M. Effect of dietary fibre on glucose control and serum lipoproteins in diabetic patients. Lancet 1980;2:447–50.

[47] King DE. Dietary fiber, inflammation, and cardiovascular disease. Mol Nutr Food Res 2005;49:594–600.

[48] Wolk A, Manson JA, Stampfer MJ, Colditz GA, Hu FB, Speizer FE, Hennekens CH, Willett WC. Long-term intake of dietary fiber and decreased risk of coronary heart disease among women. JAMA 1999;281:1998–2004.

[49] Gidding SS, Lichtenstein AH, Faith MS, Karpyn A, Mennella JA, Popkin B, Rowe J, Van Horn L, Whitsel L. Implementing American Heart Association pediatric and adult nutrition guidelines: a scientific statement from the American Heart Association Nutrition Committee of the Council on Nutrition, Physical Activity and Metabolism, Council on Cardiovascular Disease in the Young, Council on Arteriosclerosis, Thrombosis and Vascular Biology, Council on Cardiovascular Nursing, Council on Epidemiology and Prevention, and Council for High Blood Pressure Research. Circulation 2009;119:1161–75.

[50] Kromhout D, Bosschieter EB, de Lezenne Coulander C. The inverse relation between fish consumption and 20-year mortality from coronary heart disease. N Engl J Med 1985 May 9;312(19):1205–9.

[51] Lee JH, O'Keefe JH, Lavie CJ, Marchioli R, Harris WS. Omega-3 fatty acids for cardioprotection. Mayo Clin Proc 2008;83:324–32.

[52] Baer DJ, Judd JT, Clevidence BA, Tracy RP. Dietary fatty acids affect plasma markers of inflammation in healthy men fed controlled diets: a randomized crossover study. Am J Clin Nutr 2004;79:969–73.

[53] Leon H, Shibata MC, Sivakumaran S, Dorgan M, Chatterley T, Tsuyuki RT. Effect of fish oil on arrhythmias and mortality: systematic review. BMJ 2009;338:a2931.

[54] Fernández-Real JM, Bulló M, Moreno-Navarrete JM, Ricart W, Ros E, Estruch R, Salas-Salvadó J. A Mediterranean diet enriched with olive oil is associated with higher serum total osteocalcin levels in elderly men at high cardiovascular risk. J Clin Endocrinol Metab 2012;97:3792–8.

[55] Trevisan M, Krogh V, Farinaro E, Panico S, Mancini M. Alcohol consumption, drinking pattern and blood pressure: analysis of data from the Italian National Research Council Study. Int J Epidemiol 1987;16:520–7.

[56] Stranges S, Wu T, Dorn JM, Freudenheim JL, Muti P, Farinaro E, Russell M, Nochajski TH, Trevisan M. Relationship of alcohol drinking pattern to risk of hypertension: a population-based study. Hypertension 2004;44:813–9.

[57] Mukamal KJ, Conigrave KM, Mittleman MA, Camargo CA Jr, Stampfer MJ, Willett WC, Rimm EB. Roles of drinking pattern and type of alcohol consumed in coronary heart disease in men. N Engl J Med 2003;348:109–18.

[58] de Gaetano G, De Curtis A, di Castelnuovo A, Donati MB, Iacoviello L, Rotondo S. Antithrombotic effect of polyphenols in experimental models: a mechanism of reduced vascular risk by moderate wine consumption. Ann N Y Acad Sci 2002;957:174–88.

[59] Tresserra-Rimbau A, Medina-Remón A, Lamuela-Raventós RM, Bulló M, Salas-Salvadó J, Corella D, Fitó M, Gea A, Gómez-Gracia E, Lapetra J, Arós F, Fiol M, Ros E, Serra-

Majem L, Pintó X, Muñoz MA, Estruch R; PREDIMED Study Investigators. Moderate red wine consumption is associated with a lower prevalence of the metabolic syndrome in the PREDIMED population. Br J Nutr 2015;113(Suppl 2):S121–30.

[60] Imamura G, Bertelli AA, Bertelli A, Otani H, Maulik N, Das DK. Pharmacological preconditioning with resveratrol: an insight with iNOS knockout mice. J Physiol Heart Circ Physiol 2002;282:H1996–2003.

[61] Perissinotto E, Buja A, Maggi S, Enzi G, Manzato E, Scafato E, Mastrangelo G, Frigo AC, Coin A, Crepaldi G, Sergi G. ILSA Working Group Alcohol consumption and cardiovascular risk factors in older lifelong wine drinkers: the Italian Longitudinal Study on Aging. Nutr Metab Cardiovasc Dis 2010;20:647–55.

[62] Cacciatore F, Gallo C, Ferrara N, Abete P, Paolisso G, Canonico S, Signoriello G, Terracciano C, Napoli C, Varricchio M, Rengo F. Morbidity patterns in aged population in southern Italy. A survey sampling. Arch Gerontol Geriatr 1998;26:201–13.

[63] Abete P, Della Morte D, Mazzella F, D'Ambrosio D, Galizia G, Testa G, Gargiulo G, Cacciatore F, Rengo F. Lifestyle and prevention of cardiovascular disease in the elderly: an Italian perspective. Am J Geriatr Cardiol 2006;15:28–34.

[64] Scarmeas N, Luchsinger JA, Schupf N, Brickman AM, Cosentino S, Tang MX, Stern Y. Physical activity, diet, and risk of Alzheimer disease. JAMA 2009;302:627–37.

[65] Schwingshackl L, Hoffmann G. Does a Mediterranean-type diet reduce cancer risk? Curr Nutr Rep 2016;5:9–17.

[66] Schwingshackl L, Hoffmann G. Adherence to Mediterranean diet and risk of cancer: an updated systematic review and meta-analysis of observational studies. Cancer Med 2015;4:1933–47.

[67] Knoops KT, de Groot LC, Kromhout D, Perrin AE, Moreiras-Varela O, Menotti A, van Staveren WA. Mediterranean diet, lifestyle factors, and 10-year mortality in elderly European men and women: the HALE project. JAMA 2004;292:1433–39.

[68] Buckland G, Agudo A, Travier N, Huerta JM, Cirera L, Tormo MJ, Navarro C, Chirlaque MD, Moreno-Iribas C, Ardanaz E, Barricarte A, Etxeberria J, Marin P, Quirós JR, Redondo ML, Larrañaga N, Amiano P, Dorronsoro M, Arriola L, Basterretxea M, Sanchez MJ, Molina E, González CA. Adherence to the Mediterranean diet reduces mortality in the Spanish cohort of the European Prospective Investigation into Cancer and Nutrition (EPIC-Spain). Br J Nutr 2011;106:1581–91.

[69] Hodge AM, Bassett JK, Shivappa N, Hébert JR, English DR, Giles GG, Severi G. Dietary inflammatory index, Mediterranean diet score, and lung cancer: a prospective study. Cancer Causes Control 2016;27:907–17.

[70] Vargas AJ, Neuhouser ML, George SM, Thomson CA, Ho GY, Rohan TE, Kato I, Nassir R, Hou L, Manson JE. Diet quality and colorectal cancer risk in the women's health initiative observational study. Am J Epidemiol 2016;184:23–32.

[71] Toledo E, Salas-Salvadó J, Donat-Vargas C, Buil-Cosiales P, Estruch R, Ros E, Corella D, Fitó M, Hu FB, Arós F, Gómez-Gracia E, Romaguera D, Ortega-Calvo M, Serra-Majem L, Pintó X, Schröder H, Basora J, Sorlí JV, Bulló M, Serra-Mir M, Martínez-González MA. Mediterranean diet and invasive breast cancer risk among women at high cardiovascular risk in the PREDIMED Trial: a randomized clinical trial. JAMA Intern Med 2015;175:1752–60.

[72] Byberg L, Bellavia A, Larsson SC, Orsini N, Wolk A, Michaëlsson K. Mediterranean Diet and Hip Fracture in Swedish Men and Women. J Bone Miner Res 2016;31:2098–105.

[73] Fontana L, Klein S. Aging, adiposity, and calorie restriction. JAMA 2007;297:986–94.

[74] Farinaro E, Stamler J, Upton M, Mojonnier L, Hall Y, Moss D, et al. Plasma glucose levels: long-term effect of diet in the Chicago coronary prevention evaluation program. Ann Intern Med 1977;86:147–54.

[75] Tuomilehto J, Lindström T, Eriksson JG, Valle TT, Hämäläinen H, Ilanne-Parikka P, et al for the Finnish Diabetes Prevention Study Group. Prevention of type 2 diabetes mellitus by changes in lifestyle among subjects with impaired glucose tolerance. N Engl J Med 2001;344:1343–50.

[76] Weindruch R, Walford RL. The retardation of aging and disease by dietary restriction. Springfield, IL: Charles C Thomas Publisher; 1988.

[77] Esposito K, Di Palo C, Maiorino MI, Petrizzo M, Bellastella G, Siniscalchi I, Giugliano D. Long-term effect of Mediterranean-style diet and calorie restriction on biomarkers of longevity and oxidative stress in overweight men. Cardiol Res Pract. 2011:293916, 5 pages.

[78] Fontana L. Modulating human aging and age-associated diseases. Biochim Biophys Acta 2009;1790(10):1133–8.

[79] Villareal DT, Chode S, Parimi N, Sinacore DR, Hilton T, Armamento-Villareal R, Napoli N, Qualls C, Shah K. Weight loss, exercise, or both and physical function in obese older adults. N Engl J Med 2011;364:1218–29.

[80] Rautiainen S, Manson JE, Lichtenstein AH, Sesso HD. Dietary supplements and disease prevention - a global overview. Nat Rev Endocrinol 2016;12:407–20.

[81] Assmann G, Buono P, Daniele A, Della Valle E, Farinaro E, Ferns G, Krogh V, Kromhout D, Masana L, Merino J, Misciagna G, Panico S, Riccardi G, Rivellese AA, Rozza F, Salvatore F, Salvatore V, Stranges S, Trevisan M, Trimarco B, Vetrani C. Functional foods and cardiometabolic diseases International Task Force for Prevention of Cardiometabolic Diseases. Nutr Metab Cardiovasc Dis 2014;24:1272–300.

[82] Konstantinidou V, Ruiz LAD, Ordovás JM. Personalized nutrition and cardiovascular disease prevention: from Framingham to PREDIMED. Adv Nutr 2014;5:368S–71S.

[83] Alverdy J, Gilbert J, DeFazio JR, Sadowsky MJ, Chang EB, Morowitz MJ, Teitelbaum DH. Proceedings of the 2013 A.S.P.E.N. Research workshop: the interface between nutrition and the gut microbiome: implications and applications for human health. JPEN J Parenter Enteral Nutr 2014;38(2):167–78.

Chapter 7

Isoflavones: Vegetable Sources, Biological Activity, and Analytical Methods for Their Assessment

Daniela-Saveta Popa and Marius Emil Rusu

Additional information is available at the end of the chapter

http://dx.doi.org/10.5772/66531

Abstract

Phytoestrogens are natural compounds found in various plant species and they have the ability to bind to the estrogenic receptors, exerting agonist and/or antagonist effects. The main classes of phytoestrogens are isoflavones, lignans, and coumestranes. Isoflavones are plant bioactive nonsteroidal polyphenolic metabolites with antioxidant properties. They have a very close structure with 17β-estradiol and possess estrogenic/antiestrogenic effects. The main dietary source of isoflavones is soy (*Glycine max* L.). Other legumes, such as red clover (*Trifolium pratense* L.), alfalfa (*Medicago sativa* L.), and *Genista* species, have important content in isoflavones, showing nutritional or phytotherapeutic interest. In plants, isoflavones can be found mainly as non-active glycosides which are converted after ingestion, in the corresponding aglycones (e.g., genistein, daidzein) that have pharmacological activity. Many studies have demonstrated the benefits of dietary isoflavones in menopause and multiple chronic pathologies, including cardiovascular diseases, osteoporosis, and hormonal cancers. Dietary intake of isoflavones is widespread, mainly due to the consumption of soybean products. Analytical methods applied for the quantification of isoflavones allow both assessment of dietary intake of isoflavones and highlighting natural sources with phytotherapeutic potential. Health benefits of isoflavones justify the interest for this class of functional food; therefore, further clinical and epidemiological studies are required.

Keywords: nutraceuticals, phytoestrogens, isoflavones, vegetables, analysis

1. Introduction

Phytoestrogens are natural nonsteroidal compounds able to bind to estrogenic receptors and have both estrogenic and antiestrogenic activities. They are widespread in the plant kingdom being considered ubiquitous. The main classes of phytoestrogens are isoflavones, coumestans, and lignans.

Isoflavones are plant-derived secondary metabolites with a polyphenolic structure and antioxidant properties [1]. They pertain to the flavonoid class and are found mostly in plants belonging to Fabaceae family. Soy (*Glycine max* L.) is the major natural source of isoflavones, and the benefits associated with a soy diet occur mostly because of these phytochemicals. Other natural sources of isoflavones are red clover (*Trifolium pratense* L.), alfalfa (*Medicago sativa* L.), and species of the genus *Genista*. All of these plants present phytotherapeutic and nutraceutical significance, and their by-products, herbal teas, and food supplements are often used.

Several epidemiological studies have demonstrated the benefits of dietary isoflavones in menopause and multiple chronic pathologies, including cardiovascular diseases, osteoporosis, and hormonal cancers. The main mechanisms of action of isoflavones, their benefits to human health, and the factors involved in the modulation of their bioactivity are shown in this chapter. Moreover, the analytical methods used for their quantification in plant and food samples are introduced. These are very important methods to evaluate the human exposure to isoflavones and also to assess the optimum intake for human well-being.

2. Characteristics of isoflavones

2.1. Chemistry and metabolism of isoflavones

Isoflavones (IFs) are yellow pigments derived from 3-phenyl-benzopyrone (3-phenyl-chromone) structure. They are found in plants mostly as biologically inactive glycosides: 7-*O*-β-D-glycosides, 6″-*O*-acetyl-7-*O*-β-D-glucosides, and 6″-*O*-malonyl-7-*O*-β-D-glycosides [1, 2]. After ingestion, glycosides are not bioavailable to be absorbed through enterocytes [3]. They are hydrolyzed into bioactive aglycones by both intestinal mucosa and bacterial β-glucosidases from the gut microbiota. Only these forms are absorbed into systemic circulation directly or after subsequent metabolism in the bowel by intestinal bacteria [3]. Soybeans incorporate predominantly genistin, daidzin, and glycitin as inactive glycosides, which are hydrolyzed into their corresponding biologically active aglycones: genistein, daidzein, and glycitein. Other isoflavones observed in legumes are ononin and sissotrin, with their aglycones, formononetin, and biochanin A, respectively (**Figure 1**).

The absorption of aglycones is fast and efficient. Plasmatic isoflavone levels increase up to micromolar-level values after the consumption of soy-based foods, compared to the nanomolar (≤40 nm) levels found in diets without soy [4]. First pharmacokinetic study on isolated and purified isoflavones was performed, when a single dose of 50 mg of aglycone or the equivalent dose of β-glycoside, respectively, was given to healthy adult volunteers. The plasmatic peak values (Cmax) were 341 ± 74 ng/mL for genistein and 194 ± 30.6 ng/mL for daidzein. The times when the values reached the peaks were 5.2 and 6.6 hours (tmax) in the case of direct aglycones ingestion and 9.3 and 9.0 h in the case of the ingestion of β-glycosides, genistin, and daidzin, due to the time required for their hydrolysation. The bioavailability of genistein and daidzein (based on the area under the curve in plasma concentration *versus* time graph) was higher after consumption of β-glycosides [5].

Formononetin and biochanin A can be transformed to daidzein and genistein, respectively, through 4′-*O*-demethylation by the gut microflora or in the liver [6]. Aglycones can be further metabolized through several steps: reduction, deoxygenation, hydroxylation, and C-ring

http://dx.doi.org/10.5772/66531

Aglicones	R_1	R_2	R_3	R_4
Genistein	H	H	OH	H
Daidzein	H	H	H	H
Formononetin	H	H	H	CH_3
Glycitein	H	OCH_3	H	H
Biochanin A	H	H	OH	CH_3

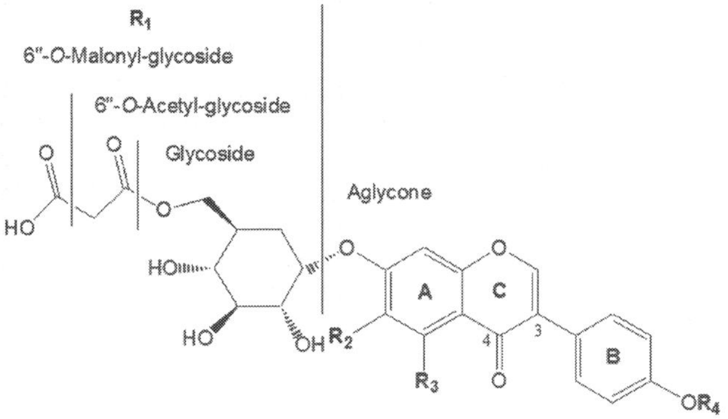

Figure 1. Chemical structure of main isoflavones.

cleavage. Daidzein forms *S*-(−)equol and *O*-desmethylangolensin (O-DMA) via dihydrodaidzein (**Figure 2**). Similarly, genistein is metabolized first as dihydrogenistein and then as 5′-hydroxy-equol and p-ethyl phenol (**Figure 2**). Another possible minor pathway is the hydroxylation of isoflavone rings at different positions, catalyzed by hepatic cytochrome P450 isoenzymes [2]. Metabolites with phenolic or polyphenolic structures are conjugated to *O*-glucuronides and sulfate esters during and after absorption through the gut barrier and more intense in the liver. The conjugated metabolites are urinary or biliary excreted and have enterohepatic circulation [4, 7].

Gut microbiota play a very important role in the isoflavone metabolism. The positive effects of a soy-rich diet derive from the existence of microorganisms in the gut capable of intense metabolization of isoflavones. It is the so-called equol producer phenotype, responsible for metabolizing daidzein to equol and identified through the equol/daidzein ratio in the 24-hour urine. Asian people (Japanese, Korean, or Chinese) and Western adult vegetarians are 50–60% equol producers, but equol producers are only 25–30% in Western population. This phenotype is rather stable and cannot be modulated through prebiotic or probiotic nutritional interventions [8]. Otherwise, there are differences between human and animal metabolism, and therefore in vivo results are not relevant to humans [9]. All tested animals had equol in urine after the ingestion of soy or clover [8]. Notably in rodents, equol constitutes 70–90% from the serum isoflavones, compared to humans where only 30% of the daidzein absorbed is metabolized as equol [4].

2.2. Isoflavone content in different sources

Isoflavones can be found in legumes [10–12], nuts, and some fruits, such as currants and raisins [13], coffee [14], and cereals [15], but the most important dietary sources are soybeans and

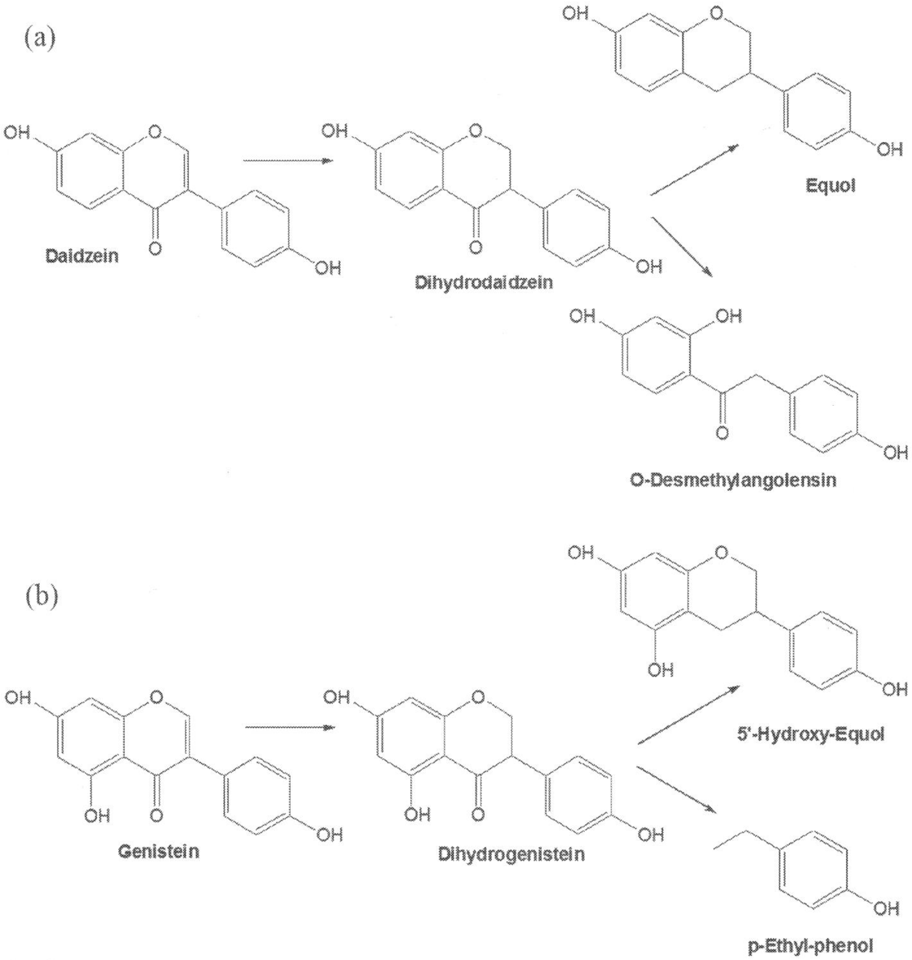

Figure 2. Metabolic pathways of daidzein and genistein.

their by-products [10, 12]. The content of isoflavones in several plants and foods is presented in **Tables 1** and **2**. Soy can be ingested as textured soy protein, as soy milk or drink, added to many fortified foods (e.g., energized bars, cereals, baby formula), or consumed as fermented soybean products, such as miso, natto, and tempeh (**Table 3**) [12]. Also, many food supplements containing soy isoflavones are on the market [16].

Isoflavone content in plants can vary greatly (up to threefold) for the same variety by growth conditions, geographical areas, years, biotic stress factors (e.g., pests), and abiotic stress factors, such as temperature, nutritional status, or drought [4]. Dietary culture has an especially

Food description	Daidzein	Genistein	Glycitein	Total IFs
Soybeans, green, mature seeds, raw	61.70	60.07	7.07	128.83
Soybeans, mature seeds, raw (the mean values from Australia, Brazil, China, Europe, Japan, Korea, Taiwan, the USA)	62.07 (27.77–78.86)	80.99 (39.78–89.32)	14.99 (9.01–22.37)	154.53 (85.68–178.81)
Soybeans, mature seeds, cooked, boiled	30.76	31.26	3.75	65.11
Beans, common, raw (*Phaseolus vulgaris*)	0.29	0.30	0.00	0.59
Beans, adzuki, mature seeds, raw	0.36	0.23	0.00	0.59
Beans, pinto, mature seeds, raw	0.01	0.17	-	0.18
Black bean, sauce	5.96	4.04	0.53	10.26
Chickpeas, mature seeds, raw	0.21	0.06	0.18	0.38
Chickpeas, mature seeds, cooked, boiled	0.00	0.02	-	0.02
Peas, green, split, mature seeds, raw	0.32	0.11	0.00	0.44

Table 1. Isoflavone content in selected legumes (mg/100 g, edible portion—the mean value derived from multiple experiments) [12].

Food description	Coumestrol	Formononetin	Biochanin A
Egg, whole, raw, fresh	0.00	0.05	0.05
Alfalfa seeds, sprouted, raw	1.60	1.43	0.04
Clover sprouts, raw	14.08	3.15	0.59
Red clover	1322.00	833.00	-
Soybeans, mature seeds, raw	0.02	8.46	0.00
Soybeans, mature seeds, sprouted, raw	0.34	0.03	0.00
Lima beans, large, mature seeds, raw	0.14	0.32	0.27
Lima beans, large, mature seeds, boiled	0.00	0.01	0.00
Chickpeas, mature seeds, raw	0.01	0.12	1.54
Chickpeas, mature seeds, canned	0.00	0.00	-

Table 2. Coumestrol, Formononetin, and Biochanin A in selected foods (mg/100 g, edible portion—the mean value derived from multiple experiments) [12].

Food description	Daidzein	Genistein	Glycitein	Total IFs
Miso	16.43	23.24	3.00	41.45
Natto	33.22	37.66	10.55	82.29
Tempeh	22.66	36.15	3.82	60.61
Tofu, raw, regular, prepared with calcium sulfate	8.56	12.99	1.98	22.73
Soybeans, green, raw (includes edamame)	20.34	22.57	7.57	48.95
Soybeans, green, cooked, boiled, drained, without salt (includes edamame)	7.41	7.06	4.60	17.92
Soybeans, mature seeds, sprouted, raw	12.86	18.77	2.88	34.39
Instant beverage, soy, powder, not reconstituted	40.07	62.18	10.90	109.51
Soy cheese, unspecified	5.79	11.14	-	25.72
Soy drink	2.75	5.10	-	7.85
Soy flour (textured)	67.69	89.42	20.02	172.55
Soy meal, defatted, raw	80.77	114.71	16.12	209.58
Soy protein drink	27.98	42.91	10.76	81.65
Soy protein isolate	30.81	57.28	8.54	91.05
Soy yogurt	13.77	16.59	2.80	33.17

Table 3. Isoflavone content in soy foods (mg/100 g, edible portion—the mean value derived from multiple experiments) [12].

big influence on isoflavone content in the diet. Asian and vegetarian diets provide 20–50 mg isoflavones/day, in some cases reaching 100 mg/day, while the Western diet contributes only 0.2–1.5 mg isoflavones/day [2]. Based on recent report of European Food Safety Authority (EFSA), in Europe the dietary isoflavone intake is usually under 1 mg/day, despite an increase in the soy food consumption [17]. The differences between the types of diets refer to the amount of isoflavone in foods, as well as the type of food consumed. In the Western diet, solid processed soy products (such as tofu) and soymilk dominate the diet, and they contain both glycosides (genistin and daidzin which are stable during processing) and aglycones. In the Asian diet, most soy products are obtained through fermentation and have higher amounts of aglycones [3]. Miso, fermented soybean paste (Japan); doenjang, fermented soybean paste (Korea); douchi, fermented soybeans (China); and tempeh, fermented soybean cake (Indonesia) are staple foods in some Asian countries. Simultaneously, health benefit probiotics are formed in these foods during the fermentation processes [18].

Besides soy, other plants in the Fabaceae family have a high content of isoflavones: species of clover, mainly red clover (*Trifolium pratense* L.), alfalfa (*Medicago sativa* L.), and hop clover (*Medicago lupulina* L.), form important part of animal feed. These plants are used in phytotherapy, as medicinal teas or as food supplements. Red clover (*Trifolium pratense* L.) incorporates mainly genistein, daidzein and formononetin, and their respective β-glycosides [19–22]. It also contains important quantities of coumestrol, a phytoestrogen part of coumestan class [19, 20], and antioxidant compounds [23, 24]. Data from scientific literature show that red clover extracts can be used as replacement for conventional hormonal therapy in menopause or hormone-dependent diseases [25]. Alfalfa (*Medicago sativa* L.) contains isoflavones (genistein, daidzein, formononetin, biochanin A) in addition to other phytoestrogens (coumestrol) and many nutrients. It is used in phytotherapy for its antianemic, antihemorrhagic, and remineralization properties [26] and for its hypocholesterolemic, antimicrobial, hypolipidemic, antioxidant, antiulcer, neuroprotective, and estrogenic properties [27]. Species of *Genista* (*G. tinctoria* L., *G. sagittalis* L.) contain essentially genistin and genistein [19, 20, 28, 29]. They are known for their hypoglycemic [30], anti-inflammatory, antiulcer, spasmolytic, antioxidant, and estrogenic properties [31]. Among these plants, *Genista tinctoria* L. show antioxidant and antitoxic activities [32, 33], protective effect against ultraviolet (UV) radiation, and in vitro melanoma cell proliferation [31].

2.3. Mechanism of estrogen-like action of isoflavones

According to the xenohormesis theory, plants synthesize phytochemicals to withstand and adapt under stress. Indeed, isoflavone biosynthesis depends on the environmental conditions in which the plant grows and is stimulated by stress. The stress-induced plant compounds have the ability to upregulate stress adaptive pathways in animals and humans. In the body, the biological effects of isoflavones are exercised by modulating pathways mediated by estrogen receptors (ERs) or various key enzymes involved in cellular signaling or metabolism and antioxidant potential [4].

3. The estrogenic/antiestrogenic effects

Isoflavones produce both estrogenic and antiestrogenic effects through several ways. Due to their structure similar to that of 17β-estradiol, they have the ability to bind to the nuclear ERs, but their affinity for these receptors is rather weak. Only genistein shows stronger affinity for ERβ to which it binds preferentially. Its relative affinity (0.87) is closer to that of the reference hormone, 17β-estradiol. Daidzein affinity for these receptors is 0.005, but equol, its metabolite, has a 5.7 times stronger affinity, thus increasing its estrogenic potential. The affinity for ERα decreases as follows: genistein > equol > daidzein, with the values of 0.04, 0.005, and 0.001, respectively. The affinities of other isoflavones are less than 0.0001 [2, 4].

Isoflavones induce agonist/antagonist effects depending on the level of the endogenous estrogen. For people with high levels of estrogen, (women premenopause, especially in the follicular phase of the menstrual cycle), the isoflavones bind to the estrogen receptors. Because of their weak estrogen potency, isoflavones exert an antagonist effect. They block the action of endogenous estrogens on their receptors. In case of low concentration of endogenous estrogens (women in menopause, after ovariectomy, or males), the estrogenic action of isoflavones becomes evident, showing additive agonist effect [34]. This is the reason why isoflavones can be used as a long-term complementary or alternative hormone therapy [35].

Isoflavones and their active metabolites can bind to the membrane ERs and induce rapid non-genomic effects by which they modulate cellular metabolism. Thus, they can change the protein kinase and lipid kinase cell signaling pathways [1]. It is believed that the activation of these signaling pathways by isoflavones causes some beneficial effects, in particular in the tissues that are not specific targets for the estrogens. At the circulatory system, the isoflavones induce vasodilation by increasing the production of nitric oxide (NO) after the activation of the endothelial NO^- synthase. At the central nervous system, they improve the cognitive function by affecting cell membrane permeability and altering the neuronal excitability. In the skeletal system, the isoflavones inhibit the tyrosine kinase causing changes in the alkaline phosphatase activity. On the other hand, they induce the apoptosis of the osteoclasts, suppress the formation of osteoclasts [34], and prevent the bone demineralization [35].

Also, isoflavones influence the activity of some of the enzymes involved in the metabolism of the sex steroid hormones. In this way they inhibit 5α-reductase (the enzyme responsible for the conversion of testosterone to 5α-dihydrotestosterone) and aromatase (involved in the conversion of testosterone to estradiol) in low concentrations, but they increase the aromatase activity at high concentrations. Isoflavones have an affinity for sex hormone-binding globulin (SHBG) and they induce its expression. Therefore, they affect the free-steroid hormone level in the systemic circulation. But these outcomes depend on many factors, including species, gender, and the hormonal status [35].

Xenoestrogens can modulate the enzyme activity of aromatase. Thus, they induce alterations in the metabolism of fats and carbohydrates through effects on ERα. The decrease of endogenous estrogen levels on ERα, aromatase inhibition or the existence of mutations affecting the enzyme activity has been correlated with visceral obesity or truncate, hyperlipidemia, glucose intolerance and insulin resistance, low physical activity, and reduced energy expenditure. Isoflavones compensate for the deficit of estrogens and have the ability to prevent the associated negative effects. Asian diets, rich in isoflavones, are correlated with low incidence of obesity and metabolic syndrome, favorable plasma profile, and a reduced body mass index in postmenopausal women [4].

4. Health benefits of isoflavones

4.1. Isoflavones and their effects on diseases

Numerous epidemiological and clinical studies have demonstrated the protective role of dietary isoflavones against development of specific menopause symptoms [36–38] and several chronic diseases, including cardiovascular diseases [39, 40], osteoporosis [38], cognitive impairment [37], and hormone-dependent cancers [41–43]. Based on human health benefits of soy diet, the Food and Drug Administration (FDA) approved the use of the following health

http://dx.doi.org/10.5772/66531

claim on the labels: "25 grams of soy protein a day, as part of a low in saturated fat and cholesterol, may reduce the risk of heart disease" [44].

Isoflavones, as all polyphenols, have a strong antioxidant activity. They can neutralize free radicals and prevent the lipid peroxydation by stopping the chain reactions. Also, isoflavones induce the antioxidant enzymes (glutathione peroxidase, catalase, and superoxide dismutase) and inhibit the expression of some enzymes, such as xanthine oxidase [1]. The antioxidant protective action of isoflavones from soy or plant extracts, such as *Trifolium pratense* L. or *Genista tinctoria* L., was proven in clinical studies [45, 46], as well as in animal models [32, 47].

4.2. Anticarcinogenic activity of isoflavones

The anticarcinogenic potential of isoflavones is based on multiple actions: binding to estrogen receptors (ERs), changing of cell signaling pathways, and inhibition of the key enzymes involved in the metabolism of sex hormones. Also, the anticarcinogenic potential of isoflavones has positive effects through independent mechanisms which do not involve ERs, such as antioxidant activity, reduction in the bioactivation of carcinogens, and stimulation of detoxification [2, 48].

Anticarcinogenic activity of genistein has been assessed more thoroughly among isoflavones. Genistein initiates apoptosis, alters cell proliferation and angiogenesis, and inhibits metastasis in many types of cancer cells [49]. It is a tyrosine kinase inhibitor. Therefore, in breast cancer cells, it slows down tumorigenesis; in the circulatory system, it prevents tumor vascularization; in the nervous system, it induces neuroprotective effects. In addition, genistein affects tumorigenesis by inhibiting DNA topoisomerases I and II [50], alteration of epigenetic regulations (both histone methylation and DNA methylation), and activating tumor suppressor genes [51]. As a polyphenol, genistein has antioxidant [1] and anti-inflammatory potential [52]. Another possible action pathway for genistein is the competitive inhibition of estrone metabolism through cytochrome P450 isoenzymes by altering the 2-hydroxy-estrone (2-OH-E_1)/16α-hydroxy-estrone (16α-OH-E_1) ratio, as noticed in vitro [53]. While 2-OH-E_1 is a weak estrogen, 16α-OH-E_1 has an important role in carcinogenesis, showing a strong estrogen effect and genotoxic properties [54]. 16α-OH-E_1 covalently binds to the estrogenic receptors and thus stimulates cell proliferation [55]. The ratio 2-OH-E_1/16α-OH-E_1 has been proposed and studied as a biomarker of breast cancer risk [55–59], but now its significance is controversial. In high concentrations, genistein decreases the hydroxylation of estrone in position 2 in favor of hydroxylation in position 16α [55]. Other studies show that genistein has no mutagenic or clastogenic activity in vivo. But in high concentration of genistein, it has clastogenic potential in vitro, explained by the topoisomerase inhibitory effect, which is known to cause chromosome damage above a certain threshold dose [60].

Anti-proliferative effects of high concentrations of genistein were demonstrated in all breast cancer cells, both ER positive and ER negative. However, there are several studies showing that genistein shows both anti-proliferative and proliferative effects, depending on the concentration, type of tumor, level of endogenous estrogens present in the tissue, or development stage. At low physiological concentrations, genistein stimulates tumorigenesis and cancels the effects of tamoxifen in ER-positive breast cancer cells [50]. Similar dual effects were observed in the case of tamoxifen and other selective estrogen receptor modulators (SERMs) [16].

In fermented soybean products (e.g., natto, miso, tempeh), aglycons can suffer changes under the effect of enzymes produced by the microorganisms involved in the fermentation process. Thus, *ortho*-hydroxygenistein (6-OHG, 8-OHG, 3′-OHG) and *ortho*-hydroxydaidzein (6-OHD, 8-OHD, 3′-OHD) were identified. These compounds are not synthesized by the plants. The hydroxylation reaction that occurs in the *ortho* position gives molecules a high antioxidant potential and a free radical scavenging activity. Moreover, several of their abilities have been proven: to suppress cell proliferation and to inhibit tyrosinase (anti-melanogenesis properties) and antimutagenic, anti-inflammatory, and hepatoprotective properties [18].

Equol has a higher estrogenic potential than daidzein, its precursor, and a preferential affinity for ERβ, as it has already been stated. This detail is of high interest for its beneficial effect in the treatment of prostate cancer, since both isomers, *S*-(−)equol and *R*-(+)equol, can bind in vivo dihydrotestosterone without having an affinity for the androgen receptor. Therefore, equol prevents the endogenous hormone to exert its stimulating effect on prostate growth. In addition, equol possesses the highest antioxidant capacity of all isoflavones tested. It causes blood vessel relaxation and modifies the inflammatory response in activated macrophages and has beneficial effects in cardiovascular and inflammatory diseases [52].

4.3. Effects of isoflavones on hormone-dependent cancers

Clinical studies show contradictory results of the efficacy of isoflavones in the treatment of breast cancer. The effects depend on a number of factors such as age, gender, hormonal status, type of isoflavones consumed (soy proteins or isolated isoflavones), dose, diet (type of food), and extent of consumption [2].

A recent meta-analysis of 35 studies shows that soy isoflavones lower the risk of breast cancer in both premenopausal and post-menopausal women. The effect is more evident in Asian women than in those living in Western countries, probably due to differences in quality (traditionally fermented foods) and quantity of the isoflavone products ingested [41]. In Asian women, a diet rich in soy food lowers breast cancer risk with 30% [61]. A higher prevalence of equol-producer phenotype in Asian population can be an essential factor. Equol-producer phenotype is associated with a substantial reduction in the risk of breast cancer. Several specific biomarkers are favorable modified, such as sex hormone-binding globulin (SHBG) and steroid hormone levels in plasma, a higher urinary 2-hydroxy-estrone/16α-hydroxy-estrone ratio, and a lower mammographic breast density [2]. However, because several studies have provided mixed or contradictory results, the general recommendation for patients diagnosed with estrogen-dependent breast cancer is to avoid consuming high quantities of products containing isoflavone. Indeed, isoflavones are selective estrogen receptor modulators (SERMs), and their effects would depend on multiple factors.

Another meta-analysis of five cohort studies that included more than 11,000 female patients diagnosed with breast cancer focused on the post-diagnostic relationship between consumption of soy foods and mortality or cancer recurrence. The study concluded that the ingestion of soy foods reduced mortality and recurrence in all types of breast cancer, especially in the ER-negative, ER-positive/PR-positive, and postmenopausal patients [42]. In women diagnosed with breast cancer under tamoxifen treatment, the consumption of plants containing isoflavones did not alter plasma levels of the drug and its metabolites [62]. Moreover, a recent study shows that a moderate intake of soy isoflavones (5–10 g soy protein/day) would have an optimal effect on tamoxifen treatment on these patients [63].

In some studies [64], excessive consumption of soy was associated with a negative impact on male fertility and reproductive hormones and the disruption of the thyroid gland function. In other studies these effects were inconsistent [65].

Isoflavones can modulate the toxicity of other xenoestrogens, but the interactions are complex and difficult to predict relying only on in vitro steroid receptor affinities [66]. In these kinds of interactions, multiple mechanisms are involved, both estrogen and non-estrogen type, such as oxidative stress [32, 47, 53]. European Food Safety Authority (EFSA) has recently conducted a systematic study of published medical literature, focusing on the correlation between the intake of soy isoflavones and the induced effects on the breast (mammographic density, proliferative marker Ki67 expression), uterus (endometrial thickness, histopathology changes), and thyroid (the thyroid hormone). Results showed that the intake of 35–150 mg isoflavones/day does not affect these organs in peri- and postmenopausal women [17]. Isoflavones have demonstrated prostate cancer efficacy in several studies: in vitro, on prostate cancer cell lines, in vivo, and in numerous clinical trials [43, 67, 68]. Conclusion of a recent meta-analysis suggests that phytoestrogen intake, mostly genistein and daidzein, can be correlated with a decreased risk of prostate cancer [69].

5. Recent advances in analytical methods of isoflavones

5.1. Isolation of isoflavones in foods and vegetable materials

In recent years, due to the health benefits provided by isoflavones, higher attention has been paid to the analytical methods that allow identification and quantification of isoflavones from different types of samples: (a) food, for dietary intake assessing [15, 70]; (b) food supplements, for standardization of nutraceuticals [5, 71]; (c) vegetable products, for phytotherapeutic evaluation [19, 20, 28]; and (d) human biological samples (plasma, urine) [5]. These analytical methods are commonly used for isoflavone bioavailability assessing and in pharmacokinetic or pharmacological studies.

Isoflavones are solubilized from food or vegetable material by refluxing or maceration, shaking, and stirring [72]. The isolation of isoflavones from the mixture can be achieved either by conventional methods, liquid-liquid extraction [11, 15, 19] or Soxhlet, or by modern ones—supercritical fluid extraction, ultrasound-assisted extraction [19, 71], pressurized fluid extraction, microwave-assisted extraction, and solid-phase extraction [5, 73] (**Table 4**).

The methods used to isolate isoflavones from food are selected function of the nature of the food, the type of the isoflavones analyzed (the total of aglycones or aglycones and

Analytes	Sample	Extraction method	Detection	Run time (min)	LOQ	References
3 IFs	Soy dry extract	Sonication/ Steam bath	HPLC-DAD	20	40–100 ng/mL	[71]
3 IFs, Cou*	10 plant species	UAE**	ULPC-PDA	4	1.97–4.08 ng/mL	[19]
12 IFs	Soybean seeds	Maceration	HPLC-UV	60	NA#	[70]
12 IFs	Soybeans, soy products	Maceration	HPLC-DAD	30	<600 nmol/L	[72]
3 IFs	Coffee	Refluxing	HPLC-DAD	35	13.7–25.0 ng/mL	[14]
17 IFs	Soymilk	Refluxing	LC-ESI(+)-MS/MS	NA#	NA#	[74]
7 IFs, Cou*	2 plant extracts	Refluxing or Maceration	LC-ESI(−)-MS/MS	18	40 ng/mL	[20]
5 IFs, Cou*	7 plant extracts	Maceration or percolation	ULPC-ESI(+)-MS/MS	5.5	5–10.78 ng/mL	[28]
5 IFs	Legumes	SPE C18	UHLPC-ESI(+)-MS/MS	18	0.1–1 ng/mL	[73]
5 IFs	Coffee	SPE C18	HPLC-ESI-MS/MS	18	0.05–1 ng/mL	[75]

* Cou, coumestrol.
** UAE, ultrasound-assisted extraction.
# NA, not available.

Table 4. HPLC and UPLC methods applied for analysis of isoflavones in different samples.

glycosides), and the instrumental method used for identification and quantification. Several examples are presented below.

Liggins et al. isolated isoflavones from cereals and derivatives after a prior sonication in a polar solvent (methanol/water 4:1, v/v), in order to break apart the cellular material, followed by filtration and evaporation of the solvent under nitrogen. In order to determine the total aglycones, glycosides were hydrolyzed in an acid medium (0.1 M acetate buffer, pH 5) by overnight incubation at 37 °C in the presence of cellulase (enzyme used for hydrolytic removal of the hydrolysis resulted carbohydrates). Aglycones were extracted into ethyl acetate and were derivatized and analyzed using GC-MS [15]. Otieno et al. analyzed isoflavones from fermented and unfermented soy milk. For the solubilization of analytes, the freeze-dried sample was refluxed in methanol for 1 hour and filtered, and after adding the internal standard, the solvent has been evaporated to dryness under nitrogen. The residue has been suspended into a buffer (10 mm ammonium acetate containing 0.1% trifluoroacetic acid) and centrifuged, and the supernatant was filtered and analyzed using high-performance liquid chromatography (HPLC) [74].

Extraction and analysis of isoflavones in soybeans can be realized through maceration of the powdered beans with 70% ethanol at room temperature, for 24 hours under constant stirring.

After centrifugation and filtering, the supernatant is analyzed directly by HPLC [70]. Also, analysis of isoflavones contained in food supplements requires a simple preparation of the samples: fine powdering of tablets, refluxing in 80% methanol for 1 hour, filtering, and injection into the HPLC system [5].

Hydroalcoholic extracts or tinctures can be prepared from either fresh or dry and pulverized vegetable materials. The hydroalcoholic extracts can be made in 70% ethanol or methanol, by refluxing and filtration; by cold maceration, pressing, and filtration [20]; by percolation [28]; or using modern methods, such as ultrasound-assisted extraction in 50% ethanol [19]. The extracts can be analyzed directly by LC-MS/MS, after an adequate dilution [20], or they can be subjected to an acid hydrolysis [19] in order to release aglycones. Further, the aglycones can be assessed directly or after liquid-liquid extraction, for a concentration of the analytes [19].

In biological samples (e.g., plasma and human urine) isoflavones can be found in different forms: as aglycones (active metabolites), aglycone derivatives (with or without bioactivity), or conjugated metabolites (β-glucuronides and sulfate esters). Isoflavone analysis can focus on individual quantification of aglycones and their metabolites or quantification of aglycones after the hydrolysis of conjugated forms. Hydrolysis of conjugated metabolites is achieved by incubation at 37 °C with a mixture of β-glucuronidase/sulfatase in the presence of a buffer (0.5 M acetate) at pH 4.5 for several hours or overnight. Isolation of free forms and/or of those freed after hydrolysis can be done by liquid-liquid extraction or solid-phase extraction [5].

5.2. Quantification of isoflavones in foods and vegetable materials

For isoflavone identification, the following chromatographic methods are used: gas chromatography coupled with mass spectrometry (GC-MS) [5, 15], high-performance liquid chromatography (HPLC) with UV detector (photodiode array, PDA) [28, 70, 71], fluorescence detector (FLD), electrochemical detector (ECD) or mass spectrometer detector (MS) [20, 74, 75], and, less often, capillary electrophoresis (CE).

Quantification of isoflavones and their derivatives can be achieved in two ways: (a) by determining the free aglycons after a prior acid hydrolysis [19, 70, 72], alkaline hydrolysis [72], or enzymatic hydrolysis [72] of the glycosides in the sample and (b) by simultaneously analyzing the glycosides and aglycones present in the sample [20, 28]. GC-MS methods are used less lately, because they require an additional step of isoflavone derivatization to the volatile compounds [5, 15]. This additional step increases both the time and the cost of the analysis and represents a potential source of error [28].

Generally, HPLC-UV is not sensitive enough (**Table 4**) for the quantification of small levels of isoflavones from plant extracts [19] or human plasma [5]. This method often requires a hydrolysis step to transform glycosides into aglycones followed by the quantification of total aglycones from the sample [71].

In order to correctly identify new isoflavones or isoflavone derivatives present in the samples analyzed, liquid chromatography coupled with mass spectrometry (LC-MS) and tandem mass spectrometry (LC-MS/MS) are the preferred methods (**Table 4**), due to the advantages: speed, selectivity, sensitivity, and robustness. In addition, mass spectrometry detection allows sure determination of the compounds based on molecular weight and ion charge. For the quantification of isoflavones, the pseudo-molecular ions or the ionic fragments resulted after fragmentation are monitored. In LC-MS/MS analysis, compound identification can be achieved even if their separation is not complete, and it is an advantage [74]. A shorter analysis can be realized by ultra-performance liquid chromatography (UPLC) [19, 28]. This method uses columns with very small size of the packing particles (1.7 μm) and consequently performs separations with superior resolution in a shorter time and a lower consumption of the mobile phase.

The isoflavones have polyphenolic structure and can easily lose a proton to form negative pseudo-molecular ions $[M-H]^-$ [20]. However, they can also be detected after ionization in positive mode to $[M + H]^+$ [74]. Isoflavones are polar compounds and they form ions in solution. For these type of compounds, electro-spray ionization (ESI) is the most commonly used source to obtain analytical ions. Atmospheric pressure chemical ionization (APCI) is the source preferred for non-polar analytes that ionize in the gas phase. The isoflavones often give poor response in this ionization source [28]. The fragmentation patterns of isoflavone glycosides (malonyl-glycosides, acetyl-glycosides, glycosides, aglycones) follow a similar trend. However each compound has a unique fragmentation pattern that allows their accurate identification (**Table 5**) [74].

Isoflavone	$[M + H]^+$	Transitions	$[M - H]^-$	Tranzitions
Daidzein	255 [28, 74]	→199 [28]	253 [20, 73, 75]	→208, 132 [73, 75]
Formononetin	269 [28]	→197 [28]	267 [20, 75]	→252, 223 [75]
Genistein	271 [28, 73, 74]	→153 [28]	269 [20, 73, 75]	→159, 133 [73, 75]
Biochanin A			283 [73, 75]	→268, 239 [73, 75]
Glycitein	285 [74]	→270, 257, 229, 196, 166 [74]	283 [20]	
Daidzin	417 [28, 73, 74]	→255 [28, 73, 74], 199 [73]	415 [20]	→253 [20]
Ononin			429 [20]	→267 [20]
Genistin	433 [28, 74, 75]	→271 [28, 73–75], 91 [73, 75]	431 [20]	→268, 269 [20]
Glycitin	447 [74]	→428, 285 [74]		
Ac-daidzin	459 [74]	→441, 255 [74]		
Ac-genistin	475 [74]	→431, 271 [74]		
Ac-glycitin	489 [74]	→471, 285 [74]		
Mal-daidzin	503 [74]	→485, 255 [74]		
Mal-genistin	519 [74]	→501, 271 [74]		

Table 5. Ions (*m*/*z*) and transitions monitored for isoflavone quantification.

6. Conclusion

Dietary intake of isoflavones is widespread, mainly due to the high consumption of soybean products. Health benefits of isoflavones justify the interest for this class of bioactive compounds, but the controversial outcomes of some clinical and epidemiological studies require further investigations. In the context of these researches, the analytical methods applied for assessment of isoflavones are very valuable. They allow for the evaluation of dietary intake of isoflavones, equating the health benefits and the circumstances in which they are exerted, and highlight the natural sources of isoflavones with phytotherapeutic potential.

Author details

Daniela-Saveta Popa[1]* and Marius Emil Rusu[2]

*Address all correspondence to: dpopa@umfcluj.ro

1 Department of Toxicology, Faculty of Pharmacy, "Iuliu Haţieganu" University of Medicine and Pharmacy Cluj-Napoca, Cluj-Napoca, Romania

2 Department of Pharmaceutical Technology and Biopharmacy, Faculty of Pharmacy, "Iuliu Haţieganu" University of Medicine and Pharmacy Cluj-Napoca, Cluj-Napoca, Romania

References

[1] Tsao R. Chemistry and biochemistry of dietary polyphenols. Nutrients. 2010;2(12):1231–1246. Doi: 10.3390/nu2121231

[2] Bolca S. Bioavailability of soy-derived isoflavones and human breast cancer. In: Watson RR, Preedy VR, Zibadi S, editors. Polyphenols in Human Health and Disease. Amsterdam: Elsevier; 2014. pp. 1241–1256. Doi: 10.1016/B978-0-12-398456-2.00094-3

[3] Setchell KD, Brown NM, Zimmer-Nechemias L, Brashear WT, Wolfe BE, Kirschner AS, Heubi JE. Evidence for lack of absorption of soy isoflavone glycosides in humans, supporting the crucial role of intestinal metabolism for bioavailability. Am J Clin Nutr. 2002;76(2):447–453.

[4] Cederroth CR, Nef S. Soy, phytoestrogens and metabolism: a review. Mol Cell Endocrinol. 2009;304(1–2):30–42. Doi: 10.1016/j.mce.2009.02.027

[5] Setchell KD, Brown NM, Desai P, Zimmer-Nechemias L, Wolfe BE, Brashear WT, Kirschner AS, Cassidy A, Heubi JE. Bioavailability of pure isoflavones in healthy humans and analysis of commercial soy isoflavone supplements. J Nutr. 2001;131(4 Suppl):1362S–1375S.

[6] Tolleson WH, Doerge DR, Churchwell MI, Marques MM, Roberts DW. Metabolism of biochanin A and formononetin by human liver microsomes *in vitro*. J Agric Food Chem. 2002;50(17):4783–4790.

[7] Gaya P, Medina M, Sánchez-Jiménez A, Landete JM. Phytoestrogen metabolism by adult human gut microbiota. Molecules. 2016;21(8),1034:1-17. Doi: 10.3390/molecules21081034

[8] Setchell KD, Clerici C. Equol: history, chemistry, and formation. J Nutr. 2010;140(7):1355S–1362S. Doi: 10.3945/jn.109.119776

[9] Douglas CC, Johnson SA, Arjmandi BH. Soy and its isoflavones: the truth behind the science in breast cancer. Anticancer Agents Med Chem. 2013;13(8):1178–1187. Doi: 10.2174/18715206113139990320

[10] Liggins J, Bluck LJ, Runswick S, Atkinson C, Coward WA, Bingham SA. Daidzein and genistein contents of vegetables. Br J Nutr. 2000;84(5):717–725.

[11] Megías C, Cortés-Giraldo I, Alaiz M, Vioque J, Girón-Calle J. Isoflavones in chickpea (Cicer arietinum) protein concentrates. J Funct Foods. 2016;21:186–192. Doi: 10.1016/j.jff.2015.12.012

[12] USDA Database for the Isoflavone Content of Selected Foods Release 2.1. 2015. https://www.ars.usda.gov/ARSUserFiles/80400525/Data/isoflav/Isoflav_R2-1.pdf.

[13] Liggins J, Bluck LJ, Runswick S, Atkinson C, Coward WA, Bingham SA. Daidzein and genistein content of fruits and nuts. J Nutr Biochem. 2000;11(6):326–331.

[14] Alves RC, Almeida IMC, Casal S, Oliveira MBPP. Method development and validation for isoflavones quantification in coffee. Food Chem. 2010;122:914–919. Doi: 10.1016/j.foodchem.2010.03.061

[15] Liggins J, Mulligan A, Runswick S, Bingham SA. Daidzein and genistein content of cereals. Eur J Clin Nutr. 2002;56(10):961–966.

[16] Patisaul HB, Jefferson W. The pros and cons of phytoestrogens. Front Neuroendocrinol. 2010;31(4):400–419. Doi: 10.1016/j.yfrne.2010.03.003

[17] EFSA ANS Panel (EFSA Panel on Food Additives and Nutrient Sources added to Food). Scientific opinion on the risk assessment for peri- and post-menopausal women taking food supplements containing isolated isoflavones. EFSA J 2015;13(10):4246. Doi: 10.2903/j.efsa.2015.4246

[18] Chang TS. Isolation, bioactivity, and production of ortho-hydroxydaidzein and ortho-hydroxygenistein. Int J Mol Sci. 2014 Apr 3;15(4):5699–5716. Doi: 10.3390/ijms15045699

[19] Kiss B, Popa DS, Hanganu D, Pop A, Loghin F. Ultra-performance liquid chromatography method for the quantification of some phytoestrogens in plant material. Rev Roum Chim. 2010;55(8):459–465.

[20] Vlase L, Popa DS, Tero-Vescan A, Olah N. New liquid chromatography: mass spectrometry assay for natural phytoestrogens from vegetable extracts. Acta Chromatogr. 2011;23:509–520.

[21] Hanganu D, Vlase L, Olah N. LC/MS analysis of isoflavones from fabaceae species extracts. Farmacia, 2010;58(2):177–183.

[22] Valls J, Millán S, Martí MP, Borràs E, Arola L. Advanced separation methods of food anthocyanins, isoflavones and flavanols. J Chromatogr A. 2009;1216(43):7143–7172. Doi: 10.1016/j.chroma.2009.07.030

[23] Kicel A, Wolbiś M. Phenolic content and DPPH radical scavenging activity of the flowers and leaves of Trifolium repens. Nat Prod Commun. 2013;8(1):99–102.

[24] Mu H, Bai YH, Wang ST, Zhu ZM, Zhang YW. Research on antioxidant effects and estrogenic effect of formononetin from Trifolium pratense (red clover). Phytomedicine. 2009 Apr;16(4):314–319. Doi: 10.1016/j.phymed.2008.07.005

[25] Beck V, Rohr U, Jungbauer A. Phytoestrogens derived from red clover: an alternative to estrogen replacement therapy? J Steroid Biochem Mol Biol. 2005;94(5):499–518. Doi: 10.1016/j.jsbmb.2004.12.038

[26] Grigorescu E, Ciulei I, Stanescu U. Index fitoterapeutic. Bucuresti: Editura Medicala; 1986. p. 252.

[27] Bora KS, Sharma A. Phytochemical and pharmacological potential of Medicago sativa: a review. Pharm Biol. 2011;49(2):211–220. Doi: 10.3109/13880209.2010.504732

[28] Kiss B, Popa DS, Păltinean R, Loghin F. A high-throughput UPLC-MS/MS for the simultaneous analysis of six phytoestrogens from *Genista tinctoria* extracts. J Liq Chromatogr Relat Technol. 2012;35(19):2735–2752.

[29] Tero-Vescan A, Imre S, Vari CE, Osan A, Dogaru MT, Csedo C. Determination of some isoflavonoids and flavonoids from *Genista tinctoria* L. by HPLC-UV. Farmacia, 2009;57(1):120–127.

[30] Rauter AP, Martins A, Lopes R, Ferreira J, Serralheiro LM, Araújo ME, Borges C, Justino J, Silva FV, Goulart M, Thomas-Oates J, Rodrigues JA, Edwards E, Noronha JP, Pinto R, Mota-Filipe H. Bioactivity studies and chemical profile of the antidiabetic plant Genista tenera. J Ethnopharmacol. 2009;122(2):384–393. Doi: 10.1016/j.jep.2008.10.011

[31] Rigano D, Cardile V, Formisano C, Maldini MT, Piacente S, Bevelacqua Y, Russo A, Senatore F. Genista sessilifolia DC. and *Genista tinctoria* L. inhibit UV light and nitric oxide-induced DNA damage and human melanoma cell growth. Chem Biol Interact. 2009;180(2):211–219. Doi: 10.1016/j.cbi.2009.02.010

[32] Popa DS, Bolfă P, Kiss B, Vlase L, Păltinean R, Pop A, Cătoi C, Crişan G, Loghin F. Influence of *Genista Tinctoria* L. or methylparaben on subchronic toxicity of bisphenol A in rats. Biomed Environ Sci. 2014;27(2):85–96. Doi: 10.3967/bes2014.021

[33] Hanganu D, Olah NK, Benedec D, Mocan A, Crisan G, Vlase L, Popica I, Oniga I. Comparative polyphenolic content and antioxidant activities of *Genista tinctoria* L. and Genistella sagittalis (L.) Gams (Fabaceae). Pak J Pharm Sci. 2016;29(1 Suppl):301–307.

[34] Hwang CS, Kwak HS, Lim HJ, Lee SH, Kang YS, Choe TB, Hur HG, Han KO. Isoflavone metabolites and their *in vitro* dual functions: they can act as an estrogenic agonist or

antagonist depending on the estrogen concentration. J Steroid Biochem Mol Biol. 2006;101(4–5):246–253. Doi: 10.1016/j.jsbmb.2006.06.020

[35] Pilšáková L, Riečanský I, Jagla F. The physiological actions of isoflavone phytoestrogens. Physiol Res. 2010;59(5):651–664.

[36] Franco OH, Chowdhury R, Troup J, Voortman T, Kunutsor S, Kavousi M, Oliver-Williams C, Muka T. Use of plant-based therapies and menopausal symptoms: a systematic review and meta-analysis. JAMA. 2016;315(23):2554–2563. Doi: 10.1001/jama.2016.8012

[37] Cheng PF, Chen JJ, Zhou XY, Ren YF, Huang W, Zhou JJ, Xie P. Do soy isoflavones improve cognitive function in postmenopausal women?. Meta Anal Menopause. 2015;22(2):198–206. Doi: 10.1097/GME.0000000000000290

[38] Pawlowski JW, Martin BR, McCabe GP, McCabe L, Jackson GS, Peacock M, Barnes S, Weaver CM. Impact of equol-producing capacity and soy-isoflavone profiles of supplements on bone calcium retention inpostmenopausal women: a randomized crossover trial. Am J Clin Nutr. 2015;102(3):695–703. Doi: 10.3945/ajcn.114.093906

[39] Cavallini DC, Manzoni MS, Bedani R, Roselino MN, Celiberto LS, Vendramini RC, de Valdez G, Abdalla DS, Pinto RA, Rosetto D, Valentini SR, Rossi EA. Probiotic soy product supplemented with isoflavones improves the lipid profile of moderately hypercholesterolemic men: a randomized controlled trial. Nutrients. 2016;8(1),52:1–18. Doi: 10.3390/nu8010052

[40] Wong JM, Kendall CW, Marchie A, Liu Z, Vidgen E, Holmes C, Jackson CJ, Josse RG, Pencharz PB, Rao AV, Vuksan V, Singer W, Jenkins DJ. Equol status and blood lipid profile in hyperlipidemia after consumption of diets containing soy foods. Am J Clin Nutr. 2012;95(3):564–571. Doi: 10.3945/ajcn.111.017418

[41] Chen M, Rao Y, Zheng Y, Wei S, Li Y, Guo T, Yin P. Association between soy isoflavone intake and breast cancer risk for pre- and post-menopausal women: a meta-analysis of epidemiological studies. PLoS One. 2014;9(2):e89288. Doi: 10.1371/journal.pone.0089288

[42] Chi F, Wu R, Zeng YC, Xing R, Liu Y, Xu ZG. Post-diagnosis soy food intake and breast cancer survival: a meta-analysis of cohort studies. Asian Pac J Cancer Prev. 2013;14(4):2407–2412. Doi: http://dx.doi.org/10.7314/APJCP.2013.14.4.2407.

[43] Zhang Q, Feng H, Qluwakemi B, Wang J, Yao S, Cheng G, Xu H, Qiu H, Zhu L, Yuan M. Phytoestrogens and risk of prostate cancer: an updated meta-analysis of epidemiologic studies. Int J Food Sci Nutr. 2016;9:1–15. Doi: 10.1080/09637486.2016.1216525

[44] US Food and Drug Administration. Guidance for Industry: A Food Labeling Guide (11. Appendix C: Health Claims). 2013. http://www.fda.gov/Food/GuidanceRegulation/GuidanceDocumentsRegulatoryInformation/LabelingNutrition/ucm064919.htm.

[45] Azadbakht L, Kimiagar M, Mehrabi Y, Esmaillzadeh A, Hu FB, Willett WC. Dietary soya intake alters plasma antioxidant status and lipid peroxidation in postmenopausal

women with the metabolic syndrome. Br J Nutr. 2007;98(4):807–813. Doi: 10.1017/S0007114507746871

[46] Cha YS, Park Y, Lee M, Chae SW, Park K, Kim Y, Lee HS. Doenjang, a Korean fermented soy food, exerts antiobesity and antioxidative activities in overweight subjects with the PPAR-γ2 C1431T polymorphism: 12-week, double-blind randomized clinical trial. J Med Food. 2014;17(1):119–127. Doi: 10.1089/jmf.2013.2877

[47] Popa DS, Hanganu D, Vlase L, Kiss B, Loghin F, Crişan G. Protective effect of *Trifolium pratense* extract on oxidative stress induced by bisphenol A in rats. Farmacia. 2014:62(2):341–349.

[48] Lockwood B. Nutraceuticals. A Guide for Healthcare Professionals. Second edition, London: Pharmaceutical Press; 2007. pp. 192–200.

[49] Spagnuolo C, Russo GL, Orhan IE, Habtemariam S, Daglia M, Sureda A, Nabavi SF, Devi KP, Loizzo MR, Tundis R, Nabavi SM. Genistein and cancer: current status, challenges, and future directions. Adv Nutr. 2015;6(4):408–419. Doi: 10.3945/an.114.008052

[50] Kwon Y. Effect of soy isoflavones on the growth of human breast tumors: findings from preclinical studies. Food Sci Nutr. 2014;2(6):613–622. Doi: 10.1002/fsn3.142.

[51] Casati L, Sendra R, Sibilia V, Celotti F. Endocrine disrupters: the new players able to affect the epigenome. Front Cell Dev Biol. 2015;3:37. Doi: 10.3389/fcell.2015.00037

[52] Setchell KD, Clerici C. Equol: pharmacokinetics and biological actions. J Nutr. 2010;140(7):1363S–1368S. Doi: 10.3945/jn.109.119784

[53] Gheldiu AM, Popa DS, Loghin F, Vlase L. Oxidative metabolism of estrone modified by genistein and bisphenol A in rat liver microsomes. Biomed Environ Sci. 2015;28(11):834–838. Doi: 10.3967/bes2015.116

[54] Lee AJ, Mills LH, Kosh JW, Conney AH, Zhu BT. NADPH-dependent metabolism of estrone by human liver microsomes. J Pharmacol Exp Ther. 2002;300(3):838–849.

[55] Eliassen AH, Missmer SA, Tworoger SS, Hankinson SE. Circulating 2-hydroxy- and 16alpha-hydroxy estrone levels and risk of breast cancer among postmenopausal women. Cancer Epidemiol Biomarkers Prev. 2008;17(8):2029–2035. Doi: 10.1158/1055-9965.EPI-08-0262

[56] Ziegler RG, Fuhrman BJ, Moore SC, Matthews CE. Epidemiologic studies of estrogen metabolism and breast cancer. Steroids. 2015;99(Pt A):67-75. Doi: 10.1016/j.steroids.2015.02.015

[57] Morimoto Y, Conroy SM, Pagano IS, Isaki M, Franke AA, Nordt FJ, Maskarinec G. Urinary estrogen metabolites during a randomized soy trial. Nutr Cancer. 2012;64(2):307–314. Doi: 10.1080/01635581.2012.648819

[58] Atkinson C, Skor HE, Dawn Fitzgibbons E, Scholes D, Chen C, Wähälä K, Schwartz SM, Lampe JW. Urinary equol excretion in relation to 2-hydroxyestrone and 16alpha-

hydroxyestrone concentrations: an observational study of young to middle-aged women. J Steroid Biochem Mol Biol. 2003;86(1):71–77.

[59] Lu LJ, Cree M, Josyula S, Nagamani M, Grady JJ, Anderson KE. Increased urinary excretion of 2-hydroxyestrone but not 16alpha-hydroxyestrone in premenopausal women during a soya diet containing isoflavones. Cancer Res. 2000;60(5):1299–1305.

[60] McClain MR, Wolz E, Davidovich A, Bausch J. Genetic toxicity studies with genistein. Food Chem Toxicol. 2006;44(1):42–55.

[61] Messina M. Impact of soy foods on the development of breast cancer and the prognosis of breast cancer patients. Forsch Komplementmed. 2016;23(2):75–80. Doi: 10.1159/000444735

[62] Hu YC, Wu CT, Lai JN, Tsai YT. Detection of a negative correlation between prescription of Chinese herbal products containing coumestrol, genistein or daidzein and risk of subsequent endometrial cancer among tamoxifen-treated female breast cancer survivors in Taiwan between 1998 and 2008: a population-based study. J Ethnopharmacol. 2015;169:356–362. Doi: 10.1016/j.jep.2015.04.028

[63] Braakhuis AJ, Campion P, Bishop KS. Reducing breast cancer recurrence: the role of dietary polyphenolics. Nutrients. 2016;8(9),547:1-15. Doi: 10.3390/nu8090547

[64] Chavarro JE, Toth TL, Sadio SM, Hauser R. Soy food and isoflavone intake in relation to semen quality parameters among men from an infertility clinic. Hum Reprod. 2008;23(11):2584–2590. Doi: 10.1093/humrep/den243

[65] D'Adamo CR, Sahin A. Soy foods and supplementation: a review of commonly perceived health benefits and risks. Altern Ther Health Med. 2014;20(Suppl 1):39–51.

[66] You L, Sar M, Bartolucci EJ, McIntyre BS, Sriperumbudur R. Modulation of mammary gland development in prepubertal male rats exposed to genistein and methoxychlor. Toxicol Sci. 2002;66(2):216–225. Doi: 10.1093/toxsci/66.2.216

[67] Hwang YW, Kim SY, Jee SH, Kim YN, Nam CM. Soy food consumption and risk of prostate cancer: a meta-analysis of observational studies. Nutr Cancer. 2009;61(5):598–606. Doi: 10.1080/01635580902825639

[68] van Die MD, Bone KM, Williams SG, Pirotta MV. Soy and soy isoflavones in prostate cancer: a systematic review and meta-analysis of randomized controlled trials. BJU Int. 2014;113(5b):E119–E130. Doi: 10.1111/bju.12435

[69] He J, Wang S, Zhou M, Yu W, Zhang Y, He X. Phytoestrogens and risk of prostate cancer: a meta-analysis of observational studies. World J Surg Oncol 2015;13:231. Doi: 10.1186/s12957-015-0648-9

[70] Sun J, Sun B, Han F, Yan S, Yang H, Kikuchi A. Rapid HPLC method for determination of 12 isoflavone components in soybean seeds. Agric Sci China. 2011;10(1):70–77. Doi: 10.1016/S1671-2927(09)60291-1

[71] César Ida C, Braga FC, Soares CD, Nunan Ede A, Pianetti GA, Condessa FA, Barbosa TA, Campos LM. Development and validation of a RP-HPLC method for quantification of isoflavone aglycones in hydrolyzed soydry extracts. J Chromatogr B Analyt Technol Biomed Life Sci. 2006;836(1–2):74–78. Doi: 10.1016/j.jchromb.2006.03.030

[72] Shao S, Duncan AM, Yang R, Marcone MF, Rajcan I, Tsao R. Systematic evaluation of pre-HPLC sample processing methods on total and individual isoflavones in soybeans and soy products. Food Res Int. 2011;44:2425–2434. Doi: 10.1016/j.foodres.2010.12.041

[73] Vila-Donat P, Caprioli G, Maggi F, Ricciutelli M, Torregiani E, Vittori S, Sagratini G. Effective clean-up and ultra high-performance liquid chromatography–tandem mass spectrometry for isoflavone determination in legumes. Food Chem. 2015;174:487–494. Doi: 10.1016/j.foodchem.2014.11.047

[74] Otieno DO, Rose H, Shah NP. Profiling and quantification of isoflavones in soymilk from soy protein isolate using extracted ion chromatography and positive ion fragmentation techniques. Food Chem. 2007;105:1642–1651. Doi: 10.1016/j.foodchem.2007.04.036

[75] Caprioli G, Navarini L, Cortese M, Ricciutelli M, Torregiani E, Vittori S, Sagratini G. Quantification of isoflavones in coffee by using solid phase extraction (SPE) and high-performance liquid chromatography-tandem mass spectrometry (HPLC-MS/MS). J Mass Spectrom. 2016;51(9):698–703. Doi: 10.1002/jms.3802

Chapter 8

Anthocyanins in Berries and Their Potential Use in Human Health

Daniela Peña-Sanhueza,

Claudio Inostroza-Blancheteau,

Alejandra Ribera-Fonseca and Marjorie Reyes-Díaz

Additional information is available at the end of the chapter

http://dx.doi.org/10.5772/67104

Abstract

Anthocyanin pigments are responsible for the red, purple, and blue colors of many fruits, vegetables, cereal grains, and flowers, increasing the interest due to their strong antioxidant capacity and their possible use to the benefit of human health. Abundant evidence is available about the preventive and therapeutic roles of anthocyanin in different kinds of chronic diseases. According to the structural differences and anthocyanin content of berries such as blackberry, blueberry, chokeberry, and others, there are different healthy properties in the treatments of circulatory disorders, cancer cell lines, and diabetes as well as antiviral and antimicrobial activities. On the other hand, molecular aspects play an important role in anthocyanin biosynthesis, making it possible to determine how biotic and abiotic factors impact its biosynthesis complex. Thus, the aim of this chapter was to describe the use of anthocyanins from berries for human health and their potential use as a pharmacological bioresource in the prevention of chronic diseases. In addition, an update of the molecular mechanisms involved in anthocyanin biosynthesis will be discussed.

Keywords: anthocyanins, berries, cancer, transcription factors

1. Introduction

The scientific evidence regarding the positive relationship between diet and health has increased consumer demand for more information related to healthy diets, including fruits and vegetables, with functional characteristics that help to delay the aging process and reduce the risk of several diseases, mainly cardiovascular diseases and cancer [1]. Berries are recognized as an important component of healthy diets due to their bioactive compounds. In this sense, commercial berry species such as blackberry (*Rubus* sp.), bilberry (*Vaccinium myrtillus* L.), blackcurrant (*Ribes rugrum* L.), chokeberry (*Aronia melanocarpa* (Michx.) Elliott.), cranberry (*V. macrocarpon* Ait.), bayberry (*Myrica* sp.), raspberry (*Rubus ideaus* L.), black raspberry (*Rubus occidentalis* L.), strawberry (*Fragaria ananassa* Duch.), highbush blueberry (*V. corymbosum* L.), maqui (*Aristotelia chilensis*), murtilla (*Ugni molinae* Turcz.), and calafate (*Berberis microphylla* G. Forst.) are particularly rich sources of antioxidants, which are usually consumed in fresh and processed products [2–5]. Higher plants, especially berry species, synthesize a diverse group of phenolic compounds such as flavonoids. These plant secondary metabolites have many biological functions, including their key role in plant-microbe interaction, plant-pathogen interaction, pollen-tube growth, UV radiation protection, tissue pigmentation, and others [6, 7]. Flavonoid compounds, which include flavonols, flavones, flavanols, flavanones, isoflavonoids, and anthocyanins, are molecules widely accumulated in vascular plants and to a lesser extent in mosses, being accumulated in all organs and tissues at different stages of development and depending on the environmental conditions [6].

Anthocyanins are natural pigments responsible for the blue, purple, red, and orange colors of many fruits and vegetables [8, 9]. Anthocyanins are a glycoside form of anthocyanidins [9], and the structural differences among them are related to the number of hydroxyl group, position, and kind and/or number of sugars linked to the molecule [10, 11]. These compounds appear to be an interesting natural resource of water-soluble dyes because they are easily incorporated in aqueous media [12]. Another important property of anthocyanins is their remarkable antioxidant activity, playing a vital role in the prevention of neuronal and cardiovascular illnesses, diabetes, cancer, etc. [11, 13]. Many reports have focused on the effect of anthocyanins in cancer prevention [14], human nutrition [15], and their biological activity [10]. Nowadays, there is an increased interest in explaining the role of anthocyanins as a natural antioxidant and their mechanism of action on human health as well as the treatment of chronic diseases and their use as a natural dye, substituting the synthetic dyes, which can be toxic to humans. This review endeavors to describe the use of anthocyanins from berries for human health and their potential use as a pharmacological bioresource in the prevention of chronic diseases. In addition, an update of the molecular mechanisms involved in anthocyanin biosynthesis will be discussed. Finally, recent clinical and preclinical studies about anthocyanin use in the prevention of human diseases are reported.

2. Anthocyanin and phenolic compounds in berries

Phenolic acid, organic acids, tannins, anthocyanins, and flavonoids are phenolic bioactive compounds with a high concentration in the berry fruits [16]. The chemical structure of phenolic compounds is characterized by one or more aromatic rings with hydroxyl groups. According to their structural characteristics, phenolic compounds are classified into five major groups: phenolic acids, stilbenes, flavonoids (flavonols or catechins, flavonols, flavones, flavonones, isoflavonoids, anthocyanins), tannins, and lignans [13]. The concentration of phenolic compounds in berry fruits is altered by many factors, such as genotype, species, agronomic management, climatic factors, ripening stage, harvesting time, and postharvest management [17, 18]. Given the plant phenol attributes of berry species, attention has largely focused on anthocyanin and flavonol antioxidant action on human health. In this way, substantial epidemiological and experimental research suggests that intakes of recognized nutritional antioxidants such as vitamin E and carotenoids can decrease the oxidative damage of proteins, lipids, and DNA *in vivo* and may reduce the incidence of developing many chronic diseases in humans [19]. The *in vitro* antioxidant effectiveness of anthocyanins and other polyphenols is due to its donation of free hydrogen atom from an aromatic hydroxyl group of the antioxidant molecules, acting as radical scavenger [20].

It has been reported that the antioxidant capacity of flavonoids is stronger than vitamins C and E [21, 22], and under *in vitro* conditions, flavonoids can prevent injury in different ways, acting as a suppressor of reactive oxygen formation, scavenging free radicals by hydrogen atom donation [22, 23], activating antioxidant enzymes [23, 24], chelating metal, reducing α-tocopheryl radicals, inhibiting oxidases, oxidative stress mitigation by nitric oxide, increasing uric acid levels, and increasing antioxidant properties of low-molecular antioxidants [22]. Anthocyanin concentration in blackberry is much higher than in raspberry and strawberry and similar to red currant blueberry, depending on the cultivar (see **Table 1**).

Anthocyanin concentration widely differs significantly among plant species, even among species of the same genus. In **Table 1**, anthocyanin and total phenolic compounds of different species and cultivars and their analysis are detailed. In blackberry, anthocyanin content is generally similar in all species, but phenolic content shows strong differences (**Table 1**). Anthocyanin content in *Rubus insularis* F. Aresch. represents 36% of the phenolic compounds, whereas in *R. fructicosus* cultivar Hull Thornless, it only represents 6.4% of the total phenolic compounds (**Table 1**). Raspberry (*R. innominatus* S. Moore) showed higher anthocyanin level, representing 41.2% of the total phenolic compounds. *R. ideaus* show high phenolic compounds; however, their high content does not necessarily represent a high anthocyanin content. *R. ideaus* Heritage cultivar has showed the highest anthocyanin percent with respect to the total phenolic compounds, representing 3.8% (**Table 1**). Additionally, blueberry cultivars showed low differences between anthocyanin and phenolic compounds, but they showed greater health benefits than other berries due to their particularly high proportion of anthocyanins. In some cases, high anthocyanin content in blueberries is related to high antioxidant capacity, but the anthocyanin contents and composition are different in each species and cultivar (**Table 1**). More specifically, the *V. corymbosum* cultivar (Duke) contains 63% anthocyanin with respect to the total phenolic compounds, followed by the cultivars CVAC5.001 and Brigitta, with 46 and 41%, respectively, and finally by Bluecrop with 27%. It is therefore necessary to evaluate the correlation between anthocyanin content and total phenolic compounds, because the ratio can exist between the two parameters, but it is not necessary to estimate in all species or among cultivars of the same genus (**Table 1**) [25–29].

Scientific name	Common name	Cultivar	Anthocyanins*	Phenolics**	References
Rubus cyri Juz.	Blackberry	Native	143	545	[25]
Rubus georgicus Focke	Blackberry	Native	89	561	[25]
Rubus insularis F. Aresch.	Blackberry	Native	170	472	[25]
Rubus ursinus (Douglas ex Hook.)	Blackberry	Native	211	629	[25]
Rubus fructicosus L.	Blackberry	Chactaw	125	1703	[26]
Rubus fructicosus	Blackberry	T. evergreen	146	2061	[26]
Rubus fructicosus	Blackberry	Hull Thornless	152	2349	[26]
Rubus idaeus L.	Raspberry	Native	65	517	[27]
Rubus innominatus S. Moore	Raspberry	Native	52	126	[25]
Rubus niveus Thunb.	Raspberry	Native	230	402	[25]
Rubus ideaus	Raspberry	Heritage	49	1280	[26]
Rubus ideaus	Raspberry	Autumm Bliss	39	2494	[26]
Rubus ideaus	Raspberry	Fallgold	3	1459	[26]
Rubus ideaus	Raspberry	Meeker	42	2116	[26]
Ribes sativum	Red currants	London Market	7.8	1115	[26]
Ribes sativum (Lam.) Mert. & Kock	Red currants	Rovada	7.5	1193	[26]
Ribes sativum	Red currants	White Versailes	1.4	657	[26]
Ribes nigrum L.	Red currants	Alagan	169	694	[25]
Ribes nigrum	Red currants	Ben Lomond	261	933	[25]
Ribes nigrum	Red currants	Ojebyn	165	830	[25]
Ribes nigrum	Red currants	Consort	411	1342	[25]
Vaccinium corymbosum L.	Blueberry	Bluecrop	84	304	[25]
Vaccinium corymbosum	Blueberry	Briggita	103	246	[25]
Vaccinium corymbosum	Blueberry	Duke	173	274	[25]
Vaccinium corymbosum	Blueberry	CVAC5.001	430	868	[25]
Vaccinium corymbosum	Blueberry	Native	62-235	181–473	[28]
Vaccinium corymbosum	Blueberry	Bluegold	206	432	[29]
Vaccinium corymbosum	Blueberry	Briggita	190	468	[29]
Vaccinium corymbosum	Blueberry	Legacy	226	570	[29]
Vaccinium angustifolium Ait.	Blueberry	Native	208	692	[25]
Vaccinium myrtillus L.	Bilberry	Native	300	525	[28]

*mg cyanidin 3-glucoside eq./100 g fresh weight.
**mg gallic acid eq./100 g fresh weight.

Table 1. Total anthocyanin and phenolic content of berry fruits.

Berry species with higher anthocyanin content are interesting for use in breeding programs for increasing their content in fruits, enhancing their antioxidant capacity, and obtaining fruit products with health properties. In addition, the understanding of the molecular network of genes involved in anthocyanin biosynthesis and how biotic and abiotic factors could affect their concentration and gene regulation are a key to use it in genetic engineering and agronomic management.

3. Molecular regulation of anthocyanin biosynthesis

Six structural genes are common in the anthocyanin pathway in all angiosperms, which are divided into two main groups. The first group is the upstream genes or early biosynthesis genes, for example, chalcone synthase (CHS), chalcone flavanone isomerase (CHI), and flavanone 3-hydroxylase (F3H), coding for enzymes that produce precursors for one or more important non-anthocyanin flavonoids. The second group is the downstream genes or late biosynthesis genes, for example, anthocyanidin synthase (ANS), dihydroflavonol-4-reductase (DFR), and UDP-glucose flavonoid 3-oxy-glucosyltransferase (UF3GT), coding for enzymes specific to anthocyanin synthesis [30–32]. In the anthocyanin pathway, L-phenylalanine is converted to naringenin by phenylalanine ammonia lyase (PAL), cinnamate 4-hydroxylase (C4H), 4-coumarate CoA ligase (4CL), chalcone synthase (CHS), and chalcone isomerase (CHI). Then, the next pathway is catalyzed by the formation of complex aglycone and anthocyanin composition by flavanone 3-hydroxylase (F3H), flavonoid 3'-hydroxylase (F3'H), dihydroflavonol 4-reductase (DFR), anthocyanidin synthase (ANS), UDP-glucoside flavonoid glucosyltransferase (UFGT), and methyl transferase (MT) [33]. It has been described that the transcription of early and late biosynthesis genes to produce anthocyanins appears to be regulated by R2R3-MYB and basic helix-loop-helix (bHLH, also known as MYC) called transcription factors in collaboration with tryptophan-aspartic acid repeat (WDR) or WD40 proteins [32, 34–37].

3.1. MYB transcription factor

The MYB transcription factors involved in the flavonoid pathway have been identified and described for several kinds of model plants, crops, and ornamental plants. The first identified and reported MYB transcription factor in plants was in *Zea mays*, which included C1 (Colorless 1) and PL1 (Purple Leaf 1) [38]. The MYB transcription factors are composed of the so-called N-terminal MYB domain, consisting of one to three imperfect repeats of almost 52 amino acids (R1, R2, and R3), beginning with R2R3, the most abundant subfamily in plants [39]. The MYB domain is involved in DNA binding and dimerization. The C-terminal region is responsible for establishing protein-protein and regulates activation or repression of gene expression [34, 40, 41]. The MYB genes are exclusive to eukaryotic organisms [42]. In animals, these genes are associated with cell proliferation and differentiation [43, 44], whereas in plants, MYB is associated with responses to different biotic and abiotic stressors (drought, cold, pathogen disease resistance), plant development (trichome formation, seed development), stomatal movement, and many other functions [34, 40, 45, 46]. Anthocyanin biosynthesis mediated by MYB transcription factors has been reported in *Arabidopsis thaliana*

(L.) Heynh. [41, 47–49], strawberry (*F. ananassa*) [50], Chilean strawberry (*Fragaria chiloensis* (L.) Duch.) [51], apple (*Malus domestica* Borkh.) [52–54], and tomato (*Solanum lycopersicum* L.) [55]. Grape (*Vitis vinifera* L.) is the main plant species studied in this way due to its agricultural and commercial importance worldwide. Thus, many MYB transcription factors have been reported in this species by different researchers. *VvMYBPA1* and *VvMYBPA2* are involved in proanthocyanidin synthesis [46, 56], while *VvMYBF1* regulates flavonol synthesis [57]. In addition, *MYBA1* and *MYBA2* genes control the last biosynthetic step of anthocyanin synthesis [58, 59]. It is reported that a glycosylation reaction mediated by the UDP-glucose flavonoid-3-O-glucosyltransferase (UFGT) enzyme produces anthocyanins in grapes [31, 39]. It is important to highlight that MYB transcription factor is conserved between different species and is one of the most important primary proteins involved in structural and biological functions. Furthermore, MYB transcription factor regulates the flavonoid pathway apparently in two ways: (a) due to variations in the C-terminal region of the protein or (b) modulating the interaction with DNA, bHLH, and WD40 protein [34, 60].

3.2. Basic helix-loop-helix (bHLH)

After MYB, bHLH proteins, also known as MYC, are the second most important family of transcription factors involved in anthocyanin biosynthesis [34, 61]. The bHLH protein domain is constituted of about 60 amino acids and is characterized by the presence of 19 conserved amino acids, five in the basic region, five in the first helix, one in the loop, and eight in the final second helix [61]. The basic region of bHLH has basic residues (5.8 on average) essential for DNA binding. In Arabidopsis, 20% of bHLH transcription factors do not have this domain and can act as a repressor because forming heterodimers are unable to bind to DNA [61]. Two cis-element boxes have been reported to bind with bHLH proteins, the E-box (5'-CANNTG-3'), and G-box (5'-CACGTG-3') elements. The G-box is the most commonly recognized sequence representing 81% of the proteins predicted to bind DNA [61, 62]. In the basic region, two amino acids conferred the property on binding DNA in Arabidopsis plants. The Glu13 and Arg16 are the E-box recognition motif [63]. Glu13 has contact with CA bases of E-box and Arg16, apparently helping Glu13 to bind and stabilize. In G-box, specific stabilization is mediated by His/Lys9, Glu13, and Arg17. The Arg17 interacts with inner G base, and His/Lys9 interacts with the last G of the G-box [61, 62]. The alpha-helix function is involved in homo- and hetero-dimerization and is formed by hydrophobic residues of isoleucine, leucine, and valine [34, 61]. Arabidopsis has been demonstrated that this residue is conserved in all bHLH proteins, indicating the importance of the basic region of the bHLH transcription factor in DNA binding [61, 63]. The second helix is involved in DNA binding through direct contact with the E-box. Finally, the loop is responsible for the three-dimensional arrangement of alpha-helices, and residues from the first helix loop junction are involved in association with bHLH proteins [34, 61, 63, 64]. Basic helix-loop-helix transcription factors in plants are involved in processes such as flower development [65, 66], hormonal response [67, 68], metal homeostasis [69], and others. Regarding bHLH and their relation to flavonoid synthesis, the first bHLH involved in this pathway was detected in maize in 1989 [70]. In this context, in *Z. mays* (ZmB, ZmR, and ZmLc), bHLH is involved in the regulation of the anthocyanin pathway [70–72], and

ZmIn1 is involved in the repression of flavonoid gene expression in maize aleurone [73]. In *A. thaliana*, it has been reported that *AtTT8* gene encodes a bHLH transcription factor involved in the control of proanthocyanidins and anthocyanins in seeds and seedlings [74]. Quatroccio et al. [75, 76]. reported PhAN1, PhJAF13 hBLH transcription factor from *Petunia hibrida* as being involved in the control of the anthocyanin pathway in flowers. For *Vitis vinifera,* VvMYCA1 (also known as bHLH) was reported as involved in promotion of anthocyanin accumulation in grape cells [37].

3.3. WDR proteins

Tryptophan-aspartic acid repeat protein (WDR) or WD40 proteins are characterized by around 44–66 amino acids, delimited by the GH dipeptide on the N-terminal size (11–24 residues from the N-terminus) and the WD dipeptide on the C-terminus [34, 77]. In Arabidopsis, WDR protein contains four (or more) tandem repeats composed of around 40 amino acids [78]. In contrast to the majority of proteins, WDR is not involved in catalytic activities such as DNA binding or gene expression regulation, mostly acting as a platform due to its capacity to interact with more than one protein at the same time [34, 78]. The work of WDR involves eukaryotic cellular process such as cell division, vesicle formation, signal transduction, RNA processing, and transcription regulation [78]. On the other hand, MYB and bHLH transcription factors have few WDR proteins involved in the flavonoid pathway, as shown in *Z. mays* (ZmPAC1), where it regulates the anthocyanin pathway in seed aleurone [79]. In Arabidopsis (AtTTG1), WDR proteins control trichomes, root hair, and seed mucilage production [80]. In petunia, AN11 regulates anthocyanin production as well as the pH of the flower vacuole [81], whereas in grape, *V. vinifera* WDR1 contributes to anthocyanin accumulation [37]. Although WDR proteins are not directly involved in the flavonoid pathway, particularly in anthocyanin synthesis, it is important to note that these proteins are highly conserved among species [34]. Nevertheless, few WDR proteins have been reported in plants, and it must be highlighted that WDR is involved in several metabolic and physiological processes [79, 80, 82]. To clarify the characteristics of WDR proteins and the complex formed with MYB and bHLH, which is involved in anthocyanin biosynthesis, species such as petunia and Arabidopsis have been used [34, 35].

3.4. MYB-bHLH-WDR (MBW complex)

MBW complex has been reported in Arabidopsis, petunia, and some varieties of grape [35, 82]. The most important function of these transcription factors is involved in the process related to DNA binding, activation of gene expression involved in the flavonoid pathway, and stabilization of the three-dimensional configuration of the complex [34]. Basic helix-loop-helix-WDR interaction is needed to WDR protein translocation into the nucleus, and this was demonstrated in onion cells using green protein fluorescent (GPF), which when expressed alone is localized in the cytosol, whereas its co-expression with PFWD and MYC-RP enables the transport and localization in the nucleus [35]. The AN11 from petunia showed the same results, being detected in the cytosol [81]. *V. vinifera* subjected to high salt concentrations showed a cultivar-dependent response for anthocyanin accumulation, which was correlated with the expression of MYBA1-2, MYCA1 and WDR1 genes [37].

4. Antioxidant capacity of anthocyanins in berries and their use in human health

The radical scavenging activity (RSA) of anthocyanins is largely due to the presence of hydroxyl groups in position 3 of ring C and also in the 3', 4', and 5' positions in ring B of the molecule. In general, RSA of anthocyanidins (aglycons) is superior to their respective anthocyanins (glycosides), and this decreases when the number of sugar increases [16]. Hanachi et al. [83] showed that fruits of *Berberis vulgaris* L. (barberry) have a high antioxidant activity, reducing the viability of cell cultures associated with liver cancer (HepG2). Furthermore, extracts of leaves and twigs of *B. vulgaris* have more antioxidants than fruits. Končić et al. [84] studied the antioxidant activity of extracts of leaves, branches, and roots of two species of *B. vulgaris* and *Berberis croatica* and demonstrated that all these organs exhibited antioxidant activity. In all cases, the activity was positively correlated with the content of phenolic acids and flavonols, and the flavonols played the main role in the total antioxidant activity of the studied species [84]. They also concluded that the antioxidant activities were significantly different (being higher in *B. croatica* than *B. vulgaris*) and among organs (being higher in leaves followed by branches and roots). The result of the anthocyanin concentration in different organs besides the fruits is interesting, because acquisition of anthocyanin in every season of the year has advantages for making new products with health properties. Thus, interesting results such as a new natural resource for promoting these compounds for human health have been reported. Končić et al. [84] suggested that studies into different species are needed to analyze all the organs of the plant, not just the fruits. Shin et al. [85] reported that in human liver cancer HepG2, cell proliferation was inhibited by strawberry extracts. Moreover, Chang et al. [86] reported that *Hibiscus sabdariffa* Linne (roselle) anthocyanin extracts mediated the apoptosis of human promyelocytic leukemia cells via the p38/Fas and Bid pathways. Research examining the use of black currant extract (BCE) with high concentrations of phenolic compounds on antiproliferative activity against gastric cancer SGC-7901 cells showed a positive antiradical activity and anticarcinogenic effects [87]. Moreover, extracts of mulberry showed an inhibition on the growth of human gastric carcinoma cells [88]. In this study, anthocyanins extracted from mulberry had notable promotive effects on the p38/jun/Fas/FasL and p38/p53/Bax signaling pathways, which accounted for its *in vitro* and *in vivo* growth-inhibitory and apoptotic responses in AGS (gastric cancer) cells. The effects of berries on diseases are shown in **Table 2**.

With respect to *in vivo* studies, Wang and Stoner [89] reported the effect of an anthocyanin-rich extract from black raspberries on the development of tumors in rat esophagus by N-nitrosomethylbenzylamine (NMBA), the most potent inducer of tumors in rat esophagus. This extract inhibited cell proliferation, inflammation, and induced apoptosis in the esophageal tissues (**Table 2**). Stoner et al. [90] compared the effect of black raspberry, red raspberry, strawberry, and blueberry anthocyanin and ellagitannins in fruit extract on the prevention of esophageal cancer induced by N-nitrosomethylbenzylamine (NMBA) in rats. Inhibition of NMBA-induced tumorigenesis in the rat esophagus was observed. The authors detected a reduction in cytokine levels in serum, interleukin 5 (IL-5), and GRO/KC, which is the rat homolog for human interleukin-8 (IL-8), and these cytokines

Disease	Scientific name	Common name	Compound	Experimental conditions	Reference
Liver cancer	*Fragaria x ananassa* Duch.	Strawberry	Crude extract	*In vitro*	[85]
Leukemia	*Hibiscus sabdariffa* L.	Rosselle	Anthocyanin rich extract	*In vitro*	[86]
Gastric cancer	*Morus alba* L.	Mulberry	Anthocyanins	*In vitro*	[88]
Gastric cancer	*Ribes nigrum* L.	Black currant	Crude extract	*In vitro*	[87]
Colon caner	*Vaccinium myrtillus* L.	Bilberry	Anthocyanin-rich extract	*In vivo* (rats)	[91]
Colon cancer	*Aronia melanocarpa* E.	Chokeberry	Anthocyanin-rich extract	*In vivo* (rats)	[91]
Colon cancer	*Vitis vinifera* L.	Grape	Anthocyanin-rich extract	*In vivo* (rats)	[91]
Esophagus cancer	*Rubus occidentalis* L.	Black raspberries	Anthocyanin-rich extract	*In vivo* (rats)	[89]
Esophagus cancer	*Rubus occidentalis*	Black raspberries	Anthocyanins and ellagitannins	*In vivo* (rats)	[90]
Esophagus cancer	*Rubus ideaus* L.	Red raspberries	Anthocyanins and ellagitannins	*In vivo* (rats)	[90]
Esophagus cancer	*Fragaria ananassa*	Strawberries	Anthocyanins and ellagitannins	*In vivo* (rats)	[90]
Esophagus cancer	*Vaccinium corymbosum* L.	Blueberries	Anthocyanins and ellagitannins	*In vivo* (rats)	[90]
Hepatic cancer	*Berberis vulgaris* Duch.	Barberries	Crude extract	*In vivo* (rats)	[83]
Liver cancer	*Vaccinium corymbosum*	Blueberries	Anthocyanin extract	*In vitro* (mice)	[94]
Liver cancer	*Berberis vulgaris*	Barberries	Crude extract	*In vitro*	[83]
Oral cancer	*Rubus occidentalis*	Black raspberries	Crude extract	*In vivo* (mice)	[96]
Mammary	*Vitis vinifera*	Grape	Crude extract	*In vivo* (rats)	[95]
Skin cancer	*Punica granatum* L.	Pomegranate	Crude extract	*In vivo* (mice)	[92, 93]

Table 2. Anticarcinogenic effects of anthocyanin/anthocyanin-rich extract from different berry species under *in vivo* and *in vitro* conditions in different chronic diseases.

were associated with an increase in serum antioxidant capacity. At molecular level, Stoner et al. [90] also reported that the use of extracts showed a differential expression in 626 and 625 genes per 4807 and 17846 of preneoplastic esophagus and esophageal papilloma genes, respectively. These genes are involved in carbohydrate and lipid metabolism, cell death and proliferation, and inflammation. These results are an important approach to estimate the relation of anthocyanin gene expression and its influence on proteins associated with cell proliferation, apoptosis, angiogenesis, and esophageal carcinogenesis. Lala et al. [91] observed an anticarcinogenic effect of anthocyanins on colon cancer induced by azoxymethane in a rat model. In that study, anthocyanin-rich extract from bilberry, chokeberry, and grapes significantly reduced azoxymethane-induced aberrant crypt foci and decreased cell proliferation and *COX-2* gene expression [91]. Delphinidin and pomegranate extracts enriched with anthocyanins, and tannins showed an inhibition in skin cancer induced by UV-B or TPA (12-O tetradecanoylphorbol-13acetate) when applied to mouse skin [92, 93]. Here, delphinidin inhibited DNA damage mediated by UV-B radiation, and pomegranate modulated the mitogen-activated protein kinase (MAPK) and nuclear factor-kappa B (NF-kB) pathways. Similarly, studies on the protective effect and antioxidant mechanism of anthocyanin extract from blueberries were conducted using a liver injury induced by CC4 in mice—the effect of which increased lipid peroxidation and reduced liver cell viability [94]. The results indicate that anthocyanin extract effectively protected mice from CC4-induced liver injury by attenuation of lipid peroxidation. In mammary adenocarcinoma induced by dimethylbenzaanthracene (DMBA) in rats, the antitumoral effect of grape juice was evaluated by Singletary et al. [95]. They demonstrated that the tumor mass was ultimately reduced by suppressing cell proliferation (**Table 2**). In general, the strong antioxidant capacity of berry species is attributed to their anthocyanin content, suggesting that it might offer potential chemopreventive properties, including the inhibition of gastric, leukemia, liver, and breast cancer cell proliferation, among others; however, the mechanism of action must be evaluated for each disease because apparently their mechanism of effects varies (inhibiting cell proliferation, activating different enzymatic activity, inducing or repressing gene expression, etc.) depending on the extract from each plant species.

5. Conclusions and future challenges

The potential use of anthocyanins from different plant species as natural compounds with a health benefit for humans opens a new trend for the prevention and alternative treatments of chronic diseases. Several reports have demonstrated that anthocyanins from berries could inhibit or decrease the growth of carcinogenic tumors by affecting cell proliferation, increasing or inhibiting enzymatic systems, and increasing expression of genes involved in cell protection. On the other hand, it is important to highlight that synthesis of anthocyanins in different tissues of plants species should be considered. In addition, the discovery and characterization of new regulatory elements of anthocyanin biosynthesis are crucial to understand and manipulate this pathway in breeding programs. Improving knowledge about increasing anthocyanin synthesis in crops of research and commercial interest, together with more animal and human model studies under *in vivo* conditions, is essential to generate better human anticarcinogenic or antichronic disease supplement products with chemopreventive effects from berries.

Acknowledgements

We are very grateful to PIA-UFRO 16-0006 and DI 16-2013 projects from the Dirección de Investigación at the Universidad de La Frontera, Temuco, Chile.

http://dx.doi.org/10.5772/67104

Author details

Daniela Peña-Sanhueza[1], Claudio Inostroza-Blancheteau[1, 2], Alejandra Ribera-Fonseca[1, 3] and Marjorie Reyes-Díaz[1, 4*]

*Address all correspondence to: marjorie.reyes@ufrontera.cl

1 Center of Plant, Soil Interaction and Natural Resources Biotechnology, Scientific and Technological Bioresource Nucleus (BIOREN), Universidad de La Frontera, Temuco, Chile

2 Núcleo de Investigación en Producción Alimentaria (NIPA-UCT), Escuela de Agronomía, Facultad de Recursos Naturales, Universidad Católica de Temuco, Temuco, Chile

3 Departamento de Producción Agropecuaria, Facultad de Ciencias Agropecurias y Forestales, Universidad de La Frontera, Temuco, Chile

4 Departamento de Ciencias Químicas y Recursos Naturales, Facultad de Ingeniería y Ciencias, Universidad de La Frontera, Temuco, Chile

References

[1] Paredes-López O, Cervantes-Ceja ML, Vigna-Pérez M, Hérnandez-Pérez T. Berries: improving human health and healthy aging, and promoting quality. A review. Plant Foods for Human Nutrition. 2010;65:299–308.

[2] Benvenuti S, Pellati F, Melegari M, Bertelli D. Polyphenols, anthocyanins, ascorbic acid, and radical scavenging activity of Rubus, Ribes, and Aronia. Journal of the Science of Food and Agriculture. 2004;69:164–169.

[3] Céspedes C, El-Hafidi H, Pavon N, Alarcon J. Antioxidant and cardioprotective activities of phenolic extracts from fruits of Chilean blackberry *Aristotelia chilensis* (Elaeocarpaceae), Maqui. Food Chemistry. 2008;107:820–829.

[4] Ruiz A, Gutierrez I, Mardones C, Vergara C, Herlitz E, Vega M, Dorau C, Winterhalter P, Von Baer D. Polyphenols and antioxidant activity of Calafate (*Berberis microphylla* G. Forst.) fruit and other native berries from southern Chile. Journal of Agricultural and Food Chemistry. 2010;58:6081–6089.

[5] Jimenez S, Guevara R, Miranda R, Feregrino A, Torres I, Vazquez M. A review: functional properties and quality characteristics of bioactive compounds in berries: biochemistry, biotechnology, and genomics. Food Research International. 2012;54:1195–1207.

[6] Koes RE, Quattrocchio F, Mol JNM. The flavonoid biosynthetic pathway in plants: function and evolution. BioEssays. 1994;16:123–132.

[7] Boss P, Davies Ch, Robinson S. Expression of anthocyanin biosynthesis pathway genes in red and white grapes. Plant Molecular Biology. 1996;32: 565–569.

[8] de Pascual-Teresa S, Moreno D, García-Viguera C. Flavanols and Anthocyanins in cardiovascular health: a review of current evidence. International Journal of Molecular Sciences. 2010;11:1679–1703.

[9] Konczak I, Zhang W. Anthocyanins-more than natures colours. Journal of Biomedicine and Biotechnology. 2004;5:239–240.

[10] Kong JM, Chia LS, Goh NK, Chia TF, Brouillard R. Analysis and biological activities of anthocyanins. Phytochemistry. 2003;64:923–933.

[11] Soriano R, Pastore G. Evaluation of the effects of anthocyanins in type 2 diabetes. Food Research International. 2012;46:378–386.

[12] Pazmino-Duran AE, Giusti MM, Wrolstad RE, Gloria BA. Anthocyanins from *Oxalis triangularis* as potential food colorants. Journal of Agricultural and Food Chemistry. 2001;75:211–216.

[13] Castaneda-Ovando A, Pacheco-Hernandez M, Paez-Hernandez M, Rodriguez J, Galan-Vidal C. Chemical studies of anthocyanins: a review. Food Chemistry. 2008;113:859–871.

[14] Nichenametla SN, Taruscio TG, Barney DL, Exon JH. A review of the effects and mechanisms of polyphenolics in cancer. Critical Reviews in Food Science and Nutrition. 2006;46:161–183.

[15] Stintzing FC, Carle R. Functional properties of anthocyanins and betalains in plants, food, and in human nutrition. Trends in Food Science and Technology. 2004;15:19–38.

[16] Szajdek A, Borowska EJ. Bioactive compounds and health promoting properties of berry fruits: review. Plant Foods for Human Nutrition. 2008;63:147–156.

[17] Castrejón ADR, Eichholz I, Rohn S, Kroh LW, Huyskens-Keil S. Phenolic profile and antioxidant activity of highbush blueberry (*Vaccinium corymbosum* L.) during fruit maturation and ripening. Food Chemistry. 2008;109:564–572.

[18] Connor AM, Luby JJ, Hancock JF, Berkheimer S, Hanson EJ. Changes in fruit antioxidant activity among blueberry cultivars during cold temperature storage. Journal Agriculture and Food Chemistry. 2002;50:893–898.

[19] Diplock AT, Charleux J-L, Crozier-Willi G, Kok FJ, Rice-Evans C, Roberfroid M. Functional food science and defence against reactive oxidative species. British Journal of Nutrition. 1998;80:77–112.

[20] Rice-Evans C, Miller NJ, Paganga G. Antioxidant properties of phenolic compounds. Trends in Plant Science. 1997;2:152–159.

[21] Prior RL, Cao G. Antioxidant phytochemicals in fruits and vegetables: diet and health implications. HortScience. 2000;35:588–592.

[22] Procházková D, Boušová I, Wilhelmov N. Antioxidant and prooxidant properties of flavonoids. Review. Fitoterapia. 2011;82:513–523.

[23] Pieta PG. Flavonoids and antioxidants. Journal of Natural Products. 2000;63:1035–1042.

[24] Nijveldt RJ, van Nood E, van Hoorn DEC, Boelens PG, van Norren K, van Leeuwen PAM. Flavonoids: a review of probable mechanisms of action and potential applications. The American Journal of Clinical Nutrition. 2001;74:418–25.

[25] Moyer R, Hummer K, Finn C, Frei B, Wrolstad R. Anthocyanins, phenolics, and antioxidant capacity in diverse small fruits: *Vaccinium, Rubus*, and *Ribes*. Journal of Agricultural and Food Chemistry. 2002;50:519–525.

[26] Pantelidis GE, Vasilakakis Manganaris GA, Diamantidis G. Antioxidant capacity, phenol, anthocyanin and ascorbic acid contents in raspberries, blackberries, red currants, gooseberries and Cornelian cherries. Journal of Agricultural and Food Chemistry. 2007;102:777–783.

[27] Wada L, Ou B. Antioxidant activity and phenolic content of Oregon caneberries. Journal of Agricultural and Food Chemistry. 2002;50:3495–3500.

[28] Prior R, Cao G, Martin A, Sofic E, McEwen J, O'Brien C, Lischner N, Ehlenfeldt M, Kalt W, Krewer G, Mainland M. Antioxidant capacity as influenced by total phenolic and anthocyanin content, maturity, and variety of *Vaccinium* species. Journal of Agricultural and Food Chemistry. 1998;46:2686–2693.

[29] Ribera AE, Reyes-Diaz M, Alberdi M, Zuñiga GE, Mora ML. Antioxidant compounds in skin and pulp of fruits change among genotypes and maturity stages in highbush blueberry (*Vaccinium corymbosum* L.) grown in southern Chile. Journal of Soil Science and Plant Nutrition. 2010;10:509–536.

[30] Rausher MD, Miler RE, Tiffin P. Patterns of evolutionary rate variation among genes oh the anthocyanin biosynthetic pathway. Molecular Biology and Evolution. 1999; 16:266–274.

[31] Matus J, Loyola R, Vega A, Peña-Neira A, Bordeu E, Arce-Jhonson P, Alcald, J. Post-veraison sunligth exposure induces MYB-mediated transcriptional regulation of anthocyanin and flavonol synthesis in berry skins of *Vitis vinifera*. Journal of Experimental Botany. 2009;60:853–867.

[32] Xi X, Zha Q, Jiang A, Tian Y. Impact of cluster thinning on transcriptional regulation of anthocyanin biosynthesis-related genes in "Summer Black" grapes. Plant Physiology and Biochemestry. 2016;104:180–187.

[33] Li Y, Baldauf S, Lim EK, Bowles DJ. Phylogenetic analysis of the UDP-glycosyltransferase multigene family of *Arabidopsis thaliana*. The Journal of Biological Chemistry. 2001;276:4338–4343.

[34] Hichri I, Barrieu F, Bogs J, Kappel Ch, Delrot S, Lauvergeat V. Recent advances in the transcriptional regulation of the biosynthetic pathway. Review paper. Journal of Experimental Botany. 2011;62:2465–2483.

[35] Payne CT, Zhang F, Lloyd AM. GL3 encodes a bHLH protein that regulates trichome development in Arabidopsis through interaction with GL1 and TTG1. Genetics. 2000;156:1349–1362.

[36] Baudry A, Caboche M, Lepiniec L. TT8 controls its own expression in a feedback regulation involving TTG1 and homologous MYB and bHLH factors, allowing a strong and cell-specific accumulation of flavonoids in *Arabidopsis thaliana.* The Plant Journal. 2006;46:768–779.

[37] Matus J, Poupin M, Cañon P, Bordeu E, Alcalde J, Arce-Johnson P. Isolation of WDR and bHLH genes related to flavonoid synthesis in grapevine (*Vitis vinifera* L.). Plant Molecular Biology. 2010;72:607–620.

[38] Paz-Ares J, Ghosal D, Wienand U, Peterson P, Saedler H. The regulatory c1 locus of *Zea mays* encodes a protein with homology to myb oncogene products and with structural similarities to transcriptional activators. The EMBO Journal. 1987;6:3553–3558.

[39] Kobayashi S, Ishimaru M, Hiraoka K, Honda C. Myb-related genes of the Kyoho grape (*Vitis labruscana*) regulate anthocyanin biosynthesis. Planta. 2002;215:924–933.

[40] Koes R, Verweij W, Quattrocchio F. Flavonoids: a colorful model for the regulation and evolution of biochemical pathways. Trends in Plant Science. 2005;10:236–242.

[41] Matus J, Aquea F, Arce-Jhonson P. Analysis of the grape MYB R2R3 subfamily reveals expanded wine quality-related clades and conserved gene structure organization across *Vitis* and *Arabidopsis* genomes. BMC Plant Biology. 2008;8:83.

[42] Yang T, Perasso R, Baroin-Tourancheau A. MYB genes in ciliates: a common origin with the MYB proto-oncogene? Protist. 2003;154:229–238.

[43] Lipsick JS. One billion years of MYB. Oncogene. 1996;13:223–235.

[44] Weston K. Myb proteins in life, death and differentiation. Current Opinion in Genetics & Development. 1998;8:76–81.

[45] Deluc L, Bogs J, Walker AR, Ferrier T, Decendit A, Merillon J-M, Robinson SP, Barrieu F. The transcription factor VvMYB5b contributes to the regulation of anthocyanin and proanthocyanidin biosynthesis in developing grape berries. Plant Physiology. 2008;147:2041–2053.

[46] Terrier N, Torregrosa L, Ageorges A, Vialet S, Verries C, Cheynier V, Romieu C. Ectopic expression of VvMybPA2 promotes proanthocyanidin biosynthesis in *Vitis vinifera* L. and suggests additional targets in the pathway. Plant Physiology. 2009;149:1028–1041.

[47] Stracke R, Werber M, Weisshaar B. The R2R3-MYB gene family in *Arabidopsis thaliana.* Current Opinion in Plant Biology. 2001;4:447–456.

[48] Mehrtens F, Kranz H, Bednarek P, Wisshaar B. The Arabidopsis transcription factor MYB12 is a flavonol-specific regulator of phenylpropanoid biosynthesis. Plant Physiology. 2005;138:1083–1096.

[49] Gonzalez A, Zhao M, Leavitt JM, Lloyd AM. Regulation of the anthocyanin biosynthetic pathway by the TTG1/bHLH/Myb transcriptional complex in Arabidopsis seedlings. The Plant Journal. 2007;53:814–827.

[50] Aharoni A, Ric De Vos CH, Wein M, Sun Z, Greco R, Kroon A, Mol JNM, O'Connell AP. The strawberry FaMYB1 transcription factor suppresses anthocyanin and flavonol accumulation in transgenic tobacco. The Plant Journal. 2001;28:319–332.

[51] Salvatierra A, Pimentel P, Moya-León MA, Herrera R. Increased accumulation of anthocyanins in *Fragaria chiloensis* fruits by transient suppression of FcMYB1 gene. Phytochemistry. 2013;90:25–36.

[52] Takos AM, Jaffé FW, Jacob SR, Bogs J, Robinson SP, Walker AM. Light-induced expression of a MYB gene regulates anthocyanin biosynthesis in red apples. Plant Physiology. 2006;142:1216–1232.

[53] Espley RV, Hellens RP, Putterill J, Stevenson DE, Kutty-Amma S, Allan AC. Red coloration in apple fruit is due to the activity of the MYB transcription factor, MdMYB10. The Plant Journal. 2007;49:414–427.

[54] Ban Y, Honda C, Hatsuyama Y, Igarashi M, Bessho H, Moriguchi T. Isolation and functional analysis of a MYB transcription factor gene that is a key regulator for the development of red coloration in apple skin. Plant Cell Physiology. 2007;48:958–970.

[55] Mathews H, Clendennen SK, Caldwell CG, Liu XL, Connors K, Matheis N, Schuster DK, Menasco DJ, Wagoner W, Lightner J, Wagner DR. Activation tagging in tomato identifies a transcriptional regulator of anthocyanin biosynthesis, modification, and transport. The Plant Cell. 2003;15:1689–1703.

[56] Bogs J, Downey MO, Harvey JS, Ashton AR, Tanner GJ, Robinson SP. Proanthocyanidin synthesis and expression of genes encoding leucanthocyanidin reductase in developing grape berries and grapevine leaves. Plant Physiology. 2005;139:652–663.

[57] Czemmel S, Stracke R, Weisshaarm B, Cordon N, Harris NN, Walker AR, Robinson SP, Bogs J. The grapevine R2R3-MYB transcription factor VvMYBF1 regulates flavonol synthesis in developing grape berries. Plant Physiology. 2009;151:1513–1530.

[58] Lijavetzky D, Ruiz-García L, Cabezas JA, De Andrés MT, Bravo G, Ibañez A, Carreño J, Cabello F, Ibañez J, Martínez-Zapater JM. Molecular genetics of berry colour variation in table grape. Molecular Genetics and Genomics. 2006;276:427–435.

[59] Walker AR, Lee E, Bogs J, McDavid DAJ, Thomas MR, Robinson SP. White grapes arose through the mutation of two similar and adjacent regulatory genes. The Plant Journal. 2007;49:772–785.

[60] Lin-Wang K, Bolitho K, Grafton K, Kortstee A, Karunairetnam A, McGhie TK, Espley RV, Hellens RP, Allan AC. An R2R3 MYB transcription factor associated with regulation of the anthocyanin biosynthetic pathway in Rosaceae. BMC Plant Biology. 2010;10:50.

[61] Toledo-Ortiz G, Huq E, Quail PH. The Arabidopsis basic/helix–loop–helix transcription factor family. The Plant Cell. 2003;15:1749–1770.

[62] Li X, Duan X, Jiang H, Sun Y, Tang Y, Yuan Z, Guo J, Liang W, Chen L, Yin J, Ma H, Wang J, Zhang D. Genome-wide analysis of basic/helix–loop–helix transcription factor family in rice and Arabidopsis. Plant Physiology. 2006;141:1167–1184.

[63] Ellenberger T, Fass D, Arnaud M, Harrison SC. Crystal structure of transcription factor E47: E-box recognition by a basic region helix–loop–helix dimer. Genes & Development. 1994;8:970–980.

[64] Heim MA, Jakoby M, Werber M, Martin C, Weisshaar B, Bailey PC. The basic helix–loop–helix transcription factor family in plants: a genome-wide study of protein structure and functional diversity. Molecular Biology and Evolution. 2003;20:735–747.

[65] Heisler MGB, Atkinson A, Bylstra YH, Walsh R, Smyth DR. SPATULA, a gene that controls development of carpel margin tissues in Arabidopsis, encodes a bHLH protein. Development. 2001;128:1089–1098.

[66] Sorensen A-M, Krober S, Unte US, Huijser P, Dekker K, Saedler H. The Arabidopsis ABORTED MICROSPORES (AMS) gene encodes a MYC class transcription factor. The Plant Journal. 2003;33:413–423.

[67] Abe H, Urao T, Ito T, Seki M, Shinozaki K, Yamaguchi-Shnizaki K. Arabidopsis AtMYC2 (bHLH) and AtMYB2 (MYB) function as transcriptional activators in abscisic acid signaling. The Plant Cell. 2003;15:63–78.

[68] Li H, Sun J, Xu Y, Jiang H, Wu X, Li C. The bHLH-type transcription factor AtAIB positively regulates ABA response in Arabidopsis. Plant Molecular Biology. 2007;65:655–665.

[69] Séguéla M, Briat JF, Vert G, Curie C. Cytokinins negatively regulate the root iron uptake machinery in Arabidopsis through a growth-dependent pathway. The Plant Journal. 2008;55:289–300.

[70] Chandler VL, Radicella JP, Robbins TP, Chen J, Turks D. Two regulatory genes of the maize anthocyanin pathway are homologous: isolation of B utilizing R genomic sequences. The Plant Cell. 1989;1:1175–1183.

[71] Ludwig SR, Habera LF, Dellaporta SL, Wessler SR. Lc, a member of the maize R gene family responsible for tissue-specific anthocyanin production, encodes a protein similar to transcription activators and contains the myc homology region. Proceedings of the National Academy of Sciences USA. 1989;86:7092–7096.

[72] Goff SA, Klein TM, Roth BA, Fromm ME, Cone KC, Radicella JP, Chandler VP. Transactivation of anthocyanin biosynthetic genes following transfer of B regulatory genes into maize tissues. EMBO Journal. 1990;9:2517–2522.

[73] Burr FA, Burr B, Scheffler BE, Blewitt M, Wienand U, Matz EC. The maize repressor-like gene intensifier shares homology with the rVb7 multigene family of transcription factors and exhibits missplicing. The Plant Cell. 1996;8:1249–1259.

[74] Nesi N, Debeaujon I, Jond C, Pelletier G, Caboche M, Lepiniec L. The TT8 gene encodes a basic helix–loop–helix domain protein required for expression of DFR and BAN genes in Arabidopsis siliques. The Plant Cell. 2000;12:1863–1878.

[75] Quattrocchio F, Wing JF, Leppen HTC, Mol JNM, Koes RE. Regulatory genes controlling anthocyanin pigmentation are functionally conserved among plant species and have distinct sets of target genes. The Plant Cell. 1993;5:1497–1512.

[76] Quattrocchio F, Wing JF, van der Woude K, Mol JNM, Koes R. Analysis of bHLH and MYB domain proteins: Species-specific regulatory differences are caused by divergent evolution of target anthocyanin genes. The Plant Journal. 1998;13:475–488.

[77] Smith TF, Gaitatzes C, Saxena K, Neer EJ. The WD repeat: a common architecture for diverse functions. Trends in Biochemical Sciences. 1999;24:181–185.

[78] Van Nocker S, Ludwig P. The WD-repeat protein superfamily in Arabidopsis: conservation and divergence in structure and function. BMC Genomics. 2003;4:50.

[79] Carey CC, Strahle JT, Selinger DA, Chandler VL. Mutations in the pale aleurone color1 regulatory gene of the *Zea mays* anthocyanin pathway have distinct phenotypes relative to the functionally similar TRANSPARENT TESTA GLABRA1 gene in *Arabidopsis thaliana*. The Plant Cell. 2004;16:450–464.

[80] Walker AR, Davison PA, Bolognesi-Winfield AC, James CM, Srinivasan N, Blundell TL, Esch JJ, Marks MD, Gray JC. The TRANSPARENT TESTA GLABRA1 locus, which regulates trichome differentiation and anthocyanin biosynthesis in Arabidopsis, encodes a WD40 repeat protein. The Plant Cell. 1999;11:1337–1350.

[81] de Vetten N, Quattrocchio F, Mol J, Koes R. The an11 locus controlling flower pigmentation in petunia encodes a novel WD-repeat protein conserved in yeast, plants, and animals. Genes & Development. 1997;11:1422–1434.

[82] Sompornpailin K, Makita Y, Yamazaki M, Saito K. A WD-repeat-containing putative regulatory protein in anthocyanin biosynthesis in *Perilla frutescens*. Plant Molecular Biology. 2002;50:485–495.

[83] Hanachi P, Kua SH, Asmah R, Motalleb G, Fauziah O. Cytotoxic effect of *Berberis vulgaris* fruit extract on the proliferation of human liver cancer cell line (HepG2) and its antioxidant properties. International Journal of Cancer Research. 2006;2:1–9.

[84] Končić MZ, Kremer D, Karlović K, Kosalec I. Evaluation of antioxidant activities and phenolic content of *Berberis vulgaris* L. and *Berberis croatica* Horvat. Food and Chemical Toxicology. 2010;48:2176–2180.

[85] Shin Y, Ryu JA, Liu RH, Nock JF, Watkins CB. Harvest maturity, storage temperature and relative humidity affect fruit quality, antioxidant contents and activity, and inhibition of cell proliferation of strawberry fruit. Postharvest Biology and Technology. 2008;49:201–209.

[86] Chang YC, Huang HP, Hsu JD, Yang SF, Wang CJ. Hibiscus anthocyanins rich extract-induced apoptotic cell death in human promyelocytic leukemia cells. Toxicology and Applied Pharmacology. 2005;205:201–212.

[87] Jia N, Xiong YL, Konga B, Liua Q, Xia X. Radical scavenging activity of black currant (*Ribes nigrum L.*) extract and its inhibitory effect on gastric cancer cell proliferation via induction of apoptosis. Journal of Functional Foods. 2012;4:382–390.

[88] Huang H, Chang Y, Wu Ch, Hung Ch, Wang Ch. Anthocyanin-rich Mulberry extract inhibit the gastric cancer cell growth *in vitro* and xenograft mice by inducing signals of p38/p53 and c-jun. Food Chemistry. 2011;129:1703–1709.

[89] Wang L, Stoner G. Anthocyanins and their role in cancer prevention. Cancer Letters. 2009;269:281–290.

[90] Stoner G, Wang L-S, Seguin C, Rocha C, Stoner K, Chiu S, Kinghorn AD. Multiple berry types prevent N-nitrosomethylbenzylamine-induced esophageal cancer in rats. Pharmacy Research. 2010;27:1138–1145.

[91] Lala G, Malik M, Zhao C, He J, Kwon Y, Giusti MM, Magnuson BA. Anthocyanin-rich extracts inhibit multiple biomarkers of colon cancer in rats. Nutrition Cancer. 2006;54:84–93.

[92] Afaq F, Saleem M, Krueger CG, Reed JD, Mukhtar H. Anthocyanin- and hydrolysable tannin-rich pomegranate fruit extract modulates MAPK and NF-kappaB pathways and inhibits skin tumorigenesis in CD-1 mice. International Journal of Cancer. 2005;113:423–433.

[93] Afaq F, Khan N, Syed DN, Mukhtar H. Oral feeding of pomegranate fruit extract inhibits early biomarkers UVB radiation induced carcinogenesis in SKH-1hairless mouse epidermis. Photochemestry and Photobiology. 2010;86:1318–1326.

[94] Chen K, Sun H, Sun A, Lin Q, Wang Y, Tao X. Studies pf the protective effect and antioxidant mechanism of blueberry anthocyanins in a CC14-induced liver injury model mice. Food and Agricultural Innumonoly. 2012;23:352–362.

[95] Singletary KW, Jung KJ, Giusti MM. Anthocyanin-rich grape extract blocks breast cell DNA damage. Journal of Medical Food. 2007;10:244–251.

[96] Casto BC, Kresty LA, Kraly CL. Chemoprevention of oral cancer by black raspberries. Anticancer Research. 2002;22:4005–4015.

Chapter 9

Protective Effects of *Curcumin* on Gastric Inflammation and Liver Disease

Duangporn Werawatganon

Additional information is available at the end of the chapter

http://dx.doi.org/10.5772/66119

Abstract

Curcumin (diferuloylmethane), an anti-inflammatory and antioxidant compound, is isolated from the rhizomes of the plant *Curcuma longa Linn*. Most of the anti-inflammatory effects can be explained by the efficient inhibition of nuclear factor-κB–mediated and activation of PPARγ expression. These studies have been investigating the effects of curcumin on the gastric microcirculation, cytokine production after *Helicobacter pylori*–induced gastric inflammation, gastric cancer, drug-induced liver injury, and alcoholic liver disease (ALD). The results show that curcumin prevents indomethacin-induced gastropathy via decreased leukocyte-endothelium interaction at postcapillary venule, decreased ICAM-1 and TNF-α level, and improved gastric microcirculation. Curcumin attenuated gastric inflammation and gastric cancer via reduced NF-κB p65 expression, decreased vascular endothelial growth factor (VEGF) level, and macromolecular leakage in the gastric mucosa. Curcumin prevented liver injury through decreased oxidative stress, reduced liver inflammation, and restored GSH. Moreover, curcumin could decrease hepatocyte apoptosis and improved PPARγ protein expression in alcohol-induced liver injury.

Keywords: curcumin, gastric inflammation, gastric cancer, liver disease

1. Introduction

Curcumin (diferuloylmethane), the natural yellow pigment in tumeric, is isolated from the rhizomes of the plant *Curcuma longa* Linn. (*C. longa* L.). *C. longa* belongs to the Zingiberaceae family. It is a perennial herb that is distributed throughout tropical and subtropical regions of the world and is widely cultivated in Asian countries, such as India, Thailand, and China. The rhizomes are used as a traditional remedy in Nepal [1]. The powder form, called turmeric, is bright yellow and has been used as a food-coloring agent in the United States. In India, it has been used as a spice, as a food preservative, and as therapeutic agent. The current Indian medicine claims the usage of tumeric is effective against biliary disorders, anorexia, coryza, cough, diabetic wounds, hepatic disorder, rheumatism, and sinusitis [2].

In the nineteenth century, there has been considerable interest in the active compounds in tumeric called curcuminoids. Curcumin is the major curcuminoid compound that makes up approximately 90% of the curcuminoid content in tumeric, followed by demethoxycurcumin and bisdemethoxycurcumin [3]. The chemical structure of curcumin was determined by Roughley and Whiting (**Figure 1**) [4].

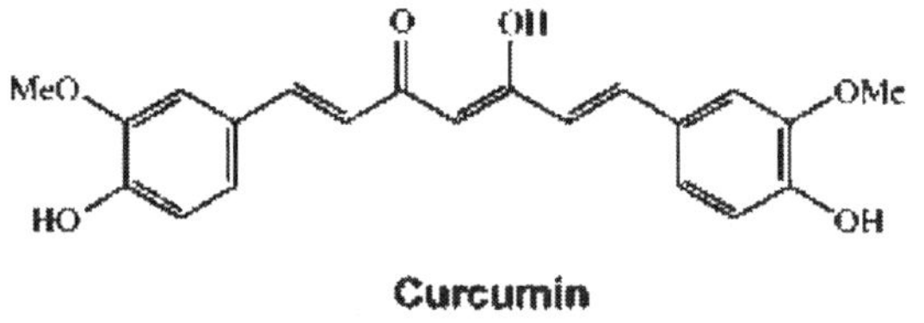

Figure 1. Chemical structure of curcumin (diferuloylmethane) [4].

Curcumin can be dissolved in organic solvents such as dimethylsulfoxide (DMSO), oil, alcohol, and petroleum agents. Interestingly, curcumin has been demonstrated to be safe for human and animals use. Human appeared to be able to tolerate high doses of curcumin without significant side-effects. A phase 1 study by Cheng et al. [5] found no adverse effects of curcumin ingestion for 3 months of dosage up to 8000 mg/day. Other human studies of curcumin included the following: a double-blinded, crossover trial in 18 patients with rheumatoid arthritis [6], a randomized, placebo-controlled trial with 45 postsurgical patients [7]. The doses of curcumin in these studies ranged from 1125 to 2500 mg/day. Only one postsurgical patient reported mild transient giddiness. No other serious adverse reactions were reported, including any changes in blood chemistry reports. Thus, curcumin appears to be safe in human even with ingestion at a high dosage.

In animals, the previous study demonstrated that curcumin is rapidly metabolized and poorly absorbed in Sprague-Dawley rats. Administrating curcumin orally was carried out by Wahlström and Blennow [8]. They demonstrated that this compound with a dose of 1–5 g/kg BW given to rats apparently did not cause any adverse effects and it was excreted about 75% in the feces, while traces found in the urine. In addition, measurements of blood plasma levels and biliary excretion showed that curcumin was poorly absorbed by the gastrointestinal (GI) tract. Curcumin could not be detected after 30 minutes when added to microsomes suspensions or hepatocyte suspensions. Furthermore, it was capable of disappearing from the blood after intravenous injection or after addition to the liver perfusion system. Moreover, oral LD_{50} was found to be 12.2 g/kg BW in rats [9]. In addition, a study in which rats were fed with curcumin

1.8 g/kg BW per day for 90 days and monkeys were fed with curcumin 0.8 mg/kg BW per day for 90 days showed no adverse effects [10].

Curcumin has been tested to demonstrate pharmacoprotective effects in various gastrointestinal (GI) and liver diseases. We highlight studies on its potential mechanism of action classified into four categories: (i) Curcumin protects against *Helicobacter pylori* infection and gastric cancer, (ii) curcumin protects against nonsteroidal anti-inflammatory drug (NSAID)-induced ulcer; (iii) curcumin protects against drug-induced liver injury; and (iv) curcumin protects against alcoholic liver disease (ALD).

2. Curcumin protects against *H. pylori* infection and gastric cancer

The discovery of *H. pylori* was first reported in 1984 by two Australian investigators, Barry Marshall and Robin Warren [11], who isolated the bacteria from mucosal biopsies of patients with chronic active gastritis. Its name was changed from *Campylobacter pyloridis, Campylobacter pylori*, and *Campylobacter*-like organism when the biochemical and genetic characterization has shown that it is in the genus *Helicobacter* [12].

H. pylori is a noninvasive, nonspore-forming, and spiral shaped gram-negative bacterium measuring approximately 3.5 × 0.5 μm. It has four to six sheathed flagella at one pole. These flagella and spiral shape of *H. pylori* help the bacterial movement into the mucus of stomach. It slowly grows in microaerophilic condition, 5% oxygen, 50% carbon dioxide at 37°C [13]. *H. pylori* is an unusual organism with a remarkably high level of genetic diversity [14], which means, it can survive in the human stomach and also multiply in high-acid environment of the stomach. When *H. pylori* infected human, it adheres on the gastric epithelial cells and induces chronic active gastritis, peptic ulcer, mucosal-associated lymphoid tissue (MALT) lymphoma, and gastric cancer.

H. pylori is highly adapted to the stomach environment. To avoid the acidic environment of the stomach lumen, *H. pylori* uses its flagella to permit entry into the mucus. It adheres to the epithelial cells by producing adhesins for attachment to epithelial cells. *H. pylori* produces a potent urease enzyme. Urease generates carbon dioxide and ammonia, which potentially buffer the surrounding microenvironment and the bacterial cytosol [15]. In addition, urea is an important source of nitrogen for the bacteria.

Powerful flagella help the bacteria to swim through the viscous mucous layer covering the gastric epithelium, where bacterial adhesion proteins mediate a close interaction with the host cells [16]. *H. pylori* can bind tightly to epithelial cells by multiple bacterial surface components. The outer-membrane protein (Hop), such as BabA, binds to the fucosylated Lewis B blood-group antigen on the gastric epithelial cells [17]. Several Hop protein families also mediate adhesion to epithelial cells. When *H. pylori* adheres on gastric epithelial cells, it releases virulent factors to immune subversion. The host response to *H. pylori* participates in the induction of gastric epithelial damage and therefore has an integral role in *H. pylori* pathogenesis. *H. pylori* adheres on the gastric epithelial cells by bacterial adhesion proteins. Then, virulent factors are delivered into host cells. Especially, cytotoxin-associated gene A (CagA) induces many pathological conditions. For example, activation of NF-κB that causes production of many inflammatory mediators inducing gastric inflammation [18].

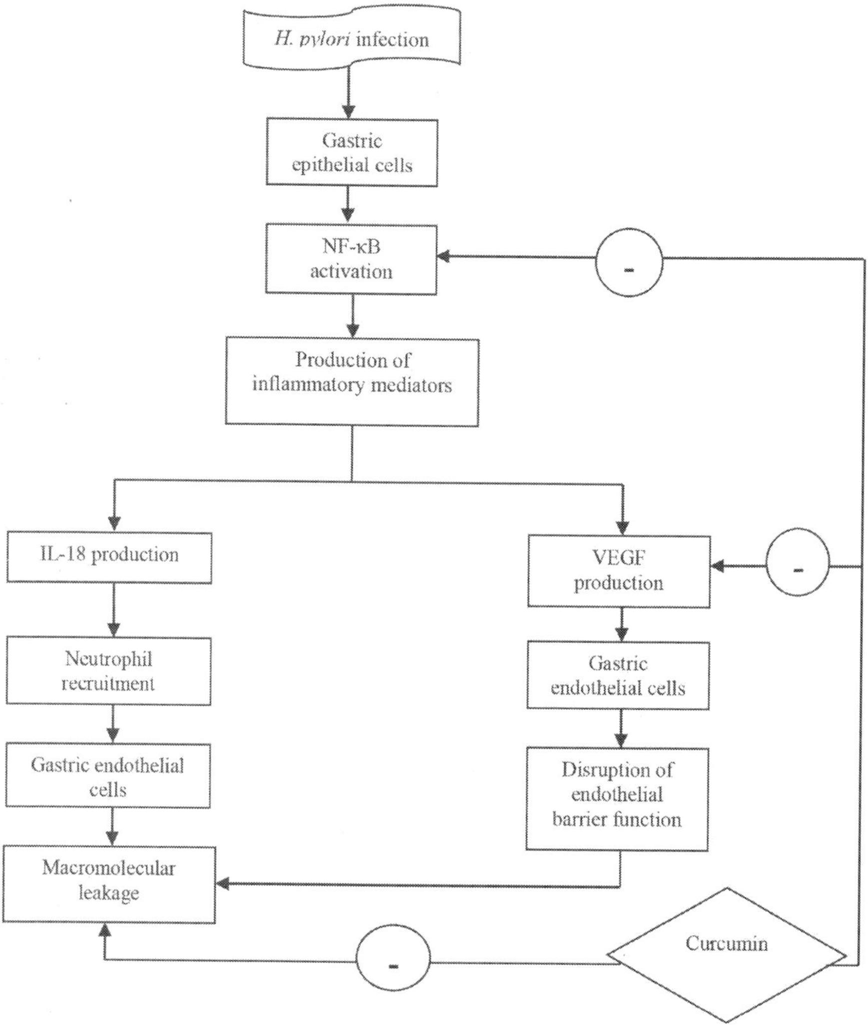

Figure 2. Protective mechanism of curcumin on *H. pylori* infection [19].

Effect of curcumin was examined by using rats [19]. The scheme of the effects was shown in **Figure 2**. Host inflammatory responses were measured by the following parameters: leakage

http://dx.doi.org/10.5772/66119

of macromolecules from gastric postcapillary venules (PCVs), serum level of vascular endothelial growth factor (VEGF), and the expression of NF-κB subunit p65.

The successful inoculation of *H. pylori* was 85%. *H. pylori* infection led to the macromolecular leakage, the NF-κB-p65 expression, and increase of VEGF level compared with control group. Curcumin alone did not significantly change baseline of these parameters. There were

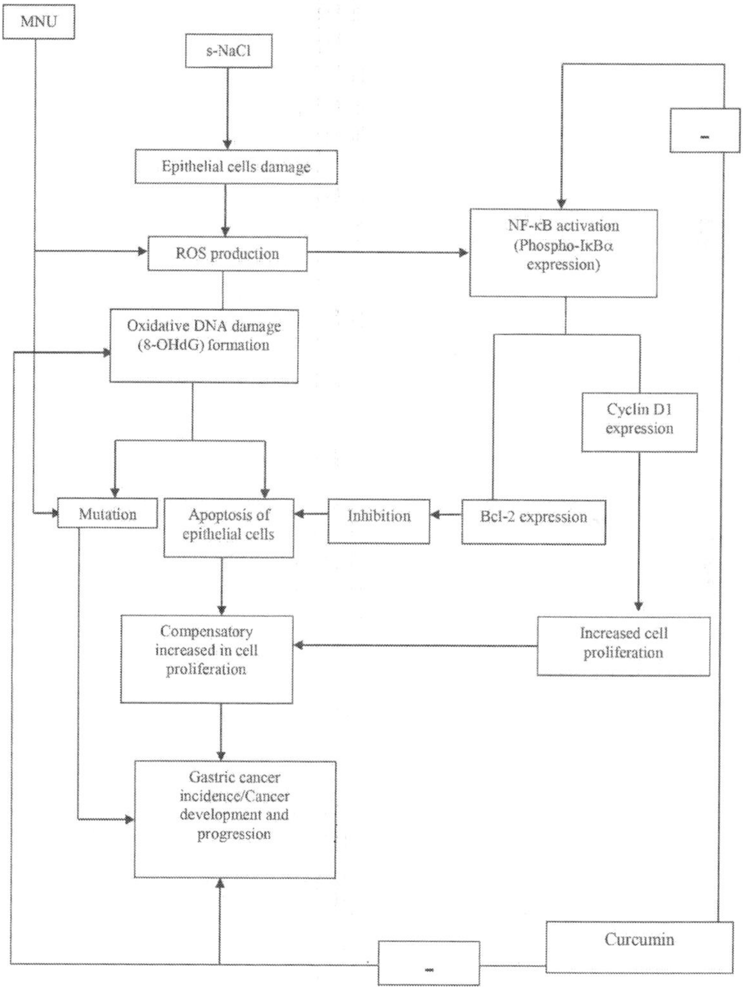

Figure 3. Protective mechanism of curcumin on N-methyl-N-nitrosourea (MNU) and saturated sodium chloride (s-NaCl)-induced gastric cancer [20].

significant decrease of macromolecular leakage and NF-κB-p65 expression ($p \leq 0.05$) in the curcumin-treated groups (curcumin 200 mg/kg and 600 mg/kg BW) compared with *H. pylori*-infected group, respectively. These results could be concluded that *H. pylori* infection increased macromolecular leakage, NF-κB-p65 expression, and serum VEGF level. Curcumin can reduce macromolecular leakage, decrease serum VEGF level, and NF-κB-p65 expression. It is implied that curcumin may have an anti-inflammatory effect on *H. pylori* infection [19].

In addition, the present study to examine the protective effect of curcumin on gastric cancer induced by N-methyl-N-nitrosourea (MNU) and saturated sodium chloride (s-NaCl) administration [20]. The scheme of the results was shown in **Figure 3**. Gastric cancer can generate in any part of the stomach. Cancers were found in the forestomach of all rats induced by MNU and s-NaCl. Curcumin supplementations showed 40–50% reduction of cancer incidence. Expressions of 8-OHdG, cyclin D1, and Bcl-2 significantly increased in rat with MNU and s-NaCl administration compared with control group. The phospho-IκBα expression had a tendency to increase in MNU and s-NaCl group compared with control group. Immunoreactive cells of 8-OHdG in curcumin supplementation significantly decreased when compared with MNU and s-NaCl group. The relative intensity of phospho-IκBα in curcumin group tended to reduce when compared with MNU and s-NaCl group. Curcumin can attenuate cancer via a reduction of phospho-IκBα and 8-OHdG expressions, which may play a promising role in gastric carcinogenesis [20].

3. Curcumin protects against NSAID-induced ulcer

Nonsteroidal anti-inflammatory drugs (NSAIDs) are one of the most commonly prescribed drugs worldwide. It is well-known that NSAIDs cause gastric mucosal damage ranging from nonspecific dyspepsia to ulceration, upper gastrointestinal (GI) bleeding, and death. These can summarize by the term "NSAIDs gastropathy." NSAIDs caused topical damage from "ion trapping" effect [21], the reduction of the hydrophobicity of the gastric mucosal surface, and uncoupling of oxidative phosphorylation [22, 23]. The systemic effect caused by inhibiting cyclo-oxygenases (COX). NSAIDs block the formation not only of proinflammatory cytokines but also of gastroprotective prostaglandins those maintain gastric mucosal blood flow and bicarbonate production [24]. The enhanced synthesis of leukotrienes may occur by shunting the arachidonic acid metabolism towards the 5-lipoxygenase pathway [25–27]. COX inhibition, enhanced synthesis of leukotrienes, contributing to gastric mucosal injury by promoting tissue ischemia and inflammation [28–31].

Mechanism of NSAID-induced gastric ulceration is a neutrophil-dependent process. NSAIDs induced neutrophil adherence to vascular endothelium [32]. Neutrophils play an important role by releasing a variety of inflammatory mediators, including neutrophil elastase and ROS caused gastric mucosal injury. Furthermore, adhesion molecules expressed on activated neutrophils, such as CD11b and CD18, play an important role in neutrophil-induced tissue injury [33–35].

http://dx.doi.org/10.5772/66119

Protective effects of curcumin on NSAIDs were examined [36]. The scheme of the effects was shown in **Figure 4**. The study demonstrates effects of curcumin on gastric microcirculation, tumor necrosis factor (TNF)-α, and intercellular adhesion molecule (ICAM)-1 levels on rat with NSAID-induced gastric injury. The stomach histopathology in NSAIDs group showed multiple erosions with mild to moderate inflammation. Serum of ICAM-1, TNF-α levels, and leukocyte-endothelium interaction increased significantly when compared with control group. Pretreatment with curcumin group resulted in decreasing the elevation serum of ICAM-1, TNF-α levels, and leukocyte-endothelium interaction. The stomach histopathology was improved in curcumin administration group. Therefore, curcumin accomplishes the protective effect on NSAID-induced gastric mucosal injury on improving gastric microcirculation and reducing inflammatory cytokines [36].

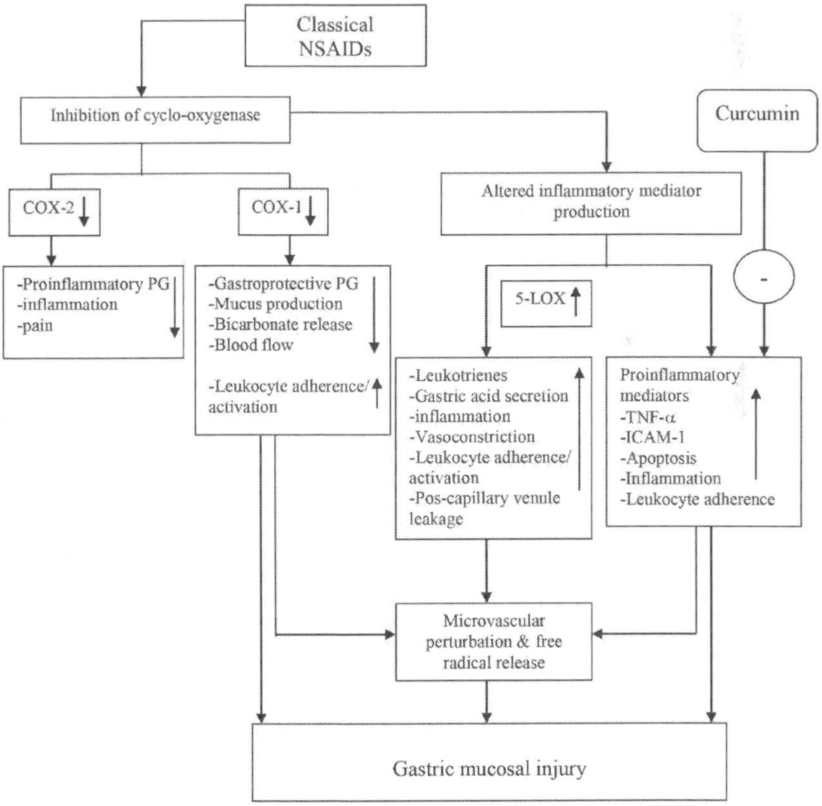

Figure 4. Protective mechanism of curcumin on NSAID-induced gastric mucosal injury [36].

4. Curcumin protects against drug-induced liver injury

N-acetyl-P-aminophenol (APAP) or paracetamol is a widely used analgesic and antipyretic drugs [37, 38]. APAP toxicity is one of the most common drug-induced liver damages worldwide, where major liver major complication is caused due to APAP overdose. APAP metabolites produced in the liver and other organs are the main contributors for the mechanism of its toxicity [39, 40].

The scheme of liver injury by drug was shown in **Figure 5**. In therapeutic doses, APAP is mainly metabolized via glucuronidation and sulfation and in conjugated forms are excreted from the body. Besides, APAP partly is metabolized by cytochrome P450 (CYP 450), to some metabolites, mainly N-acetyl-*p*-benzoquinone imine (NAPQI), which are dramatically increased in high APAP concentrations. These metabolites of APAP are detoxified by glutathione (GSH) and removed from the body. Then, in APAP overdose causes increasing of toxic metabolites. These metabolites interact with a range of cellular proteins via covalent binding, which disrupting hepatocyte function causing necrosis, apoptosis, and liver injury occurs [41, 42].

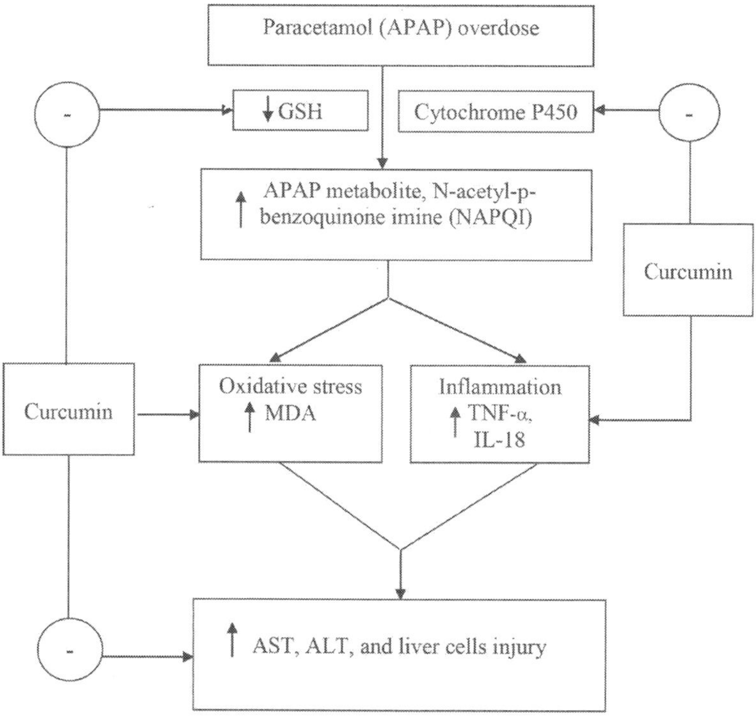

Figure 5. Protective mechanism of curcumin on paracetamol overdose-induced hepatitis [43].

The protective effects of curcumin on paracetamol overdose-induced hepatitis in mice were studied. The effect was shown in **Figure 5**. The results showed that serum transaminases, Hepatic malondialdehyde (MDA), and inflammatory cytokines (TNF-α and IL18) were increased significantly in the 400 mg/kg of APAP group compared with the control group. Curcumin treatment groups (curcumin 200 mg/kg and 600 mg/kg) were significantly decreased these parameters compared with the APAP group. The level of GSH decreased significantly in the APAP compared with the control group. Curcumin treatment groups (curcumin 200 mg/kg and 600 mg/kg) were significantly increased GSH level compared with the APAP group. The histological appearance of the liver in the control group showed normal. In the APAP group, the liver showed damage with extensive hemorrhagic hepatic necrosis at all zones. Curcumin treatment groups (curcumin 200 mg/kg and 600 mg/kg) improved the liver histopathology. In curcumin 200 mg/kg group, the liver showed mild focal necrosis and the normal architecture was well preserved in curcumin 600 mg/kg group. The results indicated that curcumin prevented APAP-induced hepatitis through decreased oxidative stress, reduced liver inflammation, and restored GSH, which caused the improvement of liver histopathology [43]

5. Curcumin protects against alcoholic liver disease

Alcoholic liver disease (ALD) represents a spectrum of clinical illness and morphological changes that range from fatty liver, hepatic inflammation, and necrosis (alcoholic hepatitis) to progressive fibrosis (alcoholic cirrhosis) [44]. Ethanol oxidation generates toxic products such as acetaldehyde, and reactive oxygen species resulted in oxidative stress that initiates apoptosis and cell injury [45–48]. More than 80–90% of heavy drinkers develop fatty liver, but only up to 20–40% of this population develops more severe forms of alcoholic liver disease (ALD), including fibrosis, alcoholic hepatitis, cirrhosis, and hepatocellular carcinoma (HCC) [49].

Pathogenic mechanisms of alcoholic liver disease were proposed. Ethanol promotes the translocation of lipopolysaccharide from the gastrointestinal lumen to the portal vein. In Kupffer cells, lipopolysaccharide binds to CD14, which combines with Toll-like receptor 4 (TLR4) which is responsible for activating the innate immune system. The increase on inflammatory cytokine production in conjunction with a decrease in signal transducer and activator of transcription (STAT) factors' expression reduces liver regeneration. Long-term alcohol consumption alters the intracellular balance of antioxidants with subsequent decrease in the release of mitochondrial cytochrome c and expression of Fas ligand, leading to hepatic apoptosis. Activated Kupffer cells and hepatocytes are suggested to be sources of free radicals (especially ROS), which are responsible for lipid peroxidation and further apoptotic damage. Activation of hepatic stellate cells also contributes to the production of cytokines, ROS and TGF-β exacerbating liver fibrosis [49].

This study demonstrated effects of curcumin attenuated inflammation and liver pathology in rats with alcoholic liver disease. The effect was shown in **Figure 6**. The results showed that the liver histopathology in ethanol group revealed moderate steatosis and necroinflammation. In ethanol group, hepatic MDA, hepatocyte apoptosis, and NF-κB activation have increased significantly when compared with control. The 400 mg/kg BW of curcumin treatment revealed the decreased of hepatocyte apoptosis, hepatic MDA, NF-κB activation. The peroxisome proliferator-activated receptor gamma (PPARγ) protein expression increased in the curcumin groups. Therefore, curcumin improved liver damage in ethanol-induced hepatitis by reduction of oxidative stress, inhibition of NF-κB activation, and restoration of PPARγ [50].

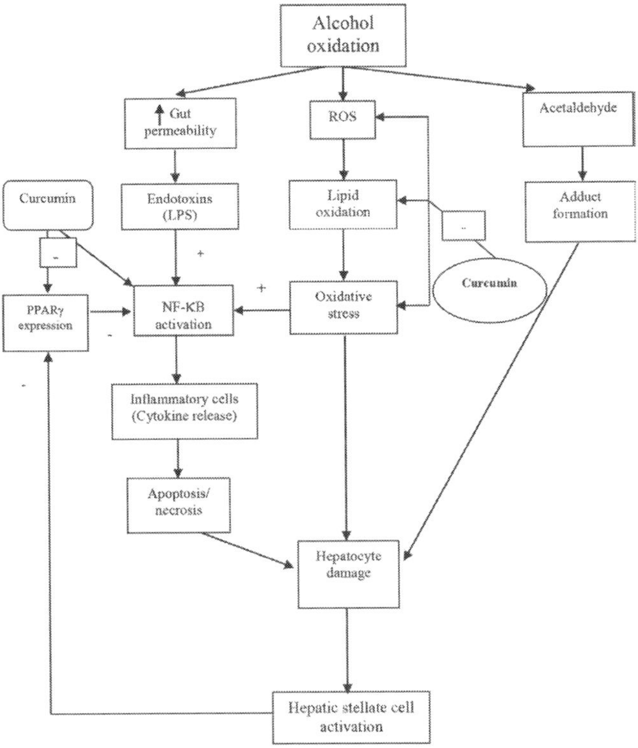

Figure 6. Protective mechanism of curcumin on changes of PPARγ expression, NF-κB activation and oxidative stress in rats with alcoholic hepatitis [50].

6. Conclusion

Curcumin prevents indomethacin-induced gastropathy by decreasing ICAM-1, TNF-α levels, and leukocyte-endothelium interaction. Curcumin reduces *H. pylori*-induced gastric inflammation and gastric cancer by reducing macromolecular leakage, decreasing serum VEGF level, and NF-κB-p65 expression. Curcumin improves liver damage caused by APAP overdose by decreasing hepatic MDA, TNF-α and IL18, and restoring GSH. Moreover, curcumin attenuates alcohol-induced liver injury by decreasing the elevation of hepatic MDA, inhibition of NF-κB activation and improving of liver pathology. Overall, the major mechanisms of curcumin are associated with reduction of oxidative stress, restoration of glutathione and PPARγ expression, inhibition the activation of NF-κB, attenuation of inflammation, and the improvement of histopathology. Therefore, these features make curcumin a very promising new therapeutic option for the treatment of gastrointestinal and liver diseases.

Author details

Duangporn Werawatganon

Address all correspondence to: dr.duangporn@gmail.com

Department of Physiology, Faculty of Medicine, Chulalongkorn University, Bangkok, Thailand

References

[1] Eigner D, Scholz D. Ferula asa-foetida and *Curcuma longa* in traditional medical treatment and diet in Nepal. *J Ethnopharmacol* 1999; 67: 1–6.

[2] Ammon HPT, Anazodo MI, Safayhi H, Dhawan BN, Srimal RC. Curcumin: a potent inhibitor of leukotriene B4 formation in rat peritoneal polymorphonuclear neutrophils (PMNL). *Planta Med* 1992; 58: 26.

[3] Ruby AJG, Kuttan KD, Babu KN, Rajasekharan R, Kuttan R. Antitumor and antioxidant activity of natural curcuminoids. *Cancer Lett* 1995; 94: 79–83.

[4] Roughley PJ, Whiting DA. Experiments in the biosynthesis of curcumin. *J Chem Soc* 1973; 20: 2379–2388.

[5] Cheng AL, Hsu CH, Lin JK, Hsu MM, Ho YF, Shen, TS. Phase I clinical trial of curcumin, a chemopreventive agent, in patients with high-risk or pre-malignant lesions. *Anti-cancer Res* 2001; 21: 2895–2900.

[6] Deodhar SD, Sethi R, Srimal RC. Preliminary studies on antirheumatic activity of curcumin. *Indian J Med Res* 1980; 71: 632–634.

[7] Satoskar RR, Shah SJ, Shenoy SG. Evaluation of anti-inflammatory property of curcumin (diferuloyl methane) in patients with postoperative inflammation. *Int J Clin Pharmacol Ther Toxicol* 1986; 24: 651–654.

[8] Wahltrom B, Blennow G. A study on the fate of curcumin in the rat. *Pharmacol Toxicol* 1978; 43: 86–92.

[9] Arora R, Basu N, Kapoor V. Anti-inflammatory studies on *Curcuma longa* (tumeric). Indian J *Med Res* 1971; 59: 1289–1295.

[10] Majeed M, Badmaev V, Shivakumar U, Rajendran R. 1995. Curcuminoids: Antioxidant Phytonutrients. Piscataway, NJ: Nutriscience Publishers, Inc.

[11] Marshall BJ, Warren JR. Unidentified curved bacilli in the stomach of patients with gastritis and peptic ulceration. *Lancet* 1984; 1: 1311–1315.

[12] Peterson WL, Graham DY. 2002. *Helicobacter pylori*. In Sleienger & Fordtran's Gastrointestinal and Liver Disease. Philadelphia: Saunders, pp. 732–746.

[13] Yamada T, Alpers DH, Yang CO, Powell DW, Silversten FE. 1991. Text Book of Gastroenterology. J.B. Lippincott, Philadelphia, pp. 1241–1250.

[14] Akopyanz N, Bukanov NO, Westblom TU, Kresovich S, Berg DE. DNA diversity among clinical isolates of *Helicobacter pylori* detected by PCR-based RAPD fingerprinting. *Nucleic Acids Res* 1992; 20: 5137–5142.

[15] Mobley HLT. 2001. *Helicobacter pylori* urease. In *Helicobacter pylori*: Molecular and Cellular Biology. Eds. M. Achtman, and S. Suerbaum Wymondham, UK: Horizon Scientific Press, pp. 155–170.

[16] Montecucco C, Rappuoli R. Living dangerously: how *Helicobacter pylori* survives in the human stomach. *Nat Rev Mol Cell Biol* 2001; 2: 457–466.

[17] Ilver D, Arnqvist A, Ogren J, Frick IM, Kersulyte D, Incecik ET. *Helicobacter pylori* adhesin binding fucosylated histo-blood group antigens revealed by retagging. *Science* 1998; 279: 373–377.

[18] Suerbaum S, Michetti P. *Helicobacter pylori* infection. *N Engl J Med* 2002; 347: 1175–1186.

[19] Sintara K, Thong-Ngam D, Patumraj S, Klaikeaw N, Chatsuwan T. Curcumin suppresses gastric NF-kappaB activation and macromolecular leakage in *Helicobacter pylori*-infected rats. *World J Gastroenterol* 2010; 16:4039–4046.

[20] Sintara K, Thong-Ngam D, Patumraj S, Klaikeaw N. Curcumin attenuates gastric cancer induced by N-methyl-N-nitrosourea and saturated sodium chloride in rats. *J Biomed Biotechnol* 2012 (2012); Article number 915380 .

[21] Davenport HW. Salicylate damage to the gastric mucosal barrier. *N Engl J Med* 1967; 276: 1307–1312.

[22] Jorgensen TG, Weis- Fogh US, Nielsen HH, Olesen HP. Salicylate and aspirin- induced uncoupling of oxidative phosphorylation in mitochondria isolated from the mucosal membrane of the stomach. *Scand J Clin Lab Invest* 1976; 36: 649–654.

[23] Lichtenberger LM. The hydrophobic barrier properties of gastrointestinal mucus. *Annu Rev Physiol* 1995; 57: 565–583.

[24] Vane JR. Inhibition of prostaglandin synthesis as a mechanism of action for aspirin-like drug. *Nat New Biol* 1971; 231: 232–235.

[25] Hudson N, Balsitis M, Everitt S, Hawkey CJ. Enhanced gastric mucosal leukotriene B4 synthesis in patients taking non- steroidal anti-inflammatory drugs. *Gut* 1993; 34: 742–747.

[26] Vaananen PM, Keenan CM, Grisham MB, Wallace JL. Pharmacological investigation of the role of leukotrienes in the pathogenesis of experimental NSAID gastropathy. *Inflammation* 1992; 16: 227–240.

[27] Martel-Pelletier J, Lajeunesse D, Reboul P, Pelletier JP. Therapeutic role of dual inhibitors of 5-LOX and COX, selective and non-selective non-steroidal anti-inflammatory drugs. *Ann Rheum Dis* 2003; 62: 501–509.

[28] Peskar BM. Role of leukotriene C4 in mucosal damage caused by necrotizing agents and indomethacin in the rat stomach. *Gastroenterology* 1991; 100: 619–626.

[29] McCafferty DM, Granger DN, Wallace JL. Indomethacin- induced gastric injury and leukocyte adherence in arthritic versus healthy rats. *Gastroenterology* 1995; 109: 1173–1180.

[30] Andrews FJ, Malcontenti-Wilson C, O' Brien PE. Effect of nonsteroidal anti-inflammatory drugs on LFA-1 and ICAM-1 expression in gastric mucosa. *Am J Physiol* 1994; 266: G657–664.

[31] Santucci L, Fiorucci S, Giansanti M, Brunori PM, Di Matteo FM, Morelli A. Penoxifylline prevents indomethacin induced acute gastric mucosal damage in rats: role of tumor necrosis factor alpha. *Gut* 1994; 35: 909–915.

[32] Wallace JL, Keenan CM, Granger DN. Gastric ulceration induced by nonsteroidal anti-inflammatory drugs is a neutrophil-dependent process. *Am J Physiol* 1990; 259: G462–67.

[33] Campbell EJ, Senior RM, McDonald JA, Cox DL. Proteolysis by neutrophils. Relative importance of cell-substrate contact and oxidative inactivation of proteinase inhibitors in vitro. *J Clin Invest* 1982; 70: 845–852.

[34] Weiss SJ, LoBuglio AF. An oxygen-dependent mechanism of neutrophil-mediated cytotoxicity. *Blood* 1980; 55: 1020–1024.

[35] Zimmerman BJ, Granger DN. Reperfusion-induced leukocyte infiltration: role of elastase. *Am J Physiol* 1990; 259: H3904.

[36] Thong-Ngam D, Choochuay S, Patumraj S, Chayanupatkul M, Klaikeaw N. Curcumin prevents indomethacin-induced gastropathy in rats. *World J Gastroenterol* 2012; 18:1479–1484.

[37] Hart SG, Beierschmitt WP, Wyand DS, Khairallah EA, Cohen SD. Acetaminophen nephrotoxicity in CD-1 mice. I. Evidence of a role for in situ activation in selective covalent binding and toxicity. *Toxicol Appl Pharmacol* 1994; 126: 267–275.

[38] Bessems JG, Vermeulen NP. Paracetamol (acetaminophen)-induced toxicity: molecular and biochemical mechanisms, analogues and protective approaches. *Crit Rev Toxicol* 2001; 31: 55–138

[39] Masubuchi Y, Suda C, Horie T. Involvement of mitochondrial permeability transition in acetaminophen-induced liver injury in mice. *J Hepatol* 2005; 42: 110–116.

[40] McCuskey RS, Bethea NW, Wong J, McCuskey MK, Abril ER, Wang X, Ito Y, DeLeve LD. Ethanol binging exacerbates sinusoidal endothelial and parenchymal injury elicited by acetaminophen. *J Hepatol* 2005; 42: 371–377.

[41] Mladenović D, Radosavljević T, Ninković M, Vucević D, Jesić-Vukićević R, Todorović V. Liver antioxidant capacity in the early phase of acute paracetamol-induced liver injury in mice. *Food Chem Toxicol* 2009; 47: 866–870.

[42] Ishibe T, Kimura A, Ishida Y, Takayasu T, Hayashi T, Tsuneyama K, Matsushima K, Sakata I, Mukaida N, Kondo T. Reduced acetaminophen-induced liver injury in mice by genetic disruption of IL-1 receptor antagonist. *Lab Invest* 2009; 89: 68–79.

[43] Somanawat K, Thong-Ngam D, Klaikeaw N. Curcumin attenuated paracetamol overdose induced hepatitis. *World Journal of Gastroenterology* 2013; 19: 1962–1967.

[44] Tome S, Lucey MR. Review article: current management of alcoholic liver disease. *Alimentary Pharmacology and Therapeutics* 2004; 19: 707–714.

[45] Das SK, Vasudevan DM. Alcohol-induced oxidative stress. *Life Sciences* 2007; 81: 177–187.

[46] Lieber CS. Pathogenesis and treatment of alcoholic liver disease: progress over the last 50 years. *Roczniki Akademii Medycznej w Białymstoku* 2005; 50: 7–20.

[47] Lieber CS. Alcohol and the liver: metabolism of alcohol and its role in hepatic and extrahepatic diseases. *Mount Sinai Journal of Medicine* 2000; 67: 84–94.

[48] Casey CA, Nanji AA, Cederbaum AI, Adachi M, Takahashi T. Alcoholic liver disease and apoptosis. *Alcoholism: Clinical and Experimental Research* 2001; 25: 49S–53S.

[49] Eric S Orman, Gemma Odena, Ramon Bataller. Alcoholic liver disease: Pathogenesis, management, and novel targets for therapy. *J Gastroenterol Hepatol* 2013; 28: 77–84.

[50] Samuhasaneeto S, Thong-Ngam D, Kulaputana O, Suyasunanont D, Klaikeaw N. Curcumin decreased oxidative stress, inhibited NF-kappaB activation, and improved liver pathology in ethanol-induced liver injury in rats. *J Biomed Biotechnol* 2009; Article number 981963.

Chapter 10

Glutamine: A Conditionally Essential Amino Acid with Multiple Biological Functions

Alberto Leguina-Ruzzi and Marcial Cariqueo

Additional information is available at the end of the chapter

http://dx.doi.org/10.5772/66488

Abstract

Glutamine (Gln) is the most abundant free amino acid (AA) in the body with concentrations fluctuating around 500–900 μmol/L. The biological functions of Gln have been widely studied, and they have opened new targets because Gln could modulate physiological functions such as immune enhancer, muscular maintainer, nitrogen transporter, neuronal mediator, pH homeostasis, gluconeogenesis, amino sugar synthesis, and insulin release modulation. In 1990, it was identified that Gln is a conditionally essential AA, meaning that in hypercatabolic or stress conditions, the body suffers depletion in its circulating levels. Moreover, this condition is an independent risk factor of mortality, has been correlated with increase in infection rates, and length of hospital stay in intensive care units (ICU) patients. This characteristic confers the option of Gln use, meaning that through its targets, it could improve the outcome of patients who are suffering a hypercatabolic or hypermetabolic condition.

Keywords: glutamine, parenteral nutrition, enteral nutrition, amino acid

1. Introduction

L-Glutamine (abbreviated as Gln or Q; encoded by the codons CAA and CAG) is a charge neutral, polar (at physiological pH) α-amino acid. Its free form has limited solubility and is unstable in aqueous solution. Gln solved in aqueous solution constructs the cyclized compound pyroglutamic acid (**Figure 1**), which is an uncommon amino acid associated with metabolism problems [1].

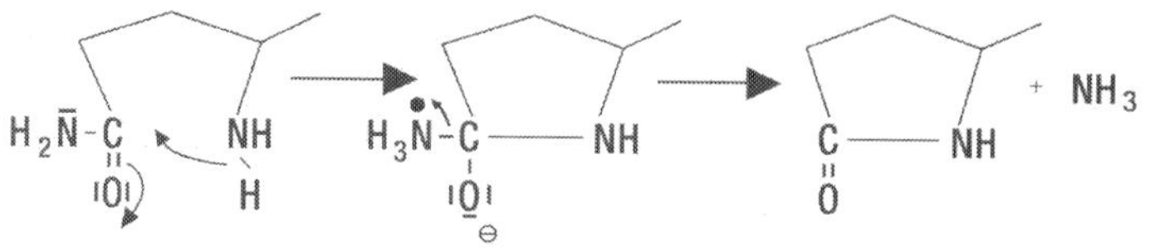

Figure 1. Suggested mechanism for the formation of pyroglutamic acid from glutamine.

Gln is commonly used as a dipetide through a conjugation with L-alanine (Ala-Gln) in the use of nutritional supplement. The dipeptide showed not only dramatically high solubility and stability (**Table 1**) but also can liberate and excrete after absorption into body (**Figure 2**) and ensure the proper excretion [2].

	Solubility (g/L H_2O at 20°C)	Stable in solution
Alanine	167.2	Yes
Cystine	0.1	Yes
Cystine-HCl	252.0	No
Bis-L-analyl-L-cystine	>500.0	Yes
Bis-glycyl-L-cystine	541.0	Yes
Tyrosine	0.4	Yes
L-alanyl-L-tyrosine	14.0	Yes
Glycyl-L-tyrosine	30.0	Yes
Glutamine	36.0	No
L-alanyl-L-glutamine	568.0	Yes
Glycyl-L-glutamine	154.0	Yes

Table 1. Solubility and stability profile of amino acids and conjugations.

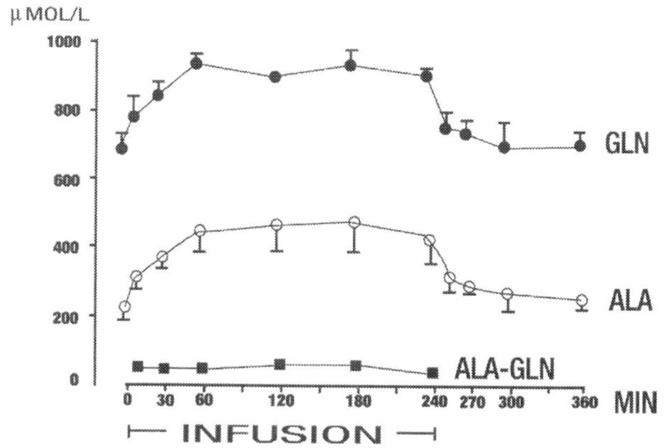

Figure 2. Plasma concentrations of alanine, glutamine, and Ala-Gln during and after continuous intravenous infusion of Ala-Gln (mean ± SD) [2].

http://dx.doi.org/10.5772/66488

Gln is the most abundant amino acid in the body, representing around 30–35% of the amino acid nitrogen in the plasma. It contains two ammonia groups: one from its precursor glutamate and the other from free ammonia in the bloodstream. Because of it, one of its classic and first described functions was as a "nitrogen shuttle," which helps to protect the body from high concentrations of ammonia: Gln behaves as a buffer, receiving excess ammonia, and then releasing it when needed to form other amino acids, amino sugars, nucleotides, and urea.

Gln is mainly distributed in the skeletal muscle (60% of the total pool), short intestine, brain, kidney, and liver. This amino acid (AA) is supplied by specific organs for its metabolic use and adequate renal excretion (**Figure 3**) [3].

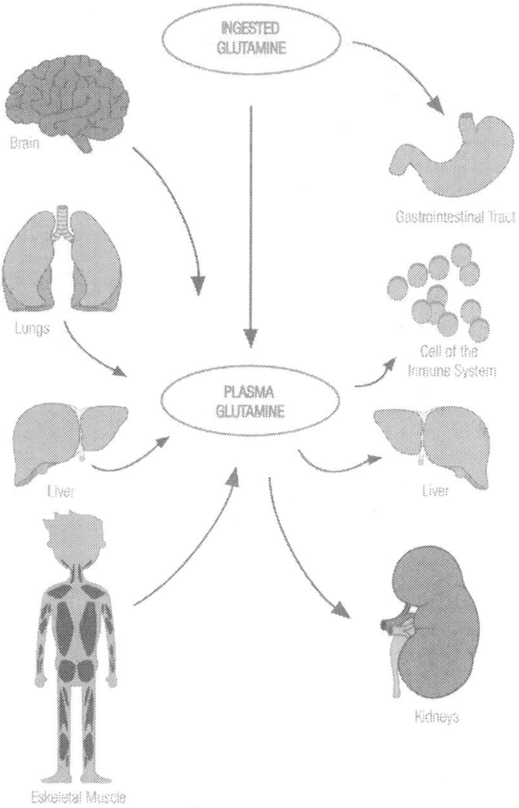

Figure 3. Glutamine distribution in the body.

The role of nutrition in the Gln metabolism has been widely studied. The initial studies on its physiological role suggested that Gln is not always necessary to be ingested from diet because it can accumulate to a high amount; however, this concept has changed in the recent years. In fact, when 5–10 g\day of Gln is consumed in the diet, the de novo synthesis of Gln is regulated by a demand to maintain a balance [4].

An extensive study that evaluated the glutamine content of a wide range of food has been performed. The results showed that the content varied from 0.01 to 9.49 g/100 g of food and a ratio of Gln contained in total protein reached around 1–33% of the intake. The most Gln-enriched foods were the one directed from beef meat, milk, tofu, white rice, corn protein, among others [5].

As mentioned above, the food containing Gln at high concentrations may be used as superfood. In this chapter, we precisely introduce the role of AA in the body, its possibility to be considered as a super nutrient, and we discuss its effectiveness in clinical practice.

2. Glutamine metabolism

Gln is considered to be a nonessential amino acid that was coined by Lacey JM and Wilmore in 1990, as human cells can readily synthesize by glutamine synthetase present in the skeletal muscles, liver, brain, and stomach tissues (**Figure 4**) [6].

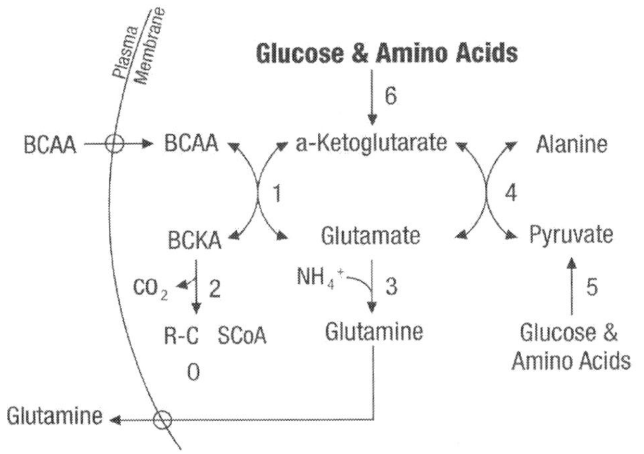

(1) BCAA transaminase; (2) BCKAD; 3) glutamine synthetase; (4) glutamate-pyruvate transaminase; (5) glucose metabolism via glycolysis and the degradation of glucogenic amino acids and (6) glucose metabolism via glycolysis and the Krebs cycle as well as the degradation of amino acids via multiple pathways including the Krebs cycle.

Figure 4. Glutamine biosynthesis process.

Gln is a highly versatile AA as shown in **Figure 5**: it can be converted to other amino acids, to glucose in the liver, and contributes amino groups to nucleotide, amino sugar, and protein biosynthesis. Moreover, it is related to multiple functions and molecular targets in physiological pathways [7].

The uptake into cellular compartments is mediated by several membrane transporters that regulate the homeostasis by coordinating the absorption, reabsorption, and delivery to tissues.

http://dx.doi.org/10.5772/66488

These redundant and ubiquitous located transporters belong to different protein families. The complex interplay between the cell polarity and types of Gln transporters have been sophistically reviewed by Pochini et al. They described the role of the glutamine transporters linked to their different transport modes and coupling with Na^+ and H^+. Most transporters share the specific transport capacity with other neutral or cationic amino acids. Na^+-dependent co-transporters efficiently accumulate glutamine, while antiporters regulate the pools of glutamine and other amino acids.

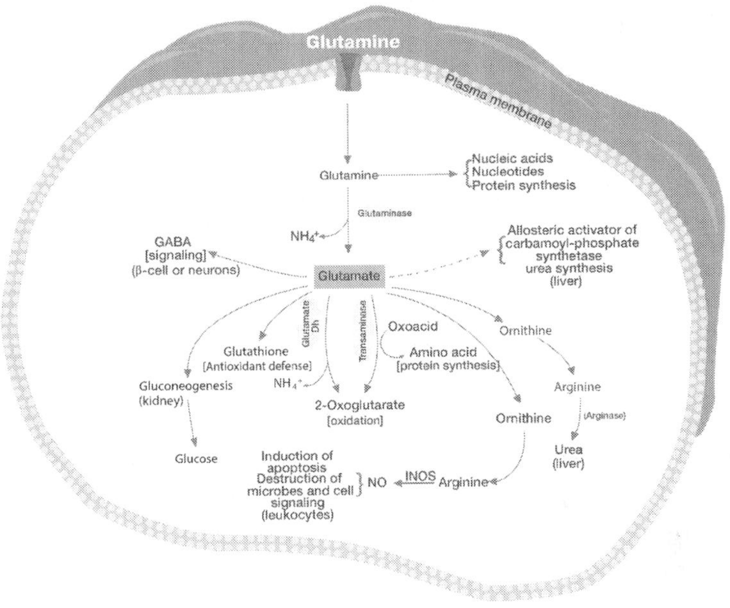

Figure 5. Glutamine intracellular metabolic pathway.

The most studied subfamilies of Gln transporters are the SLC1, 6, 7, and 38. The members involved in the homeostasis are the co-transporters B0AT1 and the SNAT members 1, 2, 3, 5, and 7; the antiporters ASCT2, LAT1 and 2 (**Figure 6**) [8].

Additionally, limited information on glycosylation and/or phosphorylation regulatory sites of the Gln transporters has been described. More studies in the field are needed to fully understand their associated mechanisms.

As shown in **Figure 7**, the metabolic pathway of Gln is a complex network of transport, and the hyperglutaminemia is a highly cytotoxic state classically reported in kidney and liver failure [9]. The proper function of these organs is to ensure the safe excretion. Gln is transformed to urea through the metabolic hepatic pathway and is excreted by the kidney. Additionally, in the intestine, muscle, and liver the Gln is converted to other compounds by chemical reactions. At those organs, Gln is degraded and converted to glutamate, aspartate, CO_2, pyruvate, lactate, alanine, and citrate, among other metabolites [10].

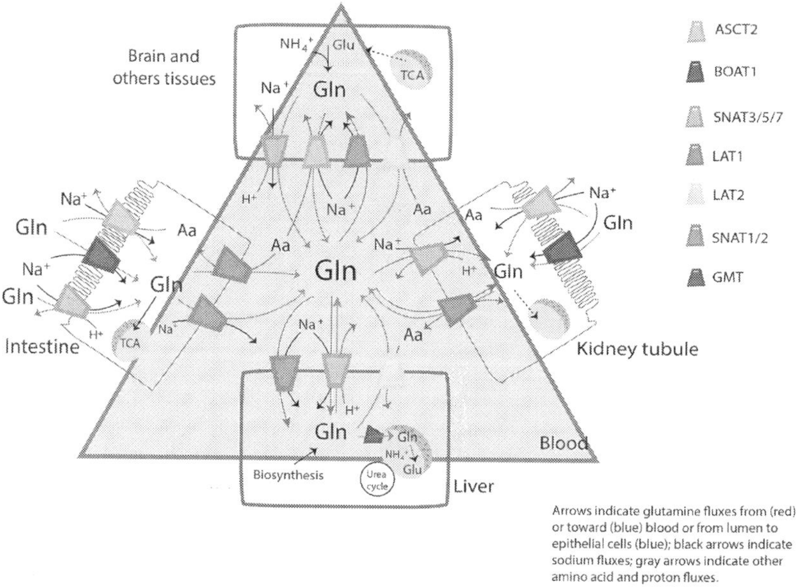

Figure 6. Glutamine transporters, modified from Pochini et al. [8].

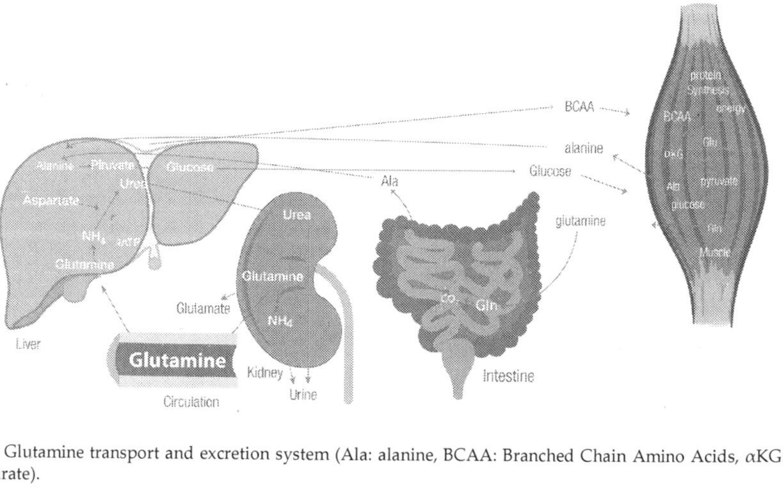

Figure 7. Glutamine transport and excretion system (Ala: alanine, BCAA: Branched Chain Amino Acids, αKG: alpha ketoglutarate).

3. Glutamine depletion in hypercatabolism

The depletion of glutamine is a generally accepted phenomenon particularly observed in the intensive care units (ICU) patients [11]. However, no clear mechanism of the onset of this alteration have been investigated. During hypercatabolic stress, proliferating lymphocytes and immune-stimulated macrophages are major glutamine consumers [12]. The increased demands of Gln in hypercatabolic stress cause high secretion of Gln from skeletal muscle leading to muscle mass loss, a prevalent feature of ICU patients [13].

Of all AA, muscular Gln levels is the most important indicator to determine the restoring forth or surviving rate in ICU patients with prolonged sepsis [14], even more it has been reported that low Gln levels are an independent mortality predictor [11]. The decreased Gln plasma levels in ICU patients are around the 50% of normal levels and negatively correlated with the severity of the pathology.

Based on the clinical observations, the reduction in Gln is associated with higher mortality, length of hospital stay (LOS), and infection [15]. The clinical experts have suggested a series of metabolic alternations that in concentration would explain the poor outcome that an ICU patient with this depletion present (**Figure 8**).

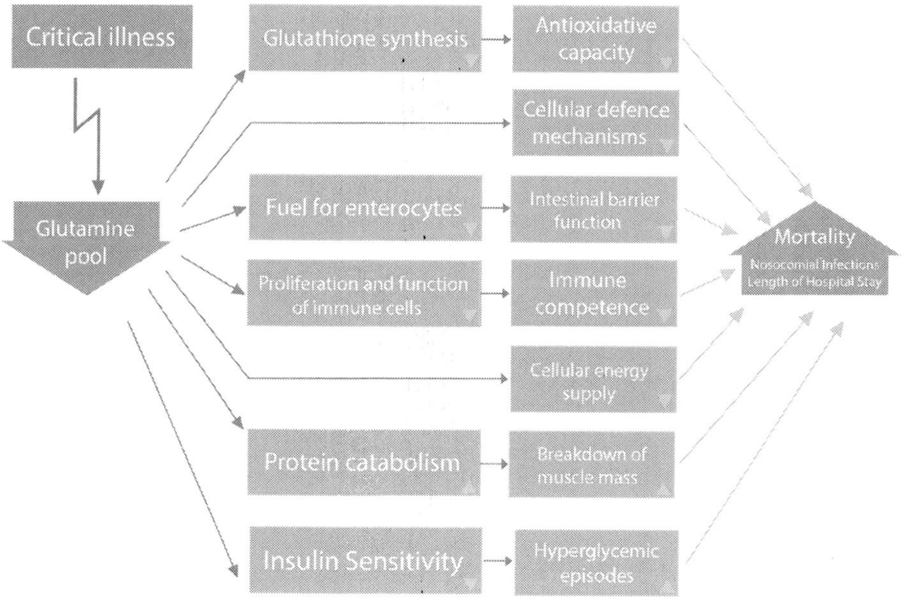

Figure 8. Consequences of glutamine depletion for the organism, modified Stehle et al. [15].

4. Molecular targets of glutamine

Several molecular actions of the Gln [16] have beneficial effects as a supplement with pharmaco logical actions. A supplement of Gln is effective for normalizing the metabolic processes that are altered in hypercatabolic patients. The last decades of biomedical research have identified the specific molecular targets in which this AA exerts its functions[1].

4.1. Glutathione biosynthesis

Glutathione (GSHis) is a reduced nonprotein thiol, which is present in all mammalian tissues and has important antioxidant and detoxification functions. Gln is a precursor of GSHis when combined with glycine and cysteine in the cytoplasmic compartment of the cell. This reduced metabolite has strong affinity to free radicals and toxins, via reaction to oxidized glutathione disulphide (GSSG). GSSG can be converted again to GSHis or be translocated to the vacuole for degradation. This conversion capacity confers its antioxidant and detoxicant effects to the cells (**Figure 9**).

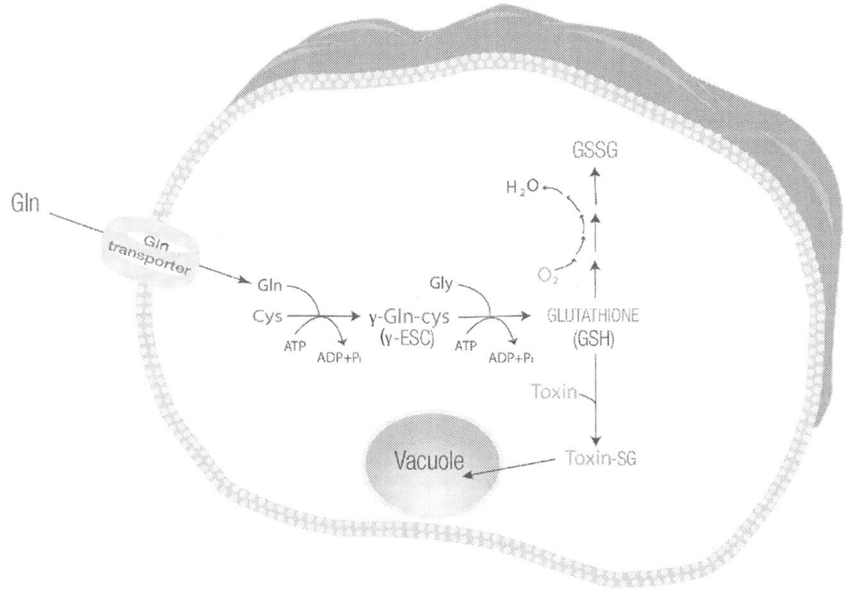

Figure 9. Antioxidant and detoxicant mechanism of glutathione (Cys: cysteine, ATP: adenosine triphosphate, ADP: adenosine diphosphate, y-ESC: gamma glutaminecystein dipeptide, Glu: glutamine, Gly: glycine, SG: glutathione disulfide).

[1] From Leguina-Ruzzi [16]. For the full compilation of references please check the cited article.

4.2. Heat Shock Proteins genetic regulation

The Heat Shock Proteins (HSP) (also known as stress response proteins) are chaperones that are highly conserved and present in all cells. The important role of HSP is to participate in protein folding, assembly, and correct transport, enabling them to act normally. It has been shown that in sepsis or in inflammatory response syndrome, there is a significant reduction in the intracellular levels of HSP70, which correlates with severity of illness and mortality. Interestingly, administration of Gln enhanced the HSP70 expression through the crosstalk with the hexosamine pathway (**Figure 10**). An *in vitro* study demonstrated that the promotion in HSP70 synthesis by Gln is accompanied by a favorable inflammatory response mediated by IL-8 and IL-10 [17].

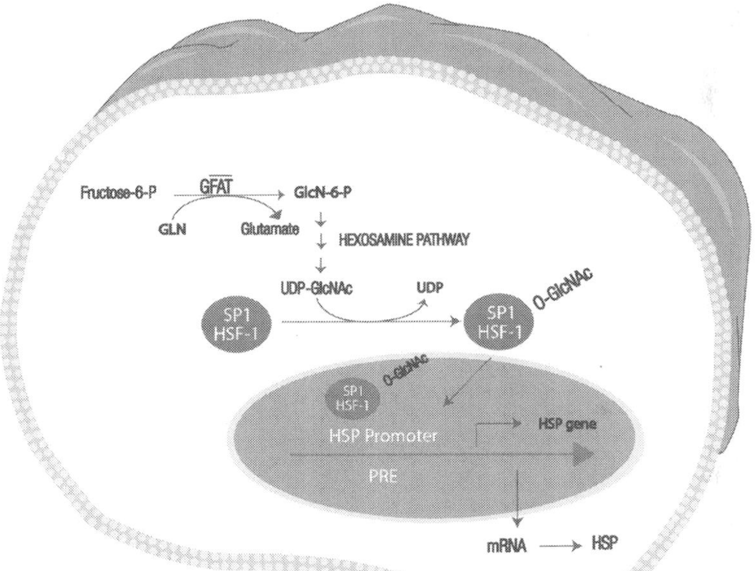

Figure 10. Enhancement of HSP synthesis by glutamine and hexosamine cross-talk (GFAT: glutamine fructose-6-phospate aminotransferase, GlcN-6-P: glucosamine-6-phospato, UDP-GlcNAc: uridine diphosphate N-acetylglucosamine, UDP: uridine di phosphate, SP1: specific protein 1, HSF-1: heat shock transcription factor 1, O-GlcNAc: O-linked N-acetylglucosamine, PRE: promotor regulatory elements).

4.3. Enterocyte integrity

The bacterial translocation (BT) is mainly occurring under pathological conditions because the passage of viable bacteria from the gastrointestinal tract to extraintestinal sites is opened in such conditions.

ICU patients are at higher risk of bacterial translocation: 15% of these patients is affected. BT is one of the main causes of sepsis and multiorgan failure.

It has been reported that the supplementation of Gln to the total parenteral nutrition (TPN) reduces the prevalence of BT and prevents inflammatory intestinal complications. *In vivo* and *in vitro* studies have reported that the enterocyte uses Gln as its principal energy source and enhances its growth and proliferation. Moreover, the enterocyte is capable of transporting Gln to and from the intestinal circulation, creating a bidirectional supply of this AA. This process uses a series of antiporters coupled with Na^+ and H^+ from the family of ASCT2. On the other hand, a deprivation of Gln facilitates BT, which suggests the importance of Gln in the intestinal barrier integrity (**Figure 11.**)

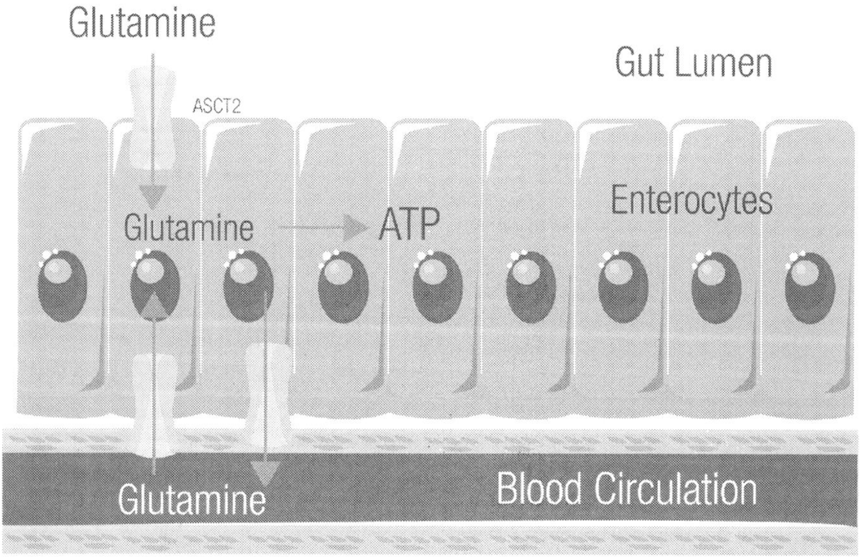

Figure 11. Glutamine transport and metabolism in the enterocyte.

4.4. Lymphocyte function

The activation of naive T cells is a pivotal process for the immune response and is a highly energetic event in which T cells require an increase in nutrient metabolism. For the series of processes required, Gln uptake is a fundamental step highly regulated by the extracellular signal-regulated kinases (ERK)/MAPK pathway (**Figure 12**).

In vitro studies have demonstrated that the Gln supplementation modulates the proliferation of the naive T cells, and the extracellular Gln is essential as a respiratory fuel. The supplementation of Gln has a prominent effect on both activations of lymphocyte and modulation of secretory functions as well as killing bacteria by neutrophils and macrophage phagocytosis.

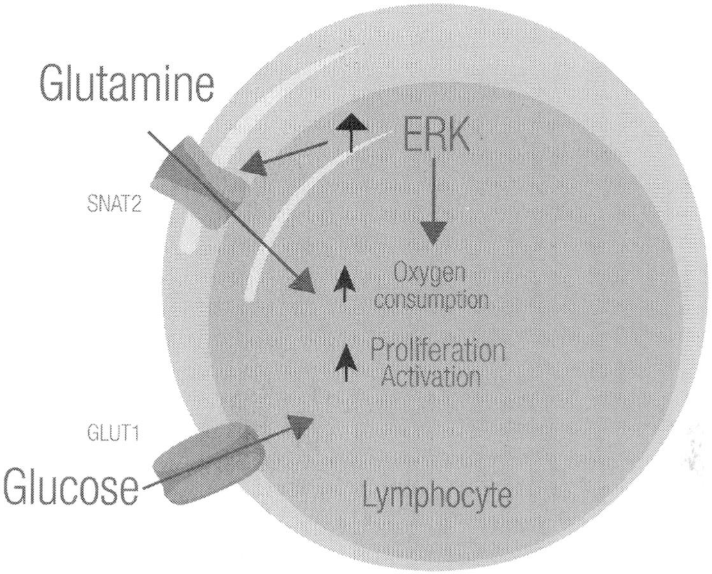

Figure 12. Metabolic processes required for lymphocyte activation (GLUT1: glucose transporter 1, ERK: extracellular signal-regulated kinases).

4.5. Nitrogen balance

The regulation of protein concentration in the body at a relatively normal level is a physiologically favorable process to maintain the cellular integrity. It is widely accepted that ICU patients with TPN present a negative balance that correlates with the clinical outcome. The capacity of Gln pool and nitrogen balance was improved by its supplementation has shown the capacity to recover the nitrogen balance in three days after supplementation. Furthermore, the Gln-supplemented diet did not affect portal ammonia concentration, showing that it does not affect the excretion pathway and did not cause anabolic effects that are associated with cardiovascular alterations [18].

Consequently, these clinical observations could be explained in part by the capacity of the Gln to act as a substrate for other AA or to construct more protein at the muscle.

4.6. Insulin release

The hyperglycemic condition is a metabolic emergency commonly associated with uncontrolled diabetes mellitus, which may result in significant morbidity and death. The prevalence in UCI patients is approximately 40% and classically was presumed to be an adaptive response to the hypercatabolism. However, more recent reports have shown that hyperglycemia is associated with unfavorable clinical outcomes.

The role of Gln in insulin sensitivity and hyperglycemia is a hot topic that has been actively studied. A recent randomized clinical trial demonstrated that the supplementation of Gln to the TPN for more than 7 days reduces significantly the hyperglycemic episodes and the insulin requirement in ICU polytrauma patients. *In vivo* and *in vitro* studies have demonstrated the Gln stimulates calcium-dependent insulin secretion and beta-cell depolarization and enhances the insulin glucose response. The improved process involves the metabolism of the gamma-glutamyl cycle, the glutathione synthesis, and the mitochondrial function (**Figure 13**).

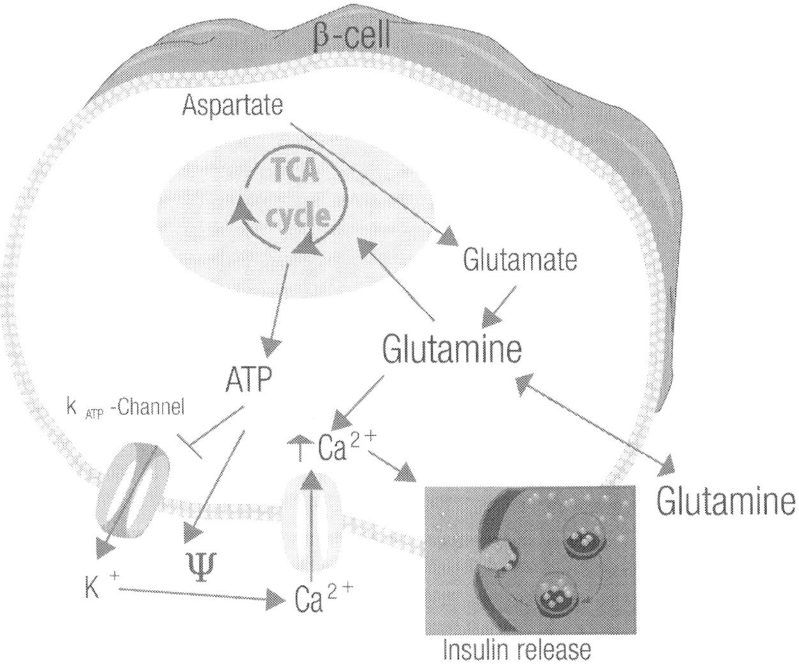

Figure 13. Glutamine-dependent insulin release (TCA: tricarboxylic acid cycle, ATP: adenosine triphosphate, K_{ATP}-channel: ATP-sensitive potassium channel).

4.7. Others beneficial effects of glutamine

The scientific community has been eager to understand the beneficial effects of Gln. Recent clinical, *in vivo* and *in vitro* studies have suggested the cardio protective role of Gln [19, 20] in ischemic heart disease and diabetic cardiomyopathy [21]. Additionally, studies suggest that the supplementation of Gln enhances the healing of the lung parenchymal injuries, reducing the air leakage [22], also regulates the pulmonary infiltration in sepsis, modulating the immunological function [23], and endogenous levels of this AA modulates the vasoactive response of the nitric oxide (NO) in pulmonary hypertension [24].

5. Enteral and parenteral glutamine supplementation

The benefits of parenteral and enteral Gln supplementation in critically ill patients have been shown in numerous clinical trials. Many authors reported in systematic reviews and meta-analyses that Gln supplementation, combined with enteral nutrition (EN) and parenteral nutrition (PN), is associated with reduced infectious morbidity and improved recovery from critical illness compared with standard [25]. The most relevant results have been observed with parenteral supplementation, a phenomenon that might be explained by highly regulated metabolism of the AA and the enterocyte participation (**Figure 11**).

Standard PN and EN formulations do not contain Gln as monopeptide due to the poor solubility and the instability in heat sterilization as described in Section 2. The solubility is limited to 35 g/l (3.5%) at 20°C, and the recommendation is not to use solutions of 2.5%, to avoid precipitation that could affect the proper nutritional administration. To solve these problems, the clinical use by parenteral formulation has been supported in the administration of the Gln dipeptides with other AA, which is more stable and allow prolonged conservation. The dipeptides are rapidly hydrolyzed by serum peptidases, allowing the utilization of 100% of Gln (**Figure 2**). The dipeptides are more soluble than Gln alone; the solubility of Gly-L-Gln is 154 g/l (15.4%) and the solubility of L-Ala-L-Gln is 568 g/l (56.8%). Importantly, parenteral formulations contain 200 g/l (20%) of dipeptide L-Ala-L-Gln which is equivalent to 134 g/l of Gln and enteral formulations contain 2–4 g/l (0.02–0.04%) [26].

Currently, the most commercially used Gln products are Dipeptiven® for PN and Reconvan® for EN produced by Fresenius Kabi Co.; however, it is important to mention that the basic enteral formulations contain Gln as a part of the protein composition and under low concentrations; for example, the Fresubin® line by Fresenius Kabi (Bad Homburg vor der Höhe, Germany) fluctuates around 3.5—9.4 g per presentation bag.

The early enteral feeding has been associated with a substantial reduction in length of hospital stay (LOS) with a significant reduction in the frequency of acquired infections. The enteral Gln supplementation has been shown to be safe and well tolerated and may help to reduce infectious complications, oxidative stress, intestinal permeability, mortality, and LOS [27]. Houdijk [28]. However, in 2015, Van Zanten et al. published a systematic review and meta-analysis with a total of 11 studies involving 1079 adult critical ill patients and enteral Gln supplementation was not associated with a reduction of hospital mortality, infectious complications, or stay in the intensive care unit. In the subset of patients with burns, there may be a significant benefit in hospital mortality [29].

The effective concentration of Gln in TPN has been suggested by different multidisciplinary medical groups. According to the ESPEN guideline, "when PN is indicated in ICU patients the amino acid solution should contain 0.2–0.4 g/kg/day of L-glutamine (e.g. 0.3–0.6 g/kg/day alanyl-glutamine dipeptide)" [30] and "glutamine should be added to a standard enteral formulation in burned patients and trauma patients" [31]. It is important to consider that the main contraindications are renal failure (Creatinine clearance <20 mL/min), metabolic acidosis, and liver failure (Liver function tests International Normalized Ratio >1.5) [32]. Interestingly,

Helling et al. evidenced the association between liver failure and high plasma Gln levels [33], a medical situation that needs to be taken into consideration at the moment to prescribe Gln supplementation.

The use of EN with Gln may not be enough to full up plasma concentration up to normal level; new remarkable data showed that enteral supplementation only is not enough to revert the Gln depletion [34]. The values and data accumulated in PN cannot be directly extrapolated to EN supplementation and, therefore, cannot be used as the base for recommendation. According to the trials, high protein enteral nutrition enriched with immuno modulator nutrients, such as Gln, did not improve infectious complications or other clinical end points compared to standard EN [35].

On the other hand, administration of TPN with Gln is effective. In 2002, Goether showed improvement in a 6-month survival in a patient with at least 9 days of parenteral Gln supplementation [28]. Until 2014, several trials showed that PN with Gln supplementation given in conjunction with nutrition support continues to be associated with a significant reduction in hospital mortality and hospital LOS. A systematic review published by Wischmeyer et al. in 2014 summarized all randomized controlled trials conducted from 1997 to 2013, showing the benefits of parenteral glutamine [36]. However, in 2013, the Reducing Deaths Due to Oxidative Stress (REDOX) study, the largest trial to date showed that the supplementation of Gln was associated with higher mortality and no beneficial effects were seen [37]. The results of this trial awaken a question to the safety and efficacy of the use of glutamine in critically ill patients, in high doses (much higher than recommended) may produce adverse effects. In this study, they used a combined enteral and IV Gln supplementation in higher doses than the recommended ones: the intervention setting was the Gln enteral 30 g/day plus parenteral 15 g/day giving around 1 g/kg/day, doses higher than the classically recommended by the clinical guidelines of the date. Furthermore, the heterogeneous enrolment included patients that fulfilled contraindication criteria for its supplementation.

A year later, the authors published a new analysis of the data (post hoc study) concluding that high doses of Gln may be associated with higher mortality in patients with multiorgan failure and particularly renal dysfunction [38]. These conclusions had a high impact not only on the pharmaceutical industry but also on the clinical practice: two of the most important guidelines changed dramatically the recommendation of the use of Gln. The Canadian Clinical Practice Guidelines (2016) [39] and ASPEN (2016) [40] downgraded the Gln supplementation based of the REDOX results and a series of studies that do not justify this.

It is important to highlight that even after these dramatic results, the research has given important steps in the years follow the REDOX. It gave not only answers to the basic questions such as the glutamine plasmatic levels in ICU patients and its safe use as supplement to TPN but also opened a whole perspective for further research and use. Pérez-Bárcena et al. [41], in 2014, showed that low doses of IV Gln for 5 days did not show beneficial effects in the ICU patients, without causing any derogative effect. Additionally, plasmatic Gln measurement in the patients showed that those with lower levels presented a worse outcome (mortality, LOS, and infections) in which a supplementation with higher doses might be necessary.

In the same year, Grintescu et al. [42] showed that Gln supplementation in trauma patients reduces hyperglycemic episodes and improves insulin response. The conclusions suggest a role for Gln as an insulin sensitizer.

Despite the controversy, when parenteral nutrition is prescribed, the Gln supplementation is still recommended in critically ill patients. There are insufficient data to generate recommendations for IV glutamine in critically ill patients who are receiving EN. New large, multicenter, prospective randomized clinical trials are needed to confirm the beneficial effects of Gln in the mortality, LOS, and infections rates as main clinical outcomes are highly relevant in the critical care unit.

6. Conclusion

The multiple functions of Gln and the possibility of its use as a pharmaco-nutrient under the hypercatabolic condition were introduced in this chapter. The studies on basic and clinical science showed the beneficial effects of Gln in the metabolism of subjects suffering from a catabolic stress condition. *In vitro* data performed under concentration ranges from 500 to 2000 μmol/L of Gln, and it could represent supplementation and supports the clinical findings. Gln pleiotropic functions make it a great candidate for its use in pathological conditions; however, important lessons have to be learned from the controversial evidence that impacts negatively on its use. Still nonconclusive data have weakened it role in oral and enteral supplementation making mandatory new research in this field.

Acknowledgements

All figures were illustrated by CatarsisCreativa.com in Santiago, Chile.

Authors contributions: A. Leguina-Ruzzi conceptualized and prepared the chapter. M. Cariqueo.

Author details

Alberto Leguina-Ruzzi[1*] and Marcial Cariqueo[2]

*Address all correspondence to: alberto@juntendo.ac.jp

1 Biochemistry Department, Research Institute for Disease of Old Age, Juntendo University, Tokyo, Japan

2 Intensive Care Unit, Clinical Hospital, University of Chile, Santiago, Chile

References

[1] Scansetti M, Hu X, McDermott BP, Lam HW: Synthesis of pyroglutamic acid derivatives via double Michael reactions of alkynones. Org Lett. 2007;9(11):2159–2162. DOI: 10.1021/ol070674f.

[2] Berg A, Rooyackers O, Norberg A, Wernerman J: Elimination kinetics of L-alanyl-L-glutamine in ICU patients. Amino Acids. 2005;29(3):221–228. DOI: 10.1007/s00726-005-0230-9.

[3] Brosnan JT: Interorgan amino acid transport and its regulation. J Nutr. 2003;133(6): 2068S–2072S.

[4] Miller AL: Therapeutic considerations of l-glutamine: a review of the literature. Altern Med Rev. 1999;4(4):239–248.

[5] Lenders CM, Liu S, Wilmore DW, Sampson L, Dougherty LW, Spiegelman D, Willett WC: Evaluation of a novel food composition database that includes glutamine and other amino acids derived from gene sequencing data. Eur J Clin Nutr. 2009;63(12):1433–1439.

[6] Lacey JM, Wilmore DW: Is glutamine a conditionally essential amino acid?. Nutr Rev. 1990;48(8):297–309.

[7] Smith RJ: Glutamine metabolism and its physiologic importance. JPEN J Parenter Enteral Nutr. 1990;14(4):40S–44S. DOI: 10.1177/014860719001400402

[8] Pochini L, Scalise M, Galluccio M, Indiveri C: Membrane transporters for the special amino acid glutamine: structure/function relationships and relevance to human health. Front Chem. 2014;eCollection 2014:61. DOI: 10.3389/fchem. 2014.00061.

[9] Helling G, Wahlin S, Smedberg M, Pettersson L, Tjäder I, Norberg Å, Rooyackers O, Wernerman: Plasma glutamine concentrations in liver failure. PLoS One. 2016;11(3):1–10. DOI: 10.1371/journal.pone.0150440.

[10] Weiner ID, Mitch WE, Sands JM: Urea and ammonia metabolism and the control of renal nitrogen excretion. Clin J Am Soc Nephrol. 2015;10(8):1444–1458. DOI: 10.2215/CJN.10311013.

[11] Oudemans-van Straaten HM, Bosman RJ, Treskes M, van der Spoel HJ, Zandstra DF: Plasma glutamine depletion and patient outcome in acute ICU admissions. Intensive Care Med. 2001;27(1):84–90. DOI: 10.1007/s001340000703.

[12] Rohde T, MacLean DA, Klarlund Pedersen B: Glutamine, lymphocyte proliferation and cytokine production. Scand J Immunol. 1996;44(6):648–650. DOI: 10.1046/j. 1365-3083.1996.d01-352.x.

[13] Askanaz Ji, Carpentier YA, Michelsen CB, Elwyn DH, Furst P, Kantrowitz LR, Gump FE, Kinney JM: Muscle and plasma amino acids following injury: influence of intercurrent infection. Ann Surg. 1980;192(1):78–85.

[14] Roth E, Funovics J, Mühlbacher F, Schemper M, Mauritz W, Sporn P, Fritsch A: Metabolic disorders in severe abdominal sepsis: glutamine deficiency in skeletal muscle. Clin Nutr. 1982;1(1):25–41. DOI: 10.1016/0261-5614(82)90004-8.

[15] Stehle P, Kuhn KS: Glutamine: an obligatory parenteral nutrition substrate in critical care therapy. Biomed Res Int. 2015. DOI: 10.1155/2015/545467.

[16] Leguina-Ruzzi A: Therapeutic targets of glutamine in parenteral nutrition: a medical science review. Int J Prev Treat. 2015;4:34–39. DOI: 10.5923/j.ijpt.20150402.03.

[17] Marino LV, Pathan N, Meyer RW, Wright VJ, Habibi P: An in vitro model to consider the effect of 2 mM glutamine and KNK437 on endotoxin-stimulated release of heat shock protein 70 and inflammatory mediators. Nutrition. 2016;32(3): 375–383. DOI: 10.1016/j.nut.2015.09.007.

[18] Carroll PV, Jackson NC, Russell-Jones DL, Treacher DF, Sönksen PH, Umpleby AM: Combined growth hormone/insulin-like growth factor I in addition to glutamine-supplemented TPN results in net protein anabolism in critical illness. Am J Physiol Endocrinol Metab. 2004;286(1):151–157. DOI: 10.1152/ajpendo.00122.2003.

[19] Kumar SHS, Anandan R: Biochemical studies on the cardioprotective effect of glutamine on tissue antioxidant defense system in isoprenaline-induced myocardial infarction in rats. J Clin Biochem Nutr. 2007;40(1):49–55. DOI: 10.3164/jcbn.40.49.

[20] Lomivorotov VV, Efremov SM, Shmirev VA, Ponomarev DN, Lomivorotov VN, Karaskov AM: Glutamine is cardioprotective in patients with ischemic heart disease following cardiopulmonary bypass. Heart Surg Forum. 2011;14(6):384–388. DOI: 10.1532/HSF98.20111074.

[21] Badole SL, Jangam GB, Chaudhari SM, Ghule AE, Zanwar AA: L-glutamine supplementation prevents the development of experimental diabetic cardiomyopathy in streptozotocin-nicotinamide induced diabetic rats. PLoS One. 2014;9(3). DOI: 10.1371/journal.pone.0092697.

[22] Sanli A, Onen A, Sarioglu S, Sis B, Guneli E, Gokcen B, Karapolat S, Acikel U: Glutamine administration enhances the healing of lung parenchymal injuries and reduces air leakage in rats. Tohoku J Exp Med. 2002;210(3):239–245. DOI: http://doi.org/10.1620/tjem.210.239.

[23] Hulsewé KW, van der Hulst RR, Ramsay G, van Berlo CL, Deutz NE, Soeters PB: Pulmonary glutamine production: effects of sepsis and pulmonary infiltrates. Intensive Care Med. 2003;29(10):1833–1836. DOI: 10.1007/s00134-003-1909-6.

[24] Xu H, Pearl RG. Effect of l-glutamine on pulmonary hypertension in the perfused rabbit lung. Pharmacology. 1994;48:260–264. DOI: 10.1159/000139188

[25] van Zanten AR, Dhaliwal R, Garrel D, Heyland DK: Enteral glutamine supplementation in critically ill patients: a systematic review and meta-analysis. Crit Care. 2015;19:294. DOI: 10.1186/s13054-015-1002-x

[26] Fürst P: New developments in glutamine delivery. J Nutr. 2001;131(9 Suppl):256–258S.

[27] Goeters C, Wenn A, Mertes N, Wempe C, Van Aken H, Stehle P, Bone HG: Parenteral L-alanyl-L-glutamine improves 6-month outcome in critically ill patients. Crit Care Med. 2002;30(9):2032–2037. DOI: 10.1097/01.CCM.0000025908.95498.A3

[28] Houdijk AP, Rijnsburger ER, Jansen J, Wesdorp RI, Weiss JK, McCamish MA, Teerlink T, Meuwissen SG, Haarman HJ, Thijs LG, van Leeuwen PA: Randomised trial of glutamine-enriched enteral nutrition on infectious morbidity in patients with multiple trauma. Lancet. 1998;352(9130):772–776. DOI: 10.1016/S0140-6736(98)02007-8

[29] van Zanten AR: Glutamine and antioxidants: status of their use in critical illness. Curr Opin Clin Nutr Metab Care. 2015;18(2):179–186. DOI: 10.1097/MCO.0000000000000152

[30] Singer P, Berger MM, Van den Berghe G, Biolo G, Calder P, Forbes A, Griffiths R, Kreyman G, Leverve X, Pichard C, ESPEN: ESPEN Guidelines on Parenteral Nutrition: intensive care. Clin Nutr. 2009;28(4):387–400. DOI: 10.1016/j.clnu.2009.04.024

[31] Kreymann KG, Berger MM, Deutz NE, Hiesmayr M, Jolliet P, Kazandjiev G, Nitenberg G, van den Berghe G, Wernerman J; DGEM (German Society for Nutritional Medicine), Ebner C, Hartl W, Heymann C, Spies C; ESPEN (European Society for Parenteral and Enteral Nutrition): ESPEN Guidelines on Enteral Nutrition: intensive care. Clin Nutr. 2006;25(2):210–223. DOI: 10.1016/j.clnu.2006.01.021

[32] Ginguay A, De Bandt JP, Cynober L: Indications and contraindications for infusing specific amino acids (leucine, glutamine, arginine, citrulline, and taurine) in critical illness. Curr Opin Clin Nutr Metab Care. 2016;19(2):161–169. DOI: 10.1097/MCO.0000000000000255

[33] Helling G, Wahlin S, Smedberg M, Pettersson L, Tjäder I, Norberg Å, Rooyackers O, Wernerman J: Plasma glutamine concentrations in liver failure. PLoS One. 2016;11(3):e0150440. DOI: 10.1371/journal.pone.0150440

[34] van Barneveld KW, Smeets BJ, Heesakkers FF, Bosmans JW, Luyer MD, Wasowicz D, Bakker JA, Roos AN, Rutten HJ, Bouvy ND, Boelens PG: Beneficial effects of early enteral nutrition after major rectal surgery: a possible role for conditionally essential amino acids? Results of a randomized clinical trial. Crit Care Med. 2016;44(6):353–361. DOI: 10.1097/CCM.0000000000001640

[35] van Zanten AR, Sztark F, Kaisers UX, Zielmann S, Felbinger TW, Sablotzki AR, De Waele JJ, Timsit JF, Honing ML, Keh D, Vincent JL, Zazzo JF, Fijn HB, Petit L, Preiser JC, van Horssen PJ, Hofman Z: High-protein enteral nutrition enriched with immune-modu-

lating nutrients vs standard high-protein enteral nutrition and nosocomial infections in the ICU: a randomized clinical trial. JAMA. 2014;312(5):514–524. DOI: 10.1001/jama.2014.7698

[36] Wischmeyer PE, Dhaliwal R, McCall M, Ziegler TR, Heyland DK: Parenteral glutamine supplementation in critical illness: a systematic review. Crit Care. 2014;18(2):R76. DOI: 10.1186/cc13836

[37] Heyland D, Muscedere J, Wischmeyer PE, Cook D, Jones G, Albert M, Elke G, Berger MM, Day AG; Canadian Critical Care Trials Group: A randomized trial of glutamine and antioxidants in critically ill patients. N Eng J Med. 2013;268(16):1489–1497. DOI: 10.1056/NEJMoa1212722

[38] Heyland DK, Elke G, Cook D, Berger MM, Wischmeyer PE, Albert M, Muscedere J, Jones G, Day AG; Canadian Critical Care Trials Group: Glutamine and antioxidants in the critically ill patient: a post hoc analysis of a large-scale randomized trial. JPEN J Parenter Enteral Nutr. 2015;39(4):401–409. DOI: 10.1177/0148607114529994

[39] Leguina-Ruzzi A: A commentary on the 2015 Canadian Clinical Practice Guidelines in glutamine supplementation to parenteral nutrition. Crit Care. 2016;8(20):7. DOI: 10.1186/s13054-015-1175-3

[40] McClave SA, Taylor BE, Martindale RG, Warren MM, Johnson DR, Braunschweig C, McCarthy MS, Davanos E, Rice TW, Cresci GA, Gervasio JM, Sacks GS, Roberts PR, Compher C; Society of Critical Care Medicine; American Society for Parenteral and Enteral Nutrition: Guidelines for the provision and assessment of nutrition support therapy in the adult critically ill patient: Society of Critical Care Medicine (SCCM) and American Society for Parenteral and Enteral Nutrition (A.S.P.E.N.). JPEN J Parenter Enteral Nutr. 2016;40(2):159–211. DOI: 10.1177/0148607115621863

[41] Pérez-Bárcena J, Marsé P, Zabalegui-Pérez A, Corral E, Herrán-Monge R, Gero-Escapa M, Cervera M, Llompart-Pou JA, Ayestarán I, Raurich JM, Oliver A, Buño A, García de Lorenzo A, Frontera G: A randomized trial of intravenous glutamine supplementation in trauma ICU patients. Intensive Care Med. 2014;40(4):539–547. DOI: 10.1007/s00134-014-3230-y

[42] Grintescu IM, Luca Vasiliu I, Cucereanu Badica I, Mirea L, Pavelescu D, Balanescu A, Grintescu IC: The influence of parenteral glutamine supplementation on glucose homeostasis in critically ill polytrauma patients—a randomized-controlled clinical study. Clin Nutr. 2014;34(3):377–382. DOI: 10.1016/j.clnu.2014.05.006

Chapter 11

The Effect of Dietary Intake of Omega-3 Polyunsaturated Fatty Acids on Cardiovascular Health: Revealing Potentials of Functional Food

Ines Drenjančević, Gordana Kralik, Zlata Kralik, Martina Mihalj, Ana Stupin, Sanja Novak and Manuela Grčević

Additional information is available at the end of the chapter

http://dx.doi.org/10.5772/67033

Abstract

Functional food is a food containing components that show beneficial effects on one or more body functions and improve general condition and health or significantly affect lowering of disease risks. This chapter is aimed to examine the effect of dietary intake of omega-3 polyunsaturated fatty acids (n3-PUFA) on cardiovascular health. This chapter presents current knowledge on functional poultry products and the reasons to consume them, omega-3 enrichment of eggs and poultry meat, and the differences in profile of fatty acids in conventional and omega-3–enriched eggs. The second part of the chapter focuses on the metabolism of fatty acids and effectiveness of n-3 PUFA in the improvement of endothelial function, improvement of elasticity of the vascular wall and the anti-inflammatory effects in patients with chronic diseases, such as metabolic syndrome, diabetes mellitus and hypercholesterolemia, and overall effect on cardiovascular health and protection. To achieve long-term protective effects, the functional food should be consumed on daily basis. There are no specific constrains in taking functional food; even more, it can be recommended to athletes and cardiovascular patients. General population can also benefit from eating functional food enriched with n-3 PUFA due to their anti-inflammatory and vascular-protective effects.

Keywords: omega-3 fatty acids, enriched eggs and poultry, cardiovascular risk

1. Introduction

Definitions of functional food differ in different parts of the world; however, they all have in common the reference toward food of natural origin that contains ingredients with beneficial effect on human health. In the United States, the definition of functional food says that functional food was "natural or processed food that contains known or unknown biologically active ingredients, which in certain, effective and non-toxic concentrations provide clinically proven and documented health benefits for prevention, treatment or healing of chronic diseases" [1]. This way of defining functional food is different from the definition in Europe, which does not mention effects of functional food in the treatment of diseases, but mainly refer to benefits in maintaining good health or reducing the risk of developing diseases. The European Commission document "Scientific Concepts of Functional Foods in Europe" states a working definition saying that food can be considered functional if it is satisfactorily shown that, in addition to appropriate nutritional effects, it has beneficial effects on one or more target functions of an organism, in a way that it is important for improving health condition and general well-being or reducing disease risks. Functional food has to be food (not in the form of pills or capsules) and it has to show its effects when consumed in normal daily amounts [2]. Regardless of different definitions [3], concluded that the main purpose of functional food had to be clear—it improves human health and well-being or general condition of the body.

Functional poultry products refer to meat and poultry eggs enriched with ingredients that have positive influence on human health. Poultry is particularly suitable for functional food products because of their ability to use the physiological and metabolic processes of their body to deposit beneficial ingredients from feed into products, that is, into meat and eggs. Meat and poultry eggs are enriched with functional ingredients (fatty acids, vitamins, and antioxidants) by feeding poultry feed supplemented with increased concentrations of those ingredients. The most common functional poultry products are meat and eggs with increased content of desirable omega-3 fatty acids, vitamin E, selenium, and carotenoids. Poultry meat is rich in protein and low in fat. As of its nutritional composition, it can be considered a dietary foodstuff. It is easily digested and especially recommended for consumption of the elderly and children. If considering all stated nutritional benefits, poultry meat enriched with functional ingredients can be considered as functional product. Chicken meat is available to wide population of consumers because of its price, which is more affordable if compared to red meat. High-quality nutritional composition of chicken meat is also one of the reasons for its frequent consumption, which is especially emphasized in recent years when consumers became more aware about the composition of foods and their effects on health.

Egg is a foodstuff that contains high-quality and easily digestible proteins, where amino acid composition is the most similar to proteins of the human body. Egg proteins are fully exploited in the human body and have greater biological value than meat proteins. Egg yolk contains essential fatty acids, vitamins, and minerals needed for proper functioning of human organism. Egg is considered a natural functional foodstuff because of its nutritional value. When compared to poultry meat, enrichment of eggs with functional ingredients is easier because of the high content of fat in egg yolk [4].

If consuming meat and eggs enriched with functional ingredients, consumers can affect the increase of the content of such functional ingredients in blood and tissue in a natural way, thus avoiding taking in some dietary supplements. The importance of functional ingredients for human health is elaborated further in the text.

2. Functional poultry products production

2.1. Metabolism of n-3 and n-6 polyunsaturated fatty acids

Fatty acids are constituent parts of fat and oil molecules. Polyunsaturated fatty acids (PUFAs) are divided into two groups of n-3 and n-6, depending on where the first double bond is found in the carbon chain, that is, where hydrogen atoms are missing. Linoleic fatty acid (LA) and arachidonic fatty acid (AA) are typical representatives of the n-6 group, and α-linolenic fatty acid (ALA), eicosapentaenoic acid (EPA), and docosahexaenoic acid (DHA) represent the n-3 group. The metabolism and the role of PUFA n-6 and n-3 may differ in living organism. Fatty acids of the n-6 and n-3 groups (LA and ALA) cannot be synthesized in an organism and are therefore called essential fatty acids (EFAs).

Importance of designed products (enriched with n-3 PUFA) is found in the fact that LA and ALA (precursors with 18 carbon atoms) may extend in human organism and desaturate into arachidonic acid and DHA. Processes are catalyzed by elongase, Δ^6- and Δ^5-desaturase [5]. The limiting factor of metabolizing n-6 PUFA and n-3 PUFA is the enzyme Δ^6-desaturase. Unfortunately, final conversion into docosapentaenoic acid (DPA) and docosahexaenoic acid is still not clear; however, the important role is attributed toΔ^4-desaturase [6]. Infante and Huszagh stated that the biosynthesis of DHA took place in the mitochondria membranes, and biosynthesis of AA, EPA, and DPA occurred in the endoplasmic reticulum [7]. Supported by enzymes of cyclooxygenase (COX) and lipoxygenase (LX) within certain processes, EFA is converted into hormone-like substances called eicosanoids. Numerous studies confirmed that linoleic, linolenic, and oleic acids during biosynthesis compete for the same Δ^6-desaturase. It was also found that linolenic acid acted as the inhibitor of n-6 PUFA metabolism. At the same time, 10 times more linoleic acids are required to inhibit metabolism of n-3 PUFA at the same level [6]. LA, ALA, and AA are essential fatty acids for poultry. The greatest importance in the composition of poultry feed should be given to those mentioned fatty acids because they are precursors to eicosapentaenoic acid and docosahexaenoic acid, both of which are also considered as essential for humans. Human organism requires daily intake of 290–390 mg of ALA and 100–200 mg of EPA and DHA. **Figure 1** depicts the metabolic pathways of fatty n-3 PUFA and n-6 PUFA.

2.2. Poultry meat and eggs enriched with n-3 PUFA

The intake of plant sources, especially linseed oil, significantly increases the content of omega-3 fatty acids in the form of ALA; however, they fail to increase the content of long-chain omega-3 fatty acids in meat and eggs. The best sources of long-chain omega-3 PUFA, EPA, and DHA are oils of sea organisms and of fish. The use of these oils is limited because of poorer organoleptic properties of final products [8]. In order to avoid unpleasant odor or taste in meat and eggs, portions of fish oil, as well as of linseed oil in feeding mixtures, must be taken into account.

In their research into effects of linseed contained in laying hens' feeding mixtures in different portions (0, 5, 10, and 15%) on the content of ALA in egg yolks, the increase of the content of

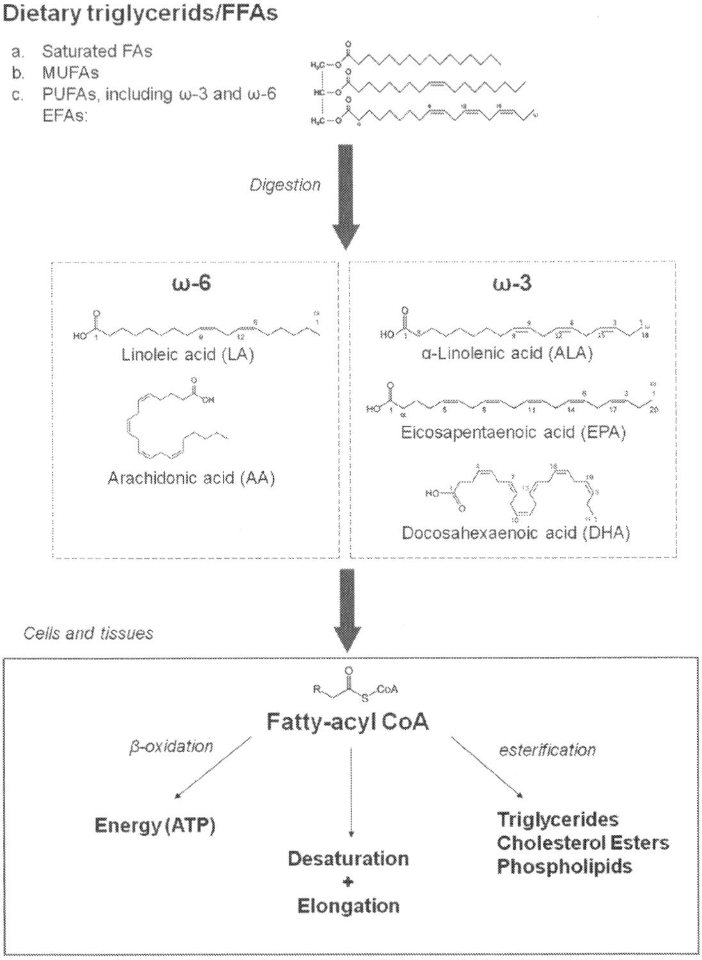

Figure 1. Fatty acids (FAs) source and metabolism.

ALA from 1.80% in the group without linseed to 7.07, 8.35, and 12.20% in the group with the highest content of linseed in feeding mixture [9] was determined. Valavan et al. recorded the increase of ALA content in egg yolks from 0.62% in the control group to 0.83, 0.93, and 1.00% in the experimental groups fed diets supplemented with linseed oil in the amounts of 1, 2, and 3% [10]. They also determined the increase in the content of EPA and DHA. Supplementation of linseed oil in the portion of 5 and 10% to laying hens' feed affected the increase of ALA portion from 0.37% to high 10.3% and 14.9% [11]. These results referring to the increase of ALA content in egg yolk correspond to the fact that linseed and linseed oil are rich in ALA.

Meluzzi et al. stated that the supplementation of 3% fish oil to laying hens' diet influenced the increase of EPA and DHA content in the egg of 19.53 and 143.70 mg/egg [12]. Gonzalez-Esquerra and Leeson pointed out that the supplementation of 6% fish oil to laying hens' diet affected the increase of EPA and DHA contents, as well as the content of total n-3 PUFA, which amounted to 246 mg on average [13]. In their paper about the production of Bio-omega-3 eggs, Imran et al. fed laying hens with mixtures supplemented with extruded linseed and determined that an increased content of extruded linseed in feeding mixtures affected the increase of DHA portion in eggs and reduced AA portion, as well as the ratio of total n-n-3 PUFA [14]. Apart from oils that can be purchased on the market, there are different commercial preparations, which can be used in poultry feeding in order to achieve the increased content of fatty acids in their products. Kralik et al. investigated the influence of Pronova Biocare EPAX 3000 (PBE), which is rich in fish oil, on the profile of fatty acids in chicken eggs [15]. The authors determined that the replacement of 3.33% of corn in chicken diet with the PBE oil resulted in the reduction of arachidonic fatty acid portion in egg yolks (C = 1.66%, E = 0.58%), and in the increase of EPA (C = 0.01%, E = 0.24%) and of DHA (C = 0.72%, and E = 1.76%). Moreover, these authors reported that the n-6/n-3 PUFA ratio was lowered from 14.88 in the control to 7.25 in the experimental group.

The possibility of altering the fatty acid composition of chicken meat is an objective in many studies. Kralik et al. emphasized more favorable ratio of total n-6/n-3 PUFA in thigh muscle lipids of chickens fed diets supplemented with linseed oil, in comparison to the control group that consumed diets with sunflower oil (2.75 and 12.23, respectively) [16]. Since poultry diet is based on corn rich in saturated fatty acids (SFAs), which are then through feed deposited into muscle tissue, feeding mixtures for chickens should be supplemented with linseed or rapeseed or their oils or with fish oil if wanting to enrich their meat with desirable n-3 fatty acids [16–18]. These authors agreed that dietary supplementation of plant oils (linseed and rapeseed oils) instead of sunflower oil affected the increase of n-3 PUFA, and the reduction of n-6 PUFA in poultry meat.

Fish oil or seafood oil are an excellent source of n-3 PUFA fatty acid, such as EPA and DHA. Mirghelenj et al. reported that the increase in the portion of fish oil contained in broiler feed influenced the increase of EPA and DHA fatty acids content in thigh and breast muscles [19]. The content of EPA in breast muscle increased from 0.014 mg/g in the group K to 0.090 mg/g in the group P4, and the content of DHA was raised from 0.046 mg/g in the group K to 0.338 mg/g in the group P4. The authors also reported the increase of EPA in thigh meat, from 0.028 mg/g in the control to 0.232 mg/g in the group P4, while the content of DHA was the lowest in the control (0.0085 mg/g), and the highest in the group P5 (0.578 mg/g). Unlike plant oils used in feeding mixtures, fish oil can negatively affect the organoleptic properties of meat [19, 20].

2.3. Specifics of fatty acid profile in conventional and n-3 PUFA-enriched products

Composition of fatty acids in eggs is influenced by many factors, such as genetic background and age of laying hens, housing system, and composition of feed [11, 21–23]. Simopoulos reported that eggs produced outdoors in the Peloponnese contained as much as 20 times more n-3 fatty acids than conventional eggs [24]. Huyghebaert et al. confirmed that feed composition significantly affected the profile of fatty acids in egg yolk [25]. The content of DHA in egg yolk is in positive correlation with the content of ALA, EPA, and DHA contained in feed, but in negative correlation with LA. Bavelaar and Beynen pointed out that EPA contained in egg yolk could be modified through laying hens' feed containing EPA, while DHA in yolk might be increased if the feed was rich in ALA or DHA [26]. **Table 1** overviews the results of the authors' own research referring to enrichment of egg yolk with n-3 PUFA.

Fatty acid	Conventional eggs	n-3 PUFA eggs	
	SFO*	SO**	MO***
ΣSFA[1]	35.34 ± 1.77	34.72 ± 1.13	31.33 ± 0.64
ΣMUFA[2]	41.21 ± 2.03	41.82 ± 2.04	41.86 ± 1.08
Σn-6 PUFA[3]	21.74 ± 1.44	20.85 ± 1.91	19.59 ± 0.83
ALA (C18:3n-3)	0.89 ± 0.18	1.17 ± 0.15	4.73 ± 0.21
ETA (C20:3n3)	0.01 ± 0.005	0.02 ± 0.01	0.03 ± 0.21
EPA (C20:5n-3)	0.01 ± 0.004	0.06 ± 0.02	0.20 ± 0.03
DHA (C22:6n-3)	0.68 ± 0.22	1.35 ± 0.22	2.37 ± 0.18
Σn-3 PUFA	1.68 ± 0.39	2.60 ± 0.18	7.32 ± 0.23
Σn-6 PUFA/Σn-3 PUFA	12.94	8.02	2.67

*SFO sunflower oil.
**SO soybean oil, fish oil.
***MO mixed oil (sunflower oil, soybean oil, linseed oil, fish oil).
[1]ΣSFA: C14:0 + C15:0 + C16:0 + C17:0 + C18:0 + C20:0 + C21:0 + C23:0.
[2]ΣMUFA: C14:1 + C16:1 + C18:1n9t + C18:1n9c + C20:1n9 + C22:1n9.
[3]Σn-6 PUFA: C18:2n6 + C18:3n6 + C20:2n6 + C20:3n6 + C20:4n6 + C22:2n6.

Table 1. Profile of fatty acids in yolk lipids of conventional and n-3 PUFA eggs (% of total fatty acids).

Eggs enriched with n-3 PUFA contain 5.3 times more ALA, 20 times more EPA, and 3.5 times more DHA compared to conventional eggs. The sum of n-3 PUFA in enriched eggs is 4.4 times higher than in conventional eggs. Samman et al. analyzed the profile of fatty acids in conventional table eggs bought in a store and omega-3 eggs [27]. They determined that the ratio of n-6/n-3 PUFA in conventional eggs was 11.03, and in eggs enriched with n-3 PUFA only 2.17. Our researches proved that lipids of omega-3 eggs contain less percentage of SFA and n-6 PUFA, and a higher percentage of ALA, EPA, and DHA than conventional eggs. The n-6/n-3 PUFA ratio in conventional eggs was 12.94 and 8.02, respectively, and in omega-3 eggs it was only 2.67. Many health organizations recommend that the n-6/n-3 PUFA ratio shall range between 3:1 and 10:1. In the USA, it is determined as 15:1, and in Japan that ratio is only 1:1 to 3:1. In Croatia, such ratio is quite wide, from 11:1 to 35:1.

Reference	Diet	ALA	EPA	DHA
		% of total FA		
Eggs				
Škrtić et al. [28]	Sunflower oil 6%	0.97	0.01	1.72
	Sunflower oil 4% + fish oil 2%	1.00	0.24	1.76
Kralik et al. [29]	Soybean oil 5%	1.17	–	0.16
	Rapeseed oil 1.5% + fish oil 3.5%	2.31	0.22	2.64
	Rapeseed oil 3.5% + fish oil 1.5%	1.21	0.10	2.23
Kralik et al. [30]	Linseed oil 1.5% + fish oil 3.5%	3.25	0.25	2.99
	Linseed oil 2.5% + fish oil 2.5%	4.33	0.20	2.35
	Linseed oil 3.5% + fish oil 1.5%	5.18	0.18	2.90
Gül et al. [31]	Soybean oil 2%	0.93	0.04	0.93
	Rapeseed oil 2%	0.51	0.68	0.71
	Rapeseed oil 4%	0.03	0.02	0.02
	Rapeseed oil 6%	0.84	1.15	1.55
Broiler breast meat				
Kralik et al. [17]	Sunflower oil 2.5%+ fish oil 2.5%	3.16	0.79	5.62
	Soybean oil 2.5% + fish oil 2.5%	2.37	0.93	6.44
	Rapeseed oil 2.5% + fish oil 2.5%	2.36	1.32	8.95
	Linseed oil 2.5% + fish oil 2.5%	6.25	1.18	5.66
Salamatdoustnobar [32]	Control	0.72	0.75	0.87
	Rapeseed oil 2%	0.37	1.18	2.03
	Rapeseed oil 4%	0.61	0.62	0.75
Galović et al. [33]	Sunflower oil 5%	1.44	0.12	0.96
	Soybean oil 5%	2.63	0.23	0.89
	Rapeseed oil 5%	2.89	0.22	0.90
	Linseed oil 5%	7.71	0.89	1.85
Gajčević [34]	Linseed oil 6%	7.09	0.77	0.90
	Linseed oil 6% + 0.3% Se	8.51	0.73	0.93
	Linseed oil 6% + 0.5% Se	6.78	0.51	0.84

Table 2. Supplementation of oils to laying hens' diet and its effect on enrichment of eggs and breast meat with n-3 PUFA.

The data presented in **Table 2** include the efficiency of enriching yolk lipids and broiler breast meat with n-3 PUFA, as reported by various authors. Most authors used feeding treatment with sunflower or soybean oil in the control groups, and for the purpose of enriching eggs with n-3 PUFA, those authors used rapeseed, linseed, and fish oils, as well as their combinations.

Results of their research showed that the most efficient enrichment of eggs with n-3 PUFA was achieved by supplementing fish oil to laying hens' diet, as well as by combining fish, rapeseed, and soybean oils. Combination of sunflower oil and fish oil was less effective in the deposition of n-3 PUFA in yolk lipids, if compared to treatments with a combination of fish oil and other plant oils. Enrichment of broiler breast meat with the n-3 PUFA was also more successful when supplementing fish oil to diets. The best deposition of EPA and DHA in breast meat (1.32 and 8.95%, respectively) was achieved in feeding treatment with 2.5% fish oil and 2.5% rapeseed oil, thus achieving 1.67 times more EPA and 1.59 times more DHA than in feeding treatment with a combination of fish oil and sunflower oil [17].

When enriching products with EPA and DHA by using particular plant oils: sunflower, soybean, rapeseed, and linseed oils supplemented in the amount of 5% in feed, and the best efficiency was proven with linseed oil, which achieved 0.89% EPA and 1.85% DHA in muscle lipids, being 7.41 and 1.92 times more than in the control group with sunflower oil [33]. Selenium supplemented to broiler feed in the amount of 3 and 6% did not have influence on enriching breast meat with n-3 PUFA.

Soybean and rapeseed oils are rich in monounsaturated fatty acids (MUFA >65%). Linseed oil is rich in n-6 polyunsaturated fatty acids (n-6 PUFA >37%) and n-3 polyunsaturated fatty acids (n-3 PUFA >28% α-linolenic acid). Fish oil is rich mostly in saturated fatty acids (39.7%) and n-3 PUFA (>31%). Our research confirmed that it was more efficient to use a combination of soybean, linseed, rapeseed, and fish oils as supplement to laying hens' diet than pure soybean oil. Results of our own research as well as of the abovementioned authors confirmed that the modification of poultry diets could influence the deposition of desirable n-3 PUFA in lipids of egg yolks and broiler meat.

3. Health consequences of n3- and n6-PUFA consumption

3.1. The fate of PUFA in human organism

Linoleic acid and alpha linolenic acid belong to the n-6 (omega-6) and n-3 (omega-3) series of polyunsaturated fatty acids, respectively. They are defined "essential" fatty acids since they are not synthesized in the human body and are mostly obtained from the diet. The best food sources of ALA and LA are the most vegetable oils, cereals and walnuts, fish meat, and fish oil. The adequate intake (AI) determined by the Food and Drug Administration (FDA) is for α-linolenic acid 1.6 g/day for men and 1.1 g/day for women, while the acceptable macronutrient distribution range (AMDR) is 0.6–1.2% of total energy [35]. The FDA has recommended that adults can safely consume a total of 3 g/day of combined DHA and EPA, with no more than 2 g/day coming from dietary supplements [36]. Linoleic acid (18:2, n-6), the shortest-chained omega-6 fatty acid, is one of many essential fatty acids. Mammalian cells lack the enzyme omega-3 desaturase and therefore cannot convert omega-6 fatty acids to omega-3 fatty acids. This outlines the importance of the proportion of omega-3 to omega-6 fatty acids in a diet [35, 36]. Omega-6 fatty acids are precursors to endocannabinoids, lipoxins, and specific eicosanoids [6, 7].

http://dx.doi.org/10.5772/67033

Arachidonic acid and EPA are precursors of different classes of pro-inflammatory or anti-inflammatory eicosanoids, respectively (**Figure 2**). AA is found in small amounts in animal food sources (e.g., eggs and meats) and can also be formed by desaturation plus elongation reactions from its precursor, LA. The long-chain n-3 PUFA, DHA and EPA, can be formed in very limited amounts in the human body, or can be consumed preformed in the diet from sources rich in DHA/EPA such as fish meat or fish oils or enriched or fortified functional foods. Dietary n-3 PUFA may counteract the inflammatory effects of AA's eicosanoids in three ways: by counteracting the effects of their AA-derived counterparts via n-3 PUFA derivatives; by displacement, since dietary n-3 PUFA decreases tissue concentrations of AA, thus less of AA-derived eicosanoids is synthesized; and by competitive inhibition with AA for the access to the cyclooxygenase and lipoxygenase enzymes [37]. AA replacement by EPA or DHA n-3 PUFA results in reduced/inhibited production of pro-inflammatory mediators such as prostaglandins, leukotrienes, and lipoxins. EPA and DHA compete with AA for the conversion by cytochrome P450 (CYP) enzymes, resulting in the formation of alternative, physiologically active, metabolites. Renal and hepatic microsomes, as well as various CYP isoforms, displayed equal or elevated activities when metabolizing EPA or DHA instead of AA. CYP2C/2J isoforms converting AA to epoxyeicosatrienoic acids (EETs) preferentially epoxidized the -3 double bond and thereby produced 17,18-epoxyeicosatetraenoic (17,18-EEQ) and 19,20-epoxydocosapentaenoic acid (19,20-EDP) from EPA and DHA. Those -3 epoxides are highly active as antiarrhythmic agents. Moreover, rats given dietary EPA/DHA supplementation exhibited substantial replacement of AA by EPA and DHA in membrane phospholipids in plasma, heart, kidney, liver, lung, and pancreas, with less pronounced changes in the brain [38]. The metabolic pathways competition of n3-PUFA and n6-PUFA is schematically represented in **Figure 2**.

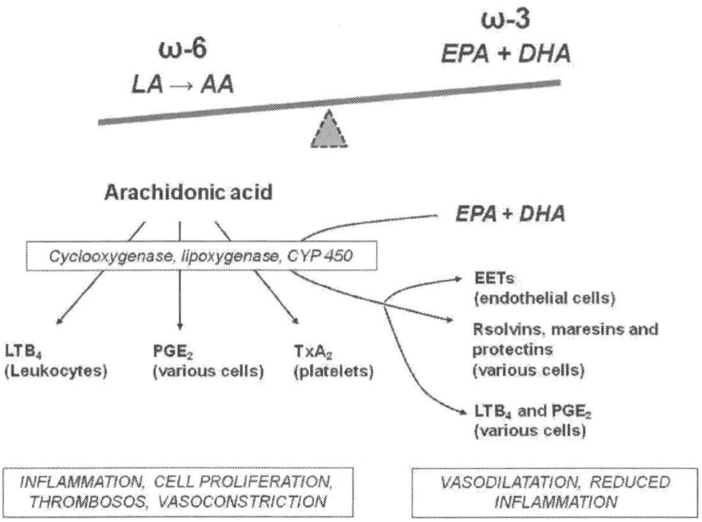

Figure 2. n3-PUFA and n6-PUFA metabolic pathways competition.

n-3 fatty acids, ALA, EPA, and DHA are especially important for good condition of heart and blood vessels, as well as for the prevention of diabetes and certain types of cancer [39]. Connor states that n-3 PUFA prevents heart diseases by preventing the occurrence of arrhythmia [40]. They have anti-inflammatory and hypolipidemic properties, they act antithrombotic, and slow down the development of atherosclerosis. Moreover, they also have a beneficial effect on digestion, improve the immune system, and reduce occurrence of allergic diseases [41]. DHA is an essential element in phospholipids of cell membranes, particularly in brain and eye retina. It is necessary for proper development and function of these organs, especially in fetuses and infants [42]. The anti-inflammatory effects of n-3 fatty acids through reduced production of pro-inflammatory mediators include reduced/inhibited leukocyte chemotaxis, adhesion molecule expression, and leukocyte-endothelial interactions. In addition, among the products of omega-3 fatty acid metabolism are the resolvins, maresins, and protectins [43, 44] which have an indispensable role in the contraction of inflammation [45].

Both n-3 and n-6 PUFA are essential in the diet; however, their ratio affects the ratio of produced pro-inflammatory and anti-inflammatory metabolites. Healthy ratios of n-6:n-3, according to some authors, is the ratio of n6 to n3 of 1:1 to 1:4 (an individual needs more omega-3 than omega-6) [46]. Typical Western diets provide ratios of n6:n3 PUFA between 10:1 and 30:1 [47].

In human, after digestion in the small intestine and transport to the blood, the n-6 and n-3 PUFAs are assimilated within tissues themselves through the body. They can be used in energy metabolism by beta-oxidation to form ATPEFAs and can also undergo esterification into cellular lipids including triglyceride, cholesterol ester, and phospholipid or can be stored in the form of triglycerides and released later by enzymatic/hydrolytic processes. EFAs can also be temporarily stored as cholesterol ester and also released to be utilized in energy metabolism. Both n-6 and n-3 PUFAs in the form of phospholipids are particularly important as they maintain both the structural integrity and the critical functioning of cellular membranes throughout the body. In addition, LA and ALA are activated to high-energy forms known as fatty-acyl CoA which provides the conversion of these dietary PUFAs into their longer-chain and more polyunsaturated products as derived by a series of desaturation plus elongation reactions which are particularly active in the liver and to a lesser extent in other tissues [48].

3.2. Metabolic effects of n-3 PUFA consumption on triacylglycerol and very low-density lipoprotein

It is known that n-3 PUFA fatty acids reduce triacylglycerol (TG) synthesis in the liver and increase very low-density lipoprotein (VLDL) clearance in the peripheral circulation. This occurs because n-3 PUFA inhibit diacylglycerol acyl transferase (DGAT), and phosphatidic acid phosphohydrolase (PA), two crucial enzymes involved in hepatic TG biosynthesis which results in decreased hepatic VLDL secretion. Furthermore, the availability of FAs for TG synthesis is decreased because of increased peroxisomal beta-oxidation of FA. Finally, due to the action of lipoprotein lipase, there is an increased plasma lipolytic activity in the peripheral circulation [49].

3.3. Anti-inflammatory effects of n-3 PUFA in chronic cardiometabolic diseases

n-3 PUFAs have numerous positive effects on immune response [50]. They are components of the plasma membrane and as such are important for cell permeability, fluidity, and flexibility [51]. Increased intake of fish oil or n-3 PUFA supplementations due to its anti-inflammatory function has beneficial effects on cardiovascular disease (CVD), metabolic syndrome, diabetes mellitus, and other diseases [52, 53].

3.4. Anti-inflammatory effect of n-3 PUFA in adiposity and glucose metabolism

Adipocytes play an important endocrine role regulating metabolism, and immune response by secreting adipokines [54]. Adipocyte in healthy subjects maintains the balance between pro- and anti-inflammatory adipokines, but in obesity secretion of inflammatory adipokines is shifted to pro-inflammatory cytokines [55], which may contribute to the pathogenesis of metabolic disorders. Accumulation of triacylglyceride in adipocytes results in adipocyte hypertrophy and dysregulation in secreting bioactive components. Obese patients are a high-risk population for developing diabetes mellitus and cardiovascular complications because they become insulin insensitive, have higher blood pressure, and heart rate (HR). It has been shown that obesity-related metabolic disorders originate from a low-grade inflammation [56].

Macrophage, lymphocyte, adipose stem cells, and preadipocytes that residue in adipose tissue also contribute to increased secretion of pro-inflammatory cytokines such as monocyte chemotactic protein (MCP)-1, IL-8, IL-6, IL-1, and tumor necrosis factor alpha (TNF-α) [57].

Anti-inflammatory effects of n-3 PUFAs have a protective effect and decrease the pro-inflammatory action of adiponectin [58, 59], as a result of the activation of AMP-activated protein kinase [60], which further regulates carbohydrate metabolism [61], and reduce the risk of developing cardiovascular diseases [62]. Although mechanisms involved in anti-inflammatory effect of n-3 PUFA are poorly understood, G protein-coupled receptor 120 (GPR120) is highly expressed on adipocytes and pro-inflammatory macrophage serves as an n-3 PUFA receptor. In mice fed with high-fat diet, supplemented with n-3 PUFA, inflammation was decreased (lower levels of TNF-α and IL-6) and systemic insulin sensitivity was enhanced, while in GPR120 knockout mice these effects were not observed [63]. They also showed that β-arrestin2 and GPR120 signaling induce the inhibition of TAB-mediated activation of transforming growth factor-β activated kinase 1 (TAK1) and inhibit toll-like receptor2/3/4 (TLR) and TNF-α pro-inflammatory-signaling pathway.

Increased intake of n-3 PUFA increases DHA and EPA in immune cells of experimental animals and human subjects [64, 65]. Since immune cells integrate more n-3 PUFAs, there is a decrease in AA content, and therefore a drop of pro-inflammatory eicosanoids secretion [50, 66]. Recent *in vitro* and *in vivo* studies show that the anti-inflammatory effects of n-3 PUFA are mediated through the inhibition of NF-κB-signaling pathway and decreased macrophage TNF-α transcription [67, 68]. Macrophages stimulated with LPS in n-3 PUFA-enriched medium significantly decreased serine 32 phosphorylation [69]. Without proper phosphorylation, NF-κB remains in the cytoplasm, inactively coupled with IκB, and in these conditions, the inflammatory response is downregulated or missing [70]. The fatty acid can also act as a ligand for peroxisome proliferator-activated receptors—PPARα and PPARγ [71]. Specifically, 8(S)-HETE and 15d-J2-PUFA metabolites are PPARs potent selective activators. PPAR receptors are a group of transcription factors regulating energy homeostasis [72] and inflammation and immunity directly inhibiting NF-κB and its downstream effects [73]. All together, these studies prove n-3 PUFA to be a potent anti-inflammatory compounds.

Besides the anti-inflammatory effect of the n-3 PUFAs, study of Mori et al. showed that the incorporation of fish into a low-fat, energy-restricted diet has decreased triglyceride level, insulin-glucose metabolism [74], and thereafter reducing the risk of developing metabolic disorders. In addition, n-3 PUFA improves glucose tolerance and insulin sensitivity in mice models of type-2 diabetes and metabolic syndrome [75]. Rats fed with high-fat diet supplemented with n-3 PUFAs showed increased insulin receptor (IR) density and increased IR and IRS1 phosphorylation, phosphatidylinositol (PI) 3′-kinase activity, and GLUT-4 content in muscles, but rats show no beneficial effect on hyperglycemia and hyperinsulinemia, indicating important role of liver in glucose metabolism [76]. Documented data seen in animal models were not always translated to human subjects [77]. Some studies show beneficial effects of n-3 PUFA on glucose metabolism, and others not. Mostad et al. showed that n-3 PUFA supplementation in type-2 diabetic and obese patients did not improve insulin sensitivity, although those patients had improved lipid metabolism [78]. On the contrary, Albert et al. showed that higher n-3 PUFA concentrations were associated with improved insulin sensitivity, lower free fatty acid, and C-reactive protein (CRP) level in a group of middle-aged overweight men [79]. Another study in women patients with type-2 diabetes showed that in 2 months of n-3 PUFA supplementation, they had decreased adiposity, significantly lower plasma triacylglycerol, but without changes in insulin sensitivity [80]. These opposing results in human insulin sensitivity may be explained by different phenotypes, sex, age, adiposity, and environmental factors of patients, but also by different n-3 PUFA dosage in studies [81].

n-3 PUFA effects on human and animals are dose and tissue dependent [82]. n-3 PUFA incorporates in the plasma membrane, and it binds to receptors as a ligand and modulates gene expression for immune and metabolic function, and by these decrease risk for cardiovascular disease. n-3 PUFA-enriched membranes have changed membrane fluidity and biophysics of lipid rafts affecting protein function and signaling events [53]. For example, n-3 PUFAs modulate the function of Na^{+} and L-type Ca^{2+} membrane ion channels, to prevent arrhythmias [83]. Altered channel function by n-3 PUFA leads to reduced myocyte excitability and cytosolic calcium fluctuation of ischemic myocardium myocyte which becomes susceptible to partial depolarization (resting inactivation) and prevents arrhythmia, while membrane potential of myocytes in the nonischemic myocardium is not drastically affected [84]. Beneficial and protective effects of n-3 PUFA are supported by many scientific studies, without notable side effect [53].

3.5. Effectiveness of n-3 polyunsaturated fatty acids in the improvement of endothelial function and improvement of elasticity of the vascular wall

A large empirical data indicate that the consumption of n-3 PUFA has beneficial effect on the risk and progression of cardiovascular diseases acting via multiple pathways and molecular mechanisms [53, 85, 86]. Since atherosclerosis is one of the main features of CVDs characterized by morphological and functional changes in blood vessel wall and its endothelium, attention has been given to numerous studies to investigate whether n-3 PUFA may prevent or delay atherosclerosis progression acting on the initial steps in its pathogenesis—endothelium and vascular wall function.

Endothelium plays a critical role in maintaining vascular tone and the term "endothelial function" is commonly used to describe its ability to release vasoactive substances, thereby regulating the blood flow [87]. Classically, endothelial dysfunction (ED) refers to reduced production and/or bioavailability of the main vasodilator nitric oxide (NO) and/or an imbalance in the relative contribution of other endothelium-derived relaxing (e.g. cyclooxygenase-1 and -2 (COX-1,2) or CYP450-epoxygenase-derived metabolites) and contracting (e.g. COX-1,-2 or CYP450-hydroxylase-derived metabolites) metabolites of AA, resulting in impaired vascular relaxation mechanisms [88]. It is considered that increased oxidative stress level is one of the main causes for ED and the development in various pathological states associated with vascular diseases such as hypertension, diabetes mellitus, hypercholesterolemia, smoking, and aging. A number of studies have shown that there is a cross-talk between the enzymes producing the vasoactive metabolites (NOS, COX-1,-2, CYP450) and reactive oxygen species (ROS), in which ROS may affect the bioavailability of NO and/or affecting other enzymes to shift their production from vasodilators to vasoconstrictors [89]. ED becomes an accepted prognostic value for future cardiovascular events in both populations at low and high cardiovascular risk and its noninvasive assessment by flow-mediated dilation (FMD) of brachial artery (gold standard) is being routinely used not only in research but in clinical practice, as well [90].

As elaborated previously, the mechanism by which n-3 PUFA may influence endothelial function is its ability to incorporate into membrane phospholipids in which signaling molecules and receptors for endothelial cell function are located [91]. Some of the possible pathways activated in this way result in increased NO production and reduced synthesis of pro-inflammatory mediators [92]. Enhanced eNOS activity/expression by n-3 PUFA administration was demonstrated in several endothelial cell cultures or experimental animal studies [93–95]. In addition, n-3 PUFAs increase NO production by directly stimulating eNOS gene and protein expression, which was reported by several studies in healthy and disease animals including atherosclerosis, diabetes mellitus, and menopause [96–102].

Taken together, these results strongly suggested that n-3 PUFA increases the bioavailability of NO acting via different molecular mechanisms. Despite that high doses of n-3 PUFA have been considered as to have a pro-oxidant effect, several studies on cell cultures and isolated blood vessels have shown that n-3 PUFA may reduce the oxidative stress level by attenuating ROS production via its direct effect on ROS formation, or reducing peroxynitrite production [97, 102, 103]. Both in vitro and in vivo experiments have demonstrated that n-3 PUFAs reduce the concentration of soluble cell adhesion molecules (sCAMs) VCAM-1 and E-selectin, as well as IL-6 and C reactive protein level resulting in the attenuation of cellular and systemic inflammation [104, 105]. It is important to emphasize that relatively high dose of n-3 PUFAs is needed to achieve this anti-inflammatory effect. Interestingly, high doses of n-3 PUFA significantly reduce triglycerides level, which indirectly also contributes to improved endothelial function in these conditions [86]. Taken together, these data suggest that n-3 PUFA has the potential to improve endothelial function by acting on the bioavailability of NO by various mechanisms, reducing oxidative stress and inflammation and thereby reducing pathological activation of the endothelium. The results of a number of functional vascular studies have been summarized in several recent meta-analyses; however, the conclusions of these meta-analyses have been a bit inconsistent. There are a few studies, both in animals and in humans which aimed to distinct the effect of n-3 PUFA on endothelium-independent vasodilation, as well, and whose results suggest that the effect of n-3 PUFA on endothelium-independent vasodilation (contribution of vascular smooth muscle cells) is negligible, as demonstrated in the meta-analysis by Wang et al. [106]. One of the main shortcomings of these functional studies was the lack of basal measurement of n-3 PUFA in the studied population. Another lack of mentioned studies is significant heterogeneity in the number of participants, inclusion criteria such as age of participants or whether participants were healthy or disease, markers of endothelial function that were measured, dose and duration of n-3 PUFA supplementation, forms of n-3 PUFA that were administered (EPA, DHA, or ALA) alone or in combination and concomitant therapy that was used. Because of the abovementioned structure heterogeneity of functional studies, conclusions that indicate that n-3 PUFA improves endothelial function are still adopted with great caution. Meta-analysis of Wang et al. from 2012 identified totally 16 eligible studies which investigated the effect of n-3 PUFA supplementation on endothelial function measured by FMD and involving 901 participants, which reported that n-3 PUFA supplementation significantly increased FMD by 2.30% at a dose range from 0.45 to 4.5 g/day during a median of 56 days. Furthermore, results of this meta-analysis suggested that the effect on n-3 PUFA on endothelial function can be modified by the health status of the participants or by the dose of n-3 PUFA supplementation [106]. A review on human intervention studies by Egert and Stehle reported that n-3 PUFA supplementation improved endothelial function in overweight DM type 2 patients with dyslipidemia; however, conflicting results were observed in CVD patients. The authors concluded that reasons for these discrepancies between studies lie in the heterogeneity in the participants' health status and age, as well as in dose, duration, and the type of n-3 PUFA supplementation [107]. A third large meta-analysis of randomized controlled trials on the fish oil supplementation on endothelial vascular function published in 2012 included 16 studies with 1385 participants involved and reported that fish oil supplementation significantly improved FMD. Furthermore, endothelial function was significantly improved particularly in normoglycemic subjects and participants with lower diastolic blood pressure [108]. But contradictory, sensitivity analysis including only double-blind, placebo-controlled studies indicated that fish oil supplementation has no significant effect on endothelial function. All together, these studies provide many indices that n-3 PUFA supplementation has beneficial effect and improves endothelial function, but large-scale and high-quality clinical trials are needed to evaluate this effect to get a definite conclusion.

Beside impairment of endothelial function, CVDs and atherosclerosis are closely linked to increased arterial wall stiffness which represents progressive deterioration in vessel elasticity [109]. Arterial wall stiffness is characterized by morphological changes in blood vessel wall structure and in mechanical properties of vascular wall, which result in changed functional possibilities of such blood vessels. The arterial stiffness, in addition to the changes in the anatomical structure of the blood vessel wall, is closely related to impaired endothelial function in promoting atherosclerosis development [86]. Therefore, the assessment of arterial stiffness became an accepted predictive factor for future cardiovascular events and mortality in patients with CVDs.

One of the most straightforward and reliable methods for large artery stiffness assessment is noninvasive measurement of pulse wave velocity (PWV), and the most commonly used methods are carotid-femoral PWV and brachial-ankle PWV. PWV presents the speed at what the so-called pulse wave (arterial pulsation produced by the ejection of blood from the heart) propagates from heart to the periphery. Higher PWV is associated with the greater blood vessel wall rigidity that is interpreted as increased arterial wall stiffness [110]. n-3 PUFA administration may influence arterial stiffness acting via passive mechanisms involving mechanical and elastic arterial wall properties, just as via active mechanisms involving cellular and molecular functions of endothelium, VMS, and extracellular matrix of blood vessel wall [86, 109]. It is well known that chronically increased blood pressure levels also increase arterial stiffness by remodeling of the artery wall itself. A large body of evidence indicates that n-3 PUFAs are able to decrease blood pressure level, and therefore act to reduce arterial stiffness as well [111–113]. A second possible link between n-3 PUFA and arterial stiffness is blood triglyceride levels, which are decreased by n-3 PUFA supplementation. Since abnormalities in lipid metabolism are considered to be one of the fundamental determinants for the atherosclerosis developments, its effect on stiffening of the arteries should be taken into account as well [114, 115]. It is considered that n-3 PUFA may act on arterial stiffness by reducing heart rate, since numerous studies in both animal model and humans reported that an increased heart rate is associated with an increased risk for CV events, and is independently associated with the progression of arterial stiffness. It has been speculated that n-3 PUFA may lower HR acting directly on cardiac electrophysiology, or through a modulation of vagal and sympathetic balance [53, 116, 117]. Therefore, beneficial effect of n-3 PUFA on arterial stiffness is multifactorial affecting both passive and active mechanisms relating to the structure and function of the arterial wall, which are very often changed and/or damaged by some major cardiovascular risk factors (such as hypertension, obesity, smoking, menopause, hyperlipidemia, etc.).

Recently, numerous studies tried to investigate the effect of n-3 PUFA supplementation on arterial stiffness in a variety of conditions associated with increased cardiovascular risk in both experimental animals and humans. Regarding results of studies in experimental animals, they have described beneficial effect of n-3 PUFA in animals with insulin resistance, hypertension, and in ovariectomized animals which presented an experimental mode for menopause [118–122]. In the meta-analysis from 2011 on the n-3 PUFA interventions to arterial stiffness which included nine studies (one on the acute effects of n-3 PUFA in healthy volunteers and others on the chronic supplementation in patients with various CVDs), all but one study reported improvement in PWV or capacitive arterial compliance compared to the controls. Furthermore, combined supplementation of EPA and DHA had greater effect on arterial stiffness improvement than that of EPA alone, while one study reported that DHA supplementation alone had no significant effect on arterial stiffness. Just as studies on the effect on n-3 PUFA on endothelial function, the disadvantage of the above mentioned experiments is heterogeneity in the sample population and supplementation dose sizes. Yet, the authors of this meta-analysis pointed out that if the different doses of n-3 PUFA supplementation acted to improve arterial stiffness in diverse populations these findings could be potentially translated to general population [109].

In conclusion, there is a growing evidence that n-3 PUFA supplementation by targeting to endothelial function and vascular wall stiffness may have beneficial effect in preventing the development and progression of atherosclerosis and incidents related to CVDs. So far, we can concisely presume that this benevolent effect of n-3 PUFA on vascular health is a sum of their actions on vasodilator mediators' bioavailability, antioxidant and anti-inflammatory capacity, modulation of lipid profile, and structural arterial remodeling. Still, stronger evidence from large clinical trials with more homogeneous experimental populations and supplementation dose is needed before n-3 PUFAs can find their place in the clinical prevention and treatment of CVDs.

4. Conclusions

The production of functional food enriched with n-3 PUFA (i.e. eggs and poultry meat) is a well-established process and the food is available at the market. Despite inconsistency in designs of the analyzed studies, up to date numerous accumulated data demonstrate beneficial effects of n-3 PUFA for human health, particularly in relation to cardiovascular and metabolic conditions. To achieve long-term protective effects, the functional food enriched with omega-3 fatty acids should be consumed on daily basis. There are no specific constrains in taking functional food; even more, it can be recommended to athletes and cardiovascular patients. General population can also benefit from eating functional food enriched with n-3 PUFA due to their anti-inflammatory and vascular-protective effects.

Author details

Ines Drenjančević[1, 2*], Gordana Kralik[1, 2], Zlata Kralik[1, 2], Martina Mihalj[1, 2], Ana Stupin[1, 2], Sanja Novak[1, 2] and Manuela Grčević[1, 2]

*Address all correspondence to: ines.drenjancevic@mefos.hr

1 Faculty of Medicine Osijek, Josip Juraj Strossmayer University of Osijek, Osijek, Croatia

2 Scientific Center of Excellence for Personalized Health Care University of Osijek, Osijek, Croatia

References

[1] Martirosyan D, Singh J. A new definition of functional food by FFC: what makes a new definition unique? Funct Foods in Health Disease. 2015;5(6):209–223.

[2] Diplock A, Aggett P, Ashwell M, Bornet F, Fern E, Roberfroid M. Scientific concepts of functional foods in Europe. Consensus document. The British Journal of Nutrition. 1999;81(Suppl 1):S1–27.

[3] Pravst, I. Functional Foods in Europe: A Focus on Health Claims. Scientific, Health and Social Aspects of the Food Industry, Valdez, B, editor: InTech; 2012. Available from: http://www.intechopen.com/books/scientific-health-and-social-aspects-of-the-food-industry/functional-foods-in-europe-a-focus-on-health-claims.

[4] Perić L, Rodić V, Milošević N. Production of poultry meat and eggs as functional food: challenges and opportunities. Biotechnology in Animal Husbandry. 2011;27(3):511–20.

[5] Meyer A, Kirsch H, Domergue F, Abbadi A, Sperling P, Bauer J, Cirpus P, Zank TK, Moreau H, Roscoe TJ, Zähringer U, Heinz E. Novel fatty acid elongases and their use for the reconstitution of docosahexaenoic acid biosynthesis. Journal of Lipid Research. 2004; 45(10):1899–1909.

[6] Holman RT. The slow discovery of the importance of omega 3 essential fatty acids in human health. The Journal of Nutrition. 1998;128(2 Suppl):427S–33S.

[7] Infante JP, Huszagh VA. Analysis of the putative role of 24-carbon polyunsaturated fatty acids in the biosynthesis of docosapentaenoic (22:5n-6) and docosahexaenoic (22:6n-3) acids. FEBS Letters. 1998;431(1):1–6.

[8] Barroeta A. Nutritive value of poultry meat: relationship between vitamin E and PUFA. World's Poultry Science Journal. 2007;63:277–84.

[9] Sari M, Aksit M, Ozdoğan M, Basmacioğlu H. Effects of addition of flaxseed of laying hens on some production characteristics, levels of yolk and serum cholesterol, and fatty acid composition of yolk. Archiv für Geflügelkunde. 2002;66(2):75–9.

[10] Valavan S, Mohan B, Chandrasekaran D, Mani K, Mohan B, Edwin S. Effects of various n-3 lipid sources on the quality characteristics and fatty acids composition of chicken egg. EPC 2006 - 12th European Poultry Conference, Verona, Italy, 10–14 September, 2006; Netherlands. Beekbergen: World's Poultry Science Association (WPSA); 2006.

[11] Grobas S, Mendez J, Lazaro R, de Blas C, Mateo GG. Influence of source and percentage of fat added to diet on performance and fatty acid composition of egg yolks of two strains of laying hens. Poultry Science. 2001;80(8):1171–9.

[12] Meluzzi A, Sirri F, Manfreda G, Tallarico N, Franchini A. Effects of dietary vitamin E on the quality of table eggs enriched with n-3 long-chain fatty acids. Poultry Science. 2000;79(4):539–45.

[13] Gonzalez-Esquerra R, Leeson S. Effect of feeding hens regular or deodorized menhaden oil on production parameters, yolk fatty acid profile, and sensory quality of eggs. Poultry Science. 2000;79(11):1597–602.

[14] Imran M, Anjum FM, Nadeem M, Ahmad N, Khan MK, Mushtaq Z, et al. Production of Bio-omega-3 eggs through the supplementation of extruded flaxseed meal in hen diet. Lipids in Health and Disease. 2015;14:126.

[15] Kralik G, Škrtić Z, Bogut I, Gajčević Z, Hanžek D. Effect of preparation rich in n-3 PUFA on yolk fatty acid profile. 1st Mediterranean Summit of WPSA "Advances and Challenges in Poultry Science", Tserveni-Goussi A, Yannnakopoulos A, Fortomaris P, Arsenos G, Sossidou E, editors; 07-10 May 2008; Porto Carras, Chalkidiki, Greece, Book of Proceedings, 663–668; 2008.

[16] Kralik G, Gajčević Z, Škrtić Z, Ivanković S, Hanžek D, editors. Fatty acid contents in chicken thigh muscles as affected by vegetable oil. 1st Mediterranean Summit of WPSA "Advances and Challenges in Poultry Science", 07–10 May 2008; 2008; Porto Carras, Chalkidiki, Greece: Book of Proceedings.

[17] Kralik G, Petrak T, Gajčević Z, Hanžek D, editors. Influence of vegetable oil on fatty acid profile of chicken meat. Proceeding of 53rd International Congress of Meat Science and Technology 5–10 Aug 2007; 2007; Beijing, China.

[18] Zuidhof MJ, Betti M, Korver DR, Hernandez FI, Schneider BL, Carney VL, et al. Omega-3-enriched broiler meat: 1. Optimization of a production system. Poultry Science. 2009;88(5):1108–20.

[19] Mirghelenj SA, Golian A, Taghizadeh V. Enrichment of chicken meat with long chain omega-3 fatty acids through dietary fish oil. Research Journal of Biological Science. 2009;4(5):604–8.

[20] Lopez-Ferrer S, Baucells MD, Barroeta AC, Grashorn MA. n-3 enrichment of chicken meat. 1. Use of very long-chain fatty acids in chicken diets and their influence on meat quality: fish oil. Poultry Science. 2001;80(6):741–52.

[21] Ahn DU, Kim SM, Shu H. Effect of egg size and strain and age of hens on the solids content of chicken eggs. Poultry Science. 1997;76(6):914–9.

[22] Nielsen H. Hen age and fatty acid composition of egg yolk lipid. British Poultry Science. 1998;39(1):53–6.

[23] Yannakopoulos A, Tserveni-Gousi A, Yannakakis S, Yamoustaris A, editors. Yolk fatty acid composition of ω-3 eggs during the laying period. XIth European Symposium on the Quality of Eggs and Egg Products; 2005; Doorwerth, The Netherlands, 23–26 May 2005.

[24] Simopoulos AP. Human requirement for N-3 polyunsaturated fatty acids. Poultry Science. 2000;79(7):961–70.

[25] Huyghebaert G, Raes K, Maertens L, Arnouts S, Delezie E, editors. The interactive impact of dietary PUFA on the deposition of DHA in the egg yolk. Proceedings of the 16th European Symposium on Poultry Nutrition; 2007; Strasbourg, France.

[26] Bavelaar FJ, Beynen AC. Relationships between the intake of n-3 polyunsaturated fatty acids by hens and the fatty acid composition of their eggs. International Journal of Poultry Science. 2004;3:690–6.

[27] Samman S, FP. K, Carter L, Foster M, Ahmad Z, Phuyal J, et al. Fatty acid composition of certified organic, conventional and omega-3 eggs. Food Chemistry. 2009;116(4):911–4 ref.34.

[28] Škrtić Z, Kralik G, Gajčević Z, Hanžek D. Effect of different source of oils on fatty acid profile and organoleptic traits of eggs. Acta agriculturae Slovenica. 2008; 92(2):213–217.

[29] Kralik G, Škrtić Z, Suchy P, Strahova E, Gajčević Z: Feeding fish oil and linseed oil to laying hens to increase the n-3 PUFA of egg yolk. Acta Veterinaria. 2008;77(4):561–568.

[30] Kralik G, Gajčević Z, Škrtić Z. The effect of different oil supplementations on laying performance and fatty acid composition of egg yolk. Italian Journal of Animal Science. 2008;7:173–183.

[31] Gül M, Yöruk MA, Aksu T, Kaya A, Kaynar Ö. The effect of Different levels of canola oil on performance, egg shell quality and fatty acid composition of laying hens. International Journal of Poultry Science. 2012;11(12):769–776.

[32] Salamatdoustnobar R, Nazeradl K, Gorbani A, Shahriar A, Fouladi P. Effect of canola oil on saturated fatty acids content in broiler meat. Journal of Animal and Veterinary Advances. 2007;6(10):1204–1208.

[33] Galović D, Kralik G, Škrtić Z, Bogut I. Fattening traits and chicken meat quality regarding the dietary oil sources. Krmiva 2009;51(6):327–33.

[34] Gajčević Z. Effect of selenium and linseed oil as dietary supplements on broiler performance and fatty acid profile in muscle tissue. Doctoral dissertation, Josip Juraj Strossmayer University of Osijek, Faculty of Agriculture in Osijek, 2011; pp. 1–124.

[35] Food and Nutrition Board Dietary Reference Intakes For Energy, Carbohydrate, Fiber, Fat, Fatty Acids, Cholesterol, Protein, and Amino Acids. Washington, D.C: Institute of Medicine of the National Academies; 2005.

[36] Bent S, Bertoglio K, Hendren RL. Omega-3 fatty acids for autistic spectrum disorder: a systematic review. Journal of Autism and Developmental Disorders. 2009;39(8):1145–54.

[37] Weylandt KH, Serini S, Chen YQ, Su HM, Lim K, Cittadini A, et al. Omega-3 polyunsaturated fatty acids: the way forward in times of mixed evidence. BioMed Research International. 2015;2015:143109.

[38] Arnold C, Markovic M, Blossey K, Wallukat G, Fischer R, Dechend R, et al. Arachidonic acid-metabolizing cytochrome P450 enzymes are targets of {omega}-3 fatty acids. The Journal of Biological Chemistry. 2010;285(43):32720–33.

[39] Barbalho SM, Goulart de Alvares R, Quesada K, Bechara MD, de Cassio Alves de Carvalho A. Inflammatory bowel disease: can omega-3 fatty acids really help? Annals of Gastroenterology. 2016;29(1):37–43.

[40] Connor W. n-3 Fatty acids and heart disease. Kritchevsky D, Carroll K, editors. Champaign, IL: American Oil Chemists' Society; 1994.

[41] N. V, Czegledi L, Javor A. Significance of functional foods of animal origin in human health. Lucrări stiińifice Zootehniesi Biotehnologii, Timisoara. 2008;41(2):263–8.

[42] Hasler CM. Functional foods: benefits, concerns and challenges—a position paper from the American Council on Science and Health. The Journal of Nutrition. 2002;132(12):3772–81.

[43] Calder PC. Polyunsaturated fatty acids and inflammatory processes: New twists in an old tale. Biochimie. 2009;91(6):791–5.

[44] Calder PC. Marine omega-3 fatty acids and inflammatory processes: Effects, mechanisms and clinical relevance. Biochimica et Biophysica Acta. 2015;1851(4):469–84.

[45] Barbalho SM, Goulart RdA, Quesada K, Bechara MD, Carvalho AdCAd. Inflammatory bowel disease: can omega-3 fatty acids really help? 2016.

[46] Lands WEM. Fish, Omega 3 and human health. Champaign, IL: AOCS Press; 2005.

[47] Hibbeln JR, Nieminen LR, Blasbalg TL, Riggs JA, Lands WE. Healthy intakes of n-3 and n-6 fatty acids: estimations considering worldwide diversity. The American Journal of Clinical Nutrition. 2006;83(6 Suppl):1483S–93S.

[48] Available from: http://www.dhaomega3.org/Overview Metabolism-of-Omega-6-and-Omega-3-Fatty-Acids-and-the-Omega-6Omega-3-Ratio.

[49] Jacobson TA. Role of n-3 fatty acids in the treatment of hypertriglyceridemia and cardiovascular disease. The American Journal of Clinical Nutrition. 2008;87(6):1981S–90S.

[50] Calder PC. n-3 fatty acids, inflammation and immunity: new mechanisms to explain old actions. The Proceedings of the Nutrition Society. 2013;72(3):326–36.

[51] Stillwell W, Wassall SR. Docosahexaenoic acid: membrane properties of a unique fatty acid. Chemistry and Physics of Lipids. 2003;126(1):1–27.

[52] Lalia AZ, Lanza IR. Insulin-sensitizing effects of omega-3 fatty acids: lost in translation? Nutrients. 2016;8(6).

[53] Mozaffarian D, Wu JH. Omega-3 fatty acids and cardiovascular disease: effects on risk factors, molecular pathways, and clinical events. Journal of the American College of Cardiology. 2011;58(20):2047–67.

[54] Conde J, Scotece M, Gomez R, Lopez V, Gomez-Reino JJ, Lago F, et al. Adipokines: biofactors from white adipose tissue. A complex hub among inflammation, metabolism, and immunity. BioFactors. 2011;37(6):413–20.

[55] Huang F, Del-Rio-Navarro BE, Perez-Ontiveros JA, Ruiz-Bedolla E, Saucedo-Ramirez OJ, Villafana S, et al. Effect of six-month lifestyle intervention on adiponectin, resistin and soluble tumor necrosis factor-alpha receptors in obese adolescents. Endocrine Journal. 2014;61(9):921–31.

[56] Das UN. Is obesity an inflammatory condition? Nutrition. 2001;17(11-12):953–66.

[57] Kalupahana NS, Claycombe KJ, Moustaid-Moussa N. (n-3) Fatty acids alleviate adipose tissue inflammation and insulin resistance: mechanistic insights. Advances in Nutrition. 2011;2(4):304–16.

[58] Flachs P, Mohamed-Ali V, Horakova O, Rossmeisl M, Hosseinzadeh-Attar MJ, Hensler M, et al. Polyunsaturated fatty acids of marine origin induce adiponectin in mice fed a high-fat diet. Diabetologia. 2006;49(2):394–7.

[59] Neschen S, Morino K, Rossbacher JC, Pongratz RL, Cline GW, Sono S, et al. Fish oil regulates adiponectin secretion by a peroxisome proliferator-activated receptor-gamma-dependent mechanism in mice. Diabetes. 2006;55(4):924–8.

[60] Yamauchi T, Kamon J, Minokoshi Y, Ito Y, Waki H, Uchida S, et al. Adiponectin stimulates glucose utilization and fatty-acid oxidation by activating AMP-activated protein kinase. Nature Medicine. 2002;8(11):1288–95.

[61] Tsitouras PD, Gucciardo F, Salbe AD, Heward C, Harman SM. High omega-3 fat intake improves insulin sensitivity and reduces CRP and IL6, but does not affect other endocrine axes in healthy older adults. Hormone and metabolic research = Hormon- und Stoffwechselforschung = Hormones et metabolisme. 2008;40(3):199–205.

[62] Stone NJ. Fish consumption, fish oil, lipids, and coronary heart disease. Circulation. 1996;94(9):2337–40.

[63] Oh DY, Talukdar S, Bae EJ, Imamura T, Morinaga H, Fan W, et al. GPR120 is an omega-3 fatty acid receptor mediating potent anti-inflammatory and insulin-sensitizing effects. Cell. 2010;142(5):687–98.

[64] Brouard C, Pascaud M. Effects of moderate dietary supplementations with n-3 fatty acids on macrophage and lymphocyte phospholipids and macrophage eicosanoid synthesis in the rat. Biochimica et Biophysica Acta. 1990;1047(1):19–28.

[65] Healy DA, Wallace FA, Miles EA, Calder PC, Newsholm P. Effect of low-to-moderate amounts of dietary fish oil on neutrophil lipid composition and function. Lipids. 2000;35(7):763–8.

[66] Peterson LD, Jeffery NM, Thies F, Sanderson P, Newsholme EA, Calder PC. Eicosapentaenoic and docosahexaenoic acids alter rat spleen leukocyte fatty acid composition and prostaglandin E2 production but have different effects on lymphocyte functions and cell-mediated immunity. Lipids. 1998;33(2):171–80.

[67] Li Z, Ren W, Han X, Liu X, Wang G, Zhang M, et al. omega3-polyunsaturated fatty acids suppress lipoprotein-associated phospholipase A2 expression in macrophages and animal models. Molecular Nutrition & Food Research. 2015;59(9):1771–9.

[68] Lo CJ, Chiu KC, Fu M, Lo R, Helton S. Fish oil decreases macrophage tumor necrosis factor gene transcription by altering the NF kappa B activity. The Journal of Surgical Research. 1999;82(2):216–21.

[69] Novak TE, Babcock TA, Jho DH, Helton WS, Espat NJ. NF-kappa B inhibition by omega -3 fatty acids modulates LPS-stimulated macrophage TNF-alpha transcription. American Journal of Physiology Lung Cellular and Molecular Physiology. 2003;284(1):L84–9.

[70] Sigal LH. Basic science for the clinician 39: NF-kappaB-function, activation, control, and consequences. Journal of Clinical Rheumatology: Practical Reports on Rheumatic & Musculoskeletal Diseases. 2006;12(4):207–11.

[71] Kliewer SA, Sundseth SS, Jones SA, Brown PJ, Wisely GB, Koble CS, et al. Fatty acids and eicosanoids regulate gene expression through direct interactions with peroxisome proliferator-activated receptors alpha and gamma. Proceedings of the National Academy of Sciences of the United States of America. 1997;94(9):4318–23.

[72] Lefebvre P, Chinetti G, Fruchart JC, Staels B. Sorting out the roles of PPAR alpha in energy metabolism and vascular homeostasis. The Journal of Clinical Investigation. 2006;116(3):571–80.

[73] Clark RB. The role of PPARs in inflammation and immunity. Journal of Leukocyte Biology. 2002;71(3):388–400.

[74] Mori TA, Bao DQ, Burke V, Puddey IB, Watts GF, Beilin LJ. Dietary fish as a major component of a weight-loss diet: effect on serum lipids, glucose, and insulin metabolism in overweight hypertensive subjects. The American Journal of Clinical Nutrition. 1999;70(5):817–25.

[75] Lombardo YB, Chicco AG. Effects of dietary polyunsaturated n-3 fatty acids on dyslipidemia and insulin resistance in rodents and humans. A review. The Journal of Nutritional Biochemistry. 2006;17(1):1–13.

[76] Taouis M, Dagou C, Ster C, Durand G, Pinault M, Delarue J. N-3 polyunsaturated fatty acids prevent the defect of insulin receptor signaling in muscle. American Journal of Physiology Endocrinology and Metabolism. 2002;282(3):E664–71.

[77] Lorente-Cebrian S, Costa AG, Navas-Carretero S, Zabala M, Martinez JA, Moreno-Aliaga MJ. Role of omega-3 fatty acids in obesity, metabolic syndrome, and cardiovascular diseases: a review of the evidence. Journal of Physiology and Biochemistry. 2013;69(3):633–51.

[78] Mostad IL, Bjerve KS, Bjorgaas MR, Lydersen S, Grill V. Effects of n-3 fatty acids in subjects with type 2 diabetes: reduction of insulin sensitivity and time-dependent alteration from carbohydrate to fat oxidation. The American Journal of Clinical Nutrition. 2006;84(3):540--50.

[79] Albert BB, Derraik JG, Brennan CM, Biggs JB, Smith GC, Garg ML, et al. Higher omega-3 index is associated with increased insulin sensitivity and more favourable metabolic profile in middle-aged overweight men. Scientific Reports. 2014;4:6697.

[80] Kabir M, Skurnik G, Naour N, Pechtner V, Meugnier E, Rome S, et al. Treatment for 2 mo with n 3 polyunsaturated fatty acids reduces adiposity and some atherogenic factors but does not improve insulin sensitivity in women with type 2 diabetes: a randomized controlled study. The American Journal of Clinical Nutrition. 2007;86(6):1670–9.

[81] Browning LM. n-3 Polyunsaturated fatty acids, inflammation and obesity-related disease. The Proceedings of the Nutrition Society. 2003;62(2):447–53.

[82] Holness MJ, Greenwood GK, Smith ND, Sugden MC. Diabetogenic impact of long-chain omega-3 fatty acids on pancreatic beta-cell function and the regulation of endogenous glucose production. Endocrinology. 2003;144(9):3958–68.

[83] Endo J, Arita M. Cardioprotective mechanism of omega-3 polyunsaturated fatty acids. Journal of Cardiology. 2016;67(1):22–7.

[84] Leaf A, Kang JX, Xiao YF, Billman GE. Clinical prevention of sudden cardiac death by n-3 polyunsaturated fatty acids and mechanism of prevention of arrhythmias by n-3 fish oils. Circulation. 2003;107(21):2646–52.

[85] Sudheendran S, Chang CC, Deckelbaum RJ. N-3 vs. saturated fatty acids: effects on the arterial wall. Prostaglandins, Leukotrienes, and Essential Fatty Acids. 2010;82(4–6):205–9.

[86] Zanetti M, Grillo A, Losurdo P, Panizon E, Mearelli F, Cattin L, et al. Omega-3 polyunsaturated fatty acids: structural and functional effects on the vascular wall. BioMed Research International. 2015;2015:791978.

[87] Ruzic A, Miletic B, Noal A, Persic V, Razov Radas M, Vcev A. Endotelna disfunkcija u "enigmatskoj slagalici" kardiovaskularnih bolesti. Med Glas 2009;6(1):2–15.

[88] Verma S, Anderson TJ. Fundamentals of endothelial function for the clinical cardiologist. Circulation. 2002;105(5):546–9.

[89] Drenjancevic-Peric I, Lombard JH. Introgression of chromosome 13 in Dahl salt-sensitive genetic background restores cerebral vascular relaxation. American Journal of Physiology Heart and Circulatory Physiology. 2004;287(2):H957–62.

[90] Green DJ, Dawson EA, Groenewoud HM, Jones H, Thijssen DH. Is flow-mediated dilation nitric oxide mediated? A meta-analysis. Hypertension. 2014;63(2):376–82.

[91] Williams JJ, Palmer TM. Cavin-1: caveolae-dependent signalling and cardiovascular disease. Biochemical Society Transactions. 2014;42(2):284–8.

[92] Layne J, Majkova Z, Smart EJ, Toborek M, Hennig B. Caveolae: a regulatory platform for nutritional modulation of inflammatory diseases. The Journal of Nutritional Biochemistry. 2011;22(9):807–11.

[93] Omura M, Kobayashi S, Mizukami Y, Mogami K, Todoroki-Ikeda N, Miyake T, et al. Eicosapentaenoic acid (EPA) induces Ca2+-independent activation and translocation of endothelial nitric oxide synthase and endothelium-dependent vasorelaxation. FEBS Letters. 2001;487(3):361–6.

[94] Stebbins CL, Stice JP, Hart CM, Mbai FN, Knowlton AA. Effects of dietary docosahexaenoic acid (DHA) on eNOS in human coronary artery endothelial cells. Journal of Cardiovascular Pharmacology and Therapeutics. 2008;13(4):261–8.

[95] Wu Y, Zhang C, Dong Y, Wang S, Song P, Viollet B, et al. Activation of the AMP-activated protein kinase by eicosapentaenoic acid (EPA, 20:5 n-3) improves endothelial function in vivo. PloS One. 2012;7(4):e35508.

[96] Casos K, Zaragoza MC, Zarkovic N, Zarkovic K, Andrisic L, Portero-Otin M, et al. A fish-oil-rich diet reduces vascular oxidative stress in apoE(-/-) mice. Free Radical Research. 2010;44(7):821–9.

[97] Gortan Cappellari G, Losurdo P, Mazzucco S, Panizon E, Jevnicar M, Macaluso L, et al. Treatment with n-3 polyunsaturated fatty acids reverses endothelial dysfunction and oxidative stress in experimental menopause. The Journal of Nutritional Biochemistry. 2013;24(1):371–9.

[98] Lopez D, Caballero C, Sanchez J, Puig-Parellada P, Mitjavila MT. Free radical production in aortic rings from rats fed a fish oil-rich diet. American Journal of Physiology Heart and Circulatory Physiology. 2001;280(6):H2929–35.

[99] Lopez D, Orta X, Casos K, Saiz MP, Puig-Parellada P, Farriol M, et al. Upregulation of endothelial nitric oxide synthase in rat aorta after ingestion of fish oil-rich diet. American Journal of Physiology Heart and Circulatory Physiology. 2004;287(2):H567–72.

[100] Matsumoto T, Nakayama N, Ishida K, Kobayashi T, Kamata K. Eicosapentaenoic acid improves imbalance between vasodilator and vasoconstrictor actions of endothelium-derived factors in mesenteric arteries from rats at chronic stage of type 2 diabetes. The Journal of Pharmacology and Experimental Therapeutics. 2009;329(1):324–34.

[101] Nyby MD, Matsumoto K, Yamamoto K, Abedi K, Eslami P, Hernandez G, et al. Dietary fish oil prevents vascular dysfunction and oxidative stress in hyperinsulinemic rats. American Journal of Hypertension. 2005;18(2 Pt 1):213–9.

[102] Zhang W, Fu F, Tie R, Liang X, Tian F, Xing W, et al. Alpha-linolenic acid intake prevents endothelial dysfunction in high-fat diet-fed streptozotocin rats and underlying mechanisms. VASA Zeitschrift fur Gefasskrankheiten. 2013;42(6):421–8.

[103] Serini S, Fasano E, Piccioni E, Cittadini AR, Calviello G. Dietary n-3 polyunsaturated fatty acids and the paradox of their health benefits and potential harmful effects. Chemical Research in Toxicology. 2011;24(12):2093–105.

[104] Miles EA, Banerjee T, Calder PC. The influence of different combinations of gamma-linolenic, stearidonic and eicosapentaenoic acids on the fatty acid composition of blood

lipids and mononuclear cells in human volunteers. Prostaglandins, Leukotrienes, and Essential Fatty Acids. 2004;70(6):529–38.

[105] Wang TM, Chen CJ, Lee TS, Chao HY, Wu WH, Hsieh SC, et al. Docosahexaenoic acid attenuates VCAM-1 expression and NF-kappaB activation in TNF-alpha-treated human aortic endothelial cells. The Journal of Nutritional Biochemistry. 2011;22(2):187–94.

[106] Wang Q, Liang X, Wang L, Lu X, Huang J, Cao J, et al. Effect of omega-3 fatty acids supplementation on endothelial function: a meta-analysis of randomized controlled trials. Atherosclerosis. 2012;221(2):536–43.

[107] Egert S, Stehle P. Impact of n-3 fatty acids on endothelial function: results from human interventions studies. Current Opinion in Clinical Nutrition and Metabolic Care. 2011;14(2):121–31.

[108] Xin W, Wei W, Li X. Effect of fish oil supplementation on fasting vascular endothelial function in humans: a meta-analysis of randomized controlled trials. PloS One. 2012;7(9):e46028.

[109] Pase MP, Grima NA, Sarris J. The effects of dietary and nutrient interventions on arterial stiffness: a systematic review. The American Journal of Clinical Nutrition. 2011;93(2):446–54.

[110] Pereira T, Correia C, Cardoso J. Novel methods for pulse wave velocity measurement. Journal of Medical and Biological Engineering. 2015;35(5):555–65.

[111] Benetos A, Laurent S, Asmar RG, Lacolley P. Large artery stiffness in hypertension. Journal of Hypertension Supplement: Official Journal of the International Society of Hypertension. 1997;15(2):S89–97.

[112] Cabo J, Alonso R, Mata P. Omega-3 fatty acids and blood pressure. The British Journal of Nutrition. 2012;107(Suppl 2):S195–200.

[113] Miller PE, Van Elswyk M, Alexander DD. Long-chain omega-3 fatty acids eicosapentaenoic acid and docosahexaenoic acid and blood pressure: a meta-analysis of randomized controlled trials. American Journal of Hypertension. 2014;27(7):885–96.

[114] Salomaa V, Riley W, Kark JD, Nardo C, Folsom AR. Non-insulin-dependent diabetes mellitus and fasting glucose and insulin concentrations are associated with arterial stiffness indexes. The ARIC Study. Atherosclerosis Risk in Communities Study. Circulation. 1995;91(5):1432–43.

[115] Task Force for the management of dyslipidaemias of the European Society of C, the European Atherosclerosis S, Catapano AL, Reiner Z, De Backer G, Graham I, et al. ESC/EAS Guidelines for the management of dyslipidaemias: the Task Force for the management of dyslipidaemias of the European Society of Cardiology (ESC) and the European Atherosclerosis Society (EAS). Atherosclerosis. 2011;217(Suppl 1):S1–44.

[116] McLennan PL. Myocardial membrane fatty acids and the antiarrhythmic actions of dietary fish oil in animal models. Lipids. 2001;36(Suppl):S111–4.

[117] Whelton SP, Blankstein R, Al-Mallah MH, Lima JA, Bluemke DA, Hundley WG, et al. Association of resting heart rate with carotid and aortic arterial stiffness: multi-ethnic study of atherosclerosis. Hypertension. 2013;62(3):477–84.

[118] Engler MM, Engler MB, Pierson DM, Molteni LB, Molteni A. Effects of docosahexaenoic acid on vascular pathology and reactivity in hypertension. Experimental Biology and Medicine. 2003;228(3):299–307.

[119] Losurdo P, Grillo A, Panizon E, Gortan Cappellari G, Fabris B, Bardelli M, et al. Supplementation of omega-3 polyunsaturated fatty acids prevents increase in arterial stiffness after experimental menopause. Journal of Cardiovascular Pharmacology and Therapeutics. 2014;19(1):114–20.

[120] Losurdo P, Grillo A, Panizon E, Zanetti M, Bardelli M, Biolo G, et al. Baroreflex sensitivity and central hemodynamics after omega-3 polyunsaturated fatty acids supplementation in an animal model of menopause. Vascular Pharmacology. 2015;71:65–9.

[121] Robbez Masson V, Lucas A, Gueugneau AM, Macaire JP, Paul JL, Grynberg A, et al. Long-chain (n-3) polyunsaturated fatty acids prevent metabolic and vascular disorders in fructose-fed rats. The Journal of Nutrition. 2008;138(10):1915–22.

[122] Sato M, Katsuki Y, Kanehiro H, Iimura M, Akada Y, Mizota M, et al. Effects of ethyl all-cis-5,8,11,14,17-icosapentaenoate on the physical properties of arterial walls in high cholesterol diet-fed rabbits. Journal of Cardiovascular Pharmacology. 1993;22(1):1–9.

Chapter 12

Evolution and Therapy of Brain by Foods Containing Unsaturated Fatty Acids

Roberto Carlos Burini,

Caroline das Neves Mendes Nunes and

Franz Homero Paganini Burini

Additional information is available at the end of the chapter

http://dx.doi.org/10.5772/67321

Abstract

About 6 million years ago, our ancestors had experienced a tremendous brain growth, widely viewed as a "major adaptive shift" in human evolution. Half of human brain composition is fat and 20% of its dry weight is long-chain polyunsaturated fatty acids (LCPUFA). Consequently, improvements in consumption of dietary fat were necessary condition for promoting encephalization. Dietary fat quantity and quality have been subjected to tremendous change over the past 10,000 years with the introduction of industrially produced *trans* fatty acids and reduced intakes of ω-3 fatty acids. The *absolute human* brain size reached its peak of approximately 90,000 years ago and has decreased by 11% since 35,000 years ago, most of it (8%) coming in the last 10,000 years. The shortfall in consumption of animal foods since the late Paleolithic and mainly consequent shortfall in consumption of preformed LCPUFA would be the plausible hypothesis for the brain size decreasing. Genetically, we are still adapted to the East African ecosystem on which our genome evolved, with some adaptations since the Out-of-Africa Diaspora. Dietary fat quantity and quality change has caused a conflict with our slowly adapting genome and this mismatch is likely to be at the basis of "typically Western" diseases. Many recommendations for the intakes of EPA + DHA have been issued, notably for prevention. However, the ultimate goal might be to return to the fat quality of our ancient diet on which our genes have evolved during the past million years of evolution.

Keywords: human encephalization, LCPUFA sources, dietary transition, LPUFA in health, W-3/W-6 LCFA and modern diseases, therapeutical W-3 LCPUFA

1. Introduction

The evolution of Homo erectus in Africa is widely viewed as a "major adaptive shift" in human evolution. Humans share a common ancestor with the chimpanzee and bonobo that probably lived in East Africa and, since some 6 million years ago our ancestors had experienced a tremendous brain growth and assumed an upright position [1].

Half of brain composition is fat, but the central nervous system contains almost a quarter of the unesterified cholesterol present in the whole human body and, long-chain polyunsaturated fatty acids (LCPUFAs) make up to 20% of brain dry weight, including 6% for arachidonic acid (AA) and 8% for docosahexaenoic acid (DHA) [2].

2. Polyunsaturated fatty acids

LCPUFAs are building blocks of the membrane phospholipids of all cells. Nevertheless, LCPUFAs are not only important structural elements of membranes, together with their highly potent metabolites (prostaglandins, thromboxanes, leukotrienes, resolvins, and (neuro)protectins), LCPUFAs are involved in the functioning of membrane-bound receptors, transporters, ion channels, and enzymes, and also in signal transduction and gene expression.

LCPUFAs are ligands of nuclear transcription factors (PPARS, SREBPs, NF-kB, and others) [3–6] that coordinate expression and repression of key enzymes and proteins participating in intermediary metabolism in glycolysis and *"de novo"* lipogenesis, thermoregulation, energy partitioning, growth and differentiation, hemostasis, and (W-3/W-6 ratio) inflammatory responses [7–11].

3. Effect of diet on brain development

The anatomic trends of human evolution (large body sizes, bigger brains, craniofacial, and intestinal changes) clearly suggest major energetic and dietary shifts [12–14].

Improvements in dietary quality and the increased consumption of dietary fat appear to have been a necessary condition for promoting encephalization in the human lineage [15].

Primitive humans with enlarging brains developed more sophisticated tool technology (including the fire cooking) and became more efficient hunter/gatherers and so gained greater access to more nutritious and easily digestible foods (e.g., fruits, nuts, and meat) [11].

Consequently, reductions of posterior tooth size (and grinding teeth) and, also the size of the face and so, no longer needed the large gastrointestinal tract. Key genetic mutations during later hominid evolution were critical to promoting the enhanced lipid metabolism necessary for subsisting on diets with greater levels of animal material [16].

In fact, associated with the evolution of our high-quality diet, humans developed distinct molecular pathways for detecting and metabolizing high-fat foods [17].

The ability to effectively detect, metabolize, and store fats likely provided tremendous selective advantages to our hominid ancestors, allowing them to expand into diverse ecosystems around the world [18].

Mammalian brain growth is dependent upon sufficient amounts of two LCPUFAs: DHA and AA [19].

It appears that mammals have a limited capacity to synthesize these fatty acids from dietary precursors. Because the composition of all mammalian brain tissue is similar with respect to these two fatty acids, species with higher levels of encephalization have greater requirements for DHA and AA. Consequently, dietary sources of DHA and AA were likely limiting nutrients that constrained the evolution of larger brain size in many mammalian lineages [18].

On average, we consume higher levels of dietary fat than other primates, and much higher levels of key LCPUFAs are critical to brain development [20, 21].

Hominins had experienced a tremendous brain growth which coincided with a change from a vegetarian to a hunting-gathering omnivore-carnivore [22–25].

Greater consumption of animal foods would have increased total dietary fat consumption in early Homo, and markedly increased the levels of key fatty acids (AA and DHA) necessary for brain development. The available evidence seems to best support a mixed dietary strategy in early Homo that involved the consumption of larger amounts of animal foods than with the australopithecines. Brain tissue is a rich source of both AA and DHA, whereas liver and muscle tissues are good sources of AA and moderate sources of DHA [18].

Dietary fat quantity and quality have been subjected to tremendous change over the past 10,000 years. Important changes are the introduction of manufactured linoleic acid (LA), trans-fatty acids, and reduced intakes of vegetal-derived alpha-linolenic acid (ALA) and fish-derived eicosapentaenoic acid (EPA) and DHA, overall leading to a reduced supplying of omega-3 fatty acids [11].

Analysis of changes in brain size in humans over the last 1.8 million years found that encephalization quotient (EQ) began reaching its peak with the first anatomically modern humans of approximately 90,000 years ago and has since remained fairly constant. Most surprisingly, however, absolute brain size has decreased by 11% since 35,000 years ago, with most of this decrease (8%) coming in just the last 10,000 years. Therefore, a genetic mutation is no more likely as an explanation for the decrease in absolute brain size. The most notorious dietary change in the last 10,000 years has been the decreased consumption of animal food (roughly from 50 to 10%) by the adventure of agriculture, followed by an increased consumption of grains. Hence, the most feasible biological hypothesis for the absolute decrease in brain size is the reduction of animal food intake with a consequent reduction of preformed long-chain fatty acids. The brain is dependent on the DHA, docosatetraenoic acid (DTA), and AA to support its growth during the formative years. These are far more plentiful in animal foods than plant. It is possible that the levels of essential fatty acids (EFAs) provided in the prehistoric diets were sufficient to support the brain expansion and evolution from prehistoric times to the present, and the current low levels of EFA intake (provided by agricultural diets) may explain the recent smaller human brain size [26].

4. Transition of diet and the development of brain disease

Dietary fat quantity and quality change have, together with other man-made changes in our environment, caused a conflict with our slowly adapting genome [11].

In fact, the EFA and other changes in our diet together with an energy intake that does not match with our current sedentary lifestyle have caused a conflict with our genome that is likely to be at the basis of typically "Western" diseases and their basis on the conflict (mismatch) caused by our current sedentary-energy rich industrialized diet, way of life with our ancestrally molded genome. The dietary composition of our ancestors has also become clear from our current (patho)physiology: epidemiological data demonstrated a negative association of fish consumption with coronary artery disease (CAD) and (postpartum) depression, while landmark trials with ALA and fish oil in CAD, and with EPA in depression and schizophrenia supported the causality of these relations.

The similarity among diseases currently associated with dietary risk factors adds the notion that there is a common insult originated from our changed environment. These effects hit different organs and systems varying in genetic susceptibility and life stages, but extremely dependent on the doses and exposure time. Low-grade inflammation might be a strong candidate for this common denominator. Low-grade inflammation can be found in metabolic syndrome and its sequelae, some psychiatric and neurodegenerative diseases. The LCPUFA, AA, EPA, and DHA are intimately related to the initiation and resolution of inflammatory responses [11].

The current balance between AA and EPA + DHA is disturbed by the dominance of AA, which originates from the diet or synthesis from LA. Higher ratio of AA/EPA+DHA might led to a pro-inflammatory condition that may precipitate a hyper-inflammatory response ("systemic inflammatory response syndrome-SIRS") with collateral damage, scarring and fibrosis and the subsequent development of an immune paralysis ("compensatory anti-inflammatory response syndrome-CARS") evidenced by debilitated host defense and secondary infection susceptibility [27, 28].

The chronic inflammation resulting from the unbalanced AA/EPA + DHA ratio might be central in the pathogenesis of the diseases of the metabolic syndrome and neurodegenerative disease, explain the relation between inflammation, depression, and dementia [29].

5. Therapy of modern diseases with polyunsaturated fatty acids

Dietary supplementation of LCFA, especially EPA and DHA, has been used during pregnancy or early postnatal life for improvement of fetal and newborn brain development, at primary and secondary CAD preventions and psychiatric diseases. Consensus has been reached that those in CAD and depression are positive but not in all others. LCPω3 supplements might especially be effective in prevention, as suggested by the outcomes of epidemiological studies on CAD and prospective studies on Alzheimer's disease, and also from the favorable effects of LCPω3 in early disease stages [30].

It takes 20 years before the human brain obtains its complex adult configuration but the most dramatic neurodevelopmental changes occur prenatally and early post-natal, including a major transformation in cortical organization 3–4 months after term and, considerable evidence indicates that prenatal and neonatal LCPUFA status is associated with neurodevelopmental outcome. Therefore, maternal and neonatal concentrations of DHA and AA are associated with improved outcomes in early infancy, and concentrations of DHA are associated with favorable neurodevelopmental outcome beyond early infancy [31].

Given the fact that LCPUFA accretion is especially abundant during the third trimester of gestation, it suggests that preterm infants would particularly profit from LCPUFA supplementation. However, studies of LCPUFA supplementation in preterm infants have not shown evidence of a positive effect on neurodevelopmental outcome. On the other hand, studies in full-term infants indicated that DHA supplementation promotes neurodevelopmental outcome in early infancy but no longer positive effects later on, being virtually absent at school age or later. Generally, the literature suggests that LCPUFA supplementation in term infants does not affect outcomes beyond the age of 4 months [31].

It is known that up to 45% of the fatty acids of synaptic membranes are EFAs [32].

There is a well-established positive correlation between depression and coronary artery disease. In fact, epidemiological studies in various countries suggest that decreased ω-3 fatty acid consumption correlates with increasing rates of depression and, adequate long-chain polyunsaturated fatty acids, particularly DHA, may reduce the development of depression just as ω-3 polyunsaturated fatty acids may reduce coronary artery disease [33].

Eight database trials that randomly assigned participants to receive ω-3 PUFAs/fish, with measured depressed mood, using human participants, came to the conclusion that trial evidence of the effects of ω-3 PUFAs on depressed mood has increased. However, the considerable heterogeneity of the studies made them difficult to summarize the results. Overall, the available evidence supports the benefit of ω-3 PUFAs in individuals with diagnosed depressive illness but no evidence of any benefit in individuals without a diagnosis of depressive illness [34].

The association between fish and meat consumption and risk of dementia in populations in developing countries was investigated in low- and middle- income countries of China, India, Cuba, the Dominican Republic, Venezuela, Mexico, and Peru. The found associations of fish and meat consumption with dementia risk to populations were consistent with mechanistic data on the neuroprotective actions of ω-3 PUFAs commonly found in fish. However, the inverse association between fish and prevalent dementia is unlikely to result from poorer dietary habits among demented individuals (reverse causality) because meat consumption was higher in those with a diagnosis of dementia. But anyway, the found beneficial effects of fish consumption on dementia provide preliminary evidence of the etiological significance of diet in dementia [35].

Given the fact that PUFAs are naturally occurring endogenous substances, present in almost all tissues and are essential components of all mammalian cells and can be taken safely for long periods of time (from few months to few years) we can conclude that PUFAs, especially ω-3 fatty acids, are useful in the prevention and treatment of Alzheimer' disease, schizophrenia, and depression [36].

The pioneering studies in Greenland Eskimos almost 30 years ago suggested that ingestion of *n*-3 fatty acids conveys protection from cardiovascular diseases [37].

Since then, many interventions have been conducted with LCPUFA, especially EPA and DHA, aiming at primary and secondary CAD preventions. From that, in most of the prospective cohort studies, *n*-3 fatty acids were found to be beneficial [38–43] but there were also exceptions with no effect [44, 45].

By comparing people who never or ate fish less than once per month, a meta-analysis of 11 prospective studies (11.8 years follow-up of more than 220 thousand subjects) showed the odds ratio for CHD mortality as 0.85 for fish consumption once per week, 0.77 for 2– 4 times/week, and 0.62 for 5 times/week. The authors calculated that each 20 g/day increase in fish intake was associated with a 7% lower risk of coronary heart disease mortality [46].

Many international organizations have made recommendations to increase the intake of EPA plus DHA, and these are summarized by the International Society for the Study of Fatty Acids and Lipids [47]. In general, these recommendations are for 200 mg/day of EPA plus DHA for all adults. The United States has also issued a Dietary Reference Intake for *n*-3 fatty acids [48].

The 2002 American Heart Association recommendations for dietary intake of *n*-3 fatty acids recommended: (1) in the absent of documented CHD it is advised to eat fish twice per week plus oils and foods rich in ALA (flaxseed, canola, soy, walnuts), this accomplishes 500 mg/day of *n*-3 fatty acids. (2) Individuals already with CHD are advised to eat 1 g/day of EPA plus DHA, preferably from oily fish, but could take EPA plus DHA supplements. (3) Individuals with hypertriglyceridemia could take 2–4 g/day of EPA plus DHA, under prescription care [49].

Several mechanisms have been proposed to explain how EPA plus DHA might beneficially influence cardiovascular disease. These include preventing arrhythmias, lowering plasma triacylglycerols, decreasing blood pressure, decreasing platelet aggregation, improving vascular reactivity, and decreasing inflammation. Overall, the therapeutic effect appears to be due to suppression of fatal arrhythmias rather than stabilization of atherosclerotic plaques [50–60].

Elevated plasma triacylglycerol concentrations have been associated with increased risk of coronary heart disease (CHD). Prospective evidence shows that nonfasting plasma triacylglycerol concentration is a strong and independent predictor of future myocardial infarction once elevated postprandial triacylglycerolemia leads to a series of metabolic reactions that reduce high-density lipoprotein (HDL)-cholesterol concentrations and promote the formation of small, dense low-density lipoprotein (LDL) particles. Metabolism of plasma triacylglycerols also influences postprandial factor VII activation [61].

EPA and DHA are *n*-3 PUFAs in fish oil which are effective hypotriacylglycerolemic agents, even when consumed at low doses (1 g *n*-3 PUFA/d). Therefore, consumption of *n*-3 PUFAs provides a realistic option for the optimization of plasma triacylglycerol metabolism [61].

Omega-3 fatty acid supplementation provided additional benefits to Lifestyle Modification Program (LSMP) in the resolution of metabolic syndrome of free living adults. The fish oil group received 3 g of fish oil per day (360 mg of DHA and 540 mg of EPA) (G2, $n = 23$)) during 20 weeks. Compared to the control group (only LSMP) the intervened group showed a significant decrease in waist circumference (1.3%) followed by metabolic syndrome reduction (29%) mainly due to normalization of blood pressure (33.3%) and triacylglycerol (27.3%). Some theories have been proposed to explain how omega-3 reduces triacylglycerol. The strongest evidence is the reduction in hepatic lipogenesis, reducing hepatic secretion of very low-density lipoprotein (VLDL). Additionally, omega-3 inhibits certain enzymes involved in the hepatic synthesis of triacylglycerol, reducing its plasma level [62].

A significant decrease of plasma oxidative-stress markers in patients with ulcerative colitis was shown when fish oil ω-3 fatty acids were used in combination with sulfasalazine [63].

Regarding type 2 diabetes mellitus (T2DM), a prospective cohort analysis of men and women showed that long-term dietary intake of long-chain omega 3 fatty acids does not decrease the risk of T2DM. Instead, a modestly but significantly higher incidence of T2DM was associated with higher fish and long-chain omega 3 fatty acid consumption [64].

Doses 3 g/day, EPA plus DHA can improve cardiovascular disease risk factors, including decreasing plasma triacylglycerols, blood pressure, platelet aggregation, and inflammation, while improving vascular reactivity [50].

By the fact that EFAs and their long-chain metabolites and other products prevent platelet aggregation, lower blood pressure, reduce LDL-cholesterol, and ameliorate the adverse actions of homocysteine, the EFAs and their metabolites show all the actions expected of the "polypill" [65].

The concept of cardiovascular "polypill" was coined [66] by the fact that, when combined, the effects of statins, aspirin, and blood pressure lowering drugs reduced the all causes mortality in CHD patients.

In conclusion, it is evident that PUFAs, especially an optimal combination of EPA, DHA, and possibly, gamma-linolenic acid (GLA), dihomo-gamma-linolenic acid (DGLA), and AA show all the qualities of the suggested "polypill", viz., aspirin-like action, inhibition of HMG-CoA and angiotensin-converting enzymes (ACEs), and possess diuretic, antihypertensive, and beta-blocker-like actions. Additionally, given the fact that PUFAs are naturally occurring endogenous substances, present in almost all tissues and are essential components of all mammalian cells and can be taken safely for long periods of time (from few months to few years), we can conclude that PUFAs, especially ω-3 fatty acids, are useful in the prevention and treatment of Alzheimer' disease, schizophrenia, and depression [36], suggesting that PUFAs have a much wider benefit compared to the "polypill" [65].

Thus, LCPω3 supplements might especially be effective in prevention, as suggested by the outcomes of epidemiological studies on CAD and prospective studies on Alzheimer's disease, and also from the favorable effects of LCPω3 in early disease stages [30]. Consensus has been reached that those interventions in CAD and depression are positive but not in all others.

6. Future directions

Genetically, we are for the greater part still adapted to the East African ecosystem on which our genome evolved, with some adaptations since the Out-of-Africa Diaspora. Dietary fat quantity and quality change have, together with other man-made changes in our environment, caused a conflict with our slowly adapting genome that is implicated in "typically Western" diseases. Fortunately, the majority of Western diseases occur typically after reproductive age. Rather than reducing our life expectancy, these diseases notably diminish our number of years in health.

Many recommendations for the intakes of saturated fat, *trans* fat, and EPA + DHA have been issued, notably for prevention. The ultimate goal might be, however, translate to the culture of the current society that our genes had evolved for million years in an entirely different dietary composition and lifestyle and therefore we must return to the fat quality of our ancient diet [11].

Author details

Roberto Carlos Burini[1]*, Caroline das Neves Mendes Nunes[2] and Franz Homero Paganini Burini[3]

*Address all correspondence to: burini@fmb.unesp.br

1 Public Health Department, UNESP Medical School, Botucatu, Brazil

2 Pathology Department, UNESP Medical School, Botucatu, Brazil

3 UNESP Medical School Clinical Hospital, Botucatu, Brazil

References

[1] White T.D., Asfaw B., DeGusta D., Gilbert H., Richards G.D., Suwa G., et al. Pleistocene Homo sapiens from middle awash, Ethiopia. Nature. 2003;423:742–747. DOI: 10.1038/nature01669

[2] Burini R.C., Nunes C.N.M., Kano H.T., Nakagaki M.S. Epygenetics of Plasma Cholesterol Variation: From an Evolutionary Genome and Environmental-Induced Frugality to Contemporary Diseases-Basis for Primary Care with Lifestyle Modifications. In: Advances in Dyslipidemia. Publisher: Avid Science. Location: Telangana, India. 2016. pp. 2–35.

[3] Clarke S.D. The multi-dimensional regulation of gene expression by fatty acids: Polyunsaturated fats as nutrient sensors. Current Opinion in Lipidology. 2004;15:13–18. DOI: 10.1097/01.mol.0000113205.97033.22

[4] Duplus E., Forest C. Is there a single mechanism for fatty acid regulation of gene transcription? Biochemical Pharmacology. 2002;64:893–901. DOI: 10.1016/S0006-2952(02)01157-7

[5] Jump D.B. Fatty acid regulation of gene transcription. Critical Reviews in Clinical Laboratory Sciences. 2004;41:41–78. DOI: 10.1080/10408360490278341

[6] Lapillonne A., Clarke S.D., Heird W.C. Polyunsaturated fatty acids and gene. Current Opinion in Clinical Nutrition & Metabolism Care. 2004;7:151–156. DOI: 10.1097/01.mco.0000120004.58274.9e

[7] Auwerx J. PPARgamma, the ultimate thrifty gene. Diabetologia. 1999;42(9):1033–1049. DOI: 10.1007/s001250051268

[8] Jump D.B. N-3 polyunsaturated fatty acid regulation of hepatic gene. Current Opinion in Lipidology. 2008;19:242–247. DOI: 10.1097/MOL.0b013e3282ffaf6a

[9] Nakamura M.T., Nara T.Y. Essential fatty acid synthesis and its regulation in. Prostaglandins, Leukotrienes and Essential Fatty Acids. 2003;68:145–150. DOI: 10.1016/S0952-3278(02)00264-8

[10] Calder P.C. N-3 polyunsaturated fatty acids, inflammation, and inflammatory. American Journal of Clinical Nutrition. 206;83:1505S–1519S.

[11] Muskiet F.A.J. Pathophysiology and Evolutionary Aspects of Dietary Fats and Long-Chain Polyunsaturated Fatty Acids across the Life Cycle. In: Montmayeur J.P., le Coutre J., editors. Fat Detection: Taste, Texture, and Post Ingestive Effects. Boca Raton, FL: CRC Press/Taylor & Francis. 2010. Chapter 2.

[12] Antón S.C. A natural history of Homo erectus. Yearbook of Physical Anthropology. 2003;46:126–170. DOI: 10.1002/ajpa.10399

[13] Leonard W.R., Robertson M.L., Snodgrass J.J., Kuzawa C.W. Metabolic correlates of hominid brain evolution. Comparative Biochemistry and Physiology. 2003;136:5–15. DOI: 10.1016/S1095-6433(03)00132-6

[14] Milton K. The critical role played by animal source foods in human (Homo) evolution. The Journal of Nutrition. 2003;133:3886S–3892S.

[15] Cordain L., Brand-Miller J., Eaton S.B., Mann N., Holt S.H.A., Speth J.D. Plant to animal subsistence ratios and macronutrient energy estimations in worldwide hunter-gatherer diets. The American Journal of Clinical Nutrition. 2010;71:682–692.

[16] Aiello L.C., Wheeler P. The expensive-tissue hypothesis: The brain and the digestive system in human and primate evolution. Current Anthropology. 1995;36:199–221. DOI: 10.1086/204350

[17] Finch C.E., Stanford C.B. Meat-adaptive genes and the evolution of slower aging in humans. The Quarterly Review of Biology. 2004;79:3–50.

[18] Leonard W.R., Snodgrass J.J., Robertson M.L. Evolutionary perspectives on fat ingestion and metabolism in humans. In: Montmayeur J.P., le Coutre J, editors. Fat Detection: Taste, Texture, and Post Ingestive Effects. Boca Raton, FL: CRC Press/Taylor & Francis. 2010. Chapter 1.

[19] Crawford M.A., Bloom M., Broadhurst C.L., Schmidt W.F., Cunnane S.C., Galli C., et al. Evidence for unique function of docosahexaenoic acid during the evolution of the modern human brain. Lipids. 1999;34:S39–S47.

[20] Popovich D.G., Jenkins D.J.A., Kendall C.W.C., Dierenfeld E.S., Carroll R.W., Tariq N., et al. The western lowland gorilla diet has implications for the health of humans and other hominoids. The Journal of Nutrition. 1997;127:2000–2005.

[21] Cordain L, Watkins BA, Mann NJ. Fatty acid composition and energy density of foods available to African hominids. World Review of Nutrition and Dietetics. 2001;90:144–161.

[22] Bobe R., Behrensmeyer A.K. Faunal change, environmental variability and late Pliocene hominin evolution. Journal of Human Evolution. 2002;42:475–497. DOI: 10.1006/jhev.2001.0535

[23] de Menocal P.B. African climate change and faunal evolution during the Pliocene–Pleistocene. Earth and Planetary Science Letters. 2004;220:3–24. DOI: 10.1016/S0012-821X(04)00003-2

[24] Wynn J.G. Influence of Plio-Pleistocene aridification on human evolution: Evidence from paleosols from the Turkana Basin, Kenya. American Journal of Physical Anthropology. 2004;123:106–118. DOI: 10.1002/ajpa.10317

[25] Reed K. Early hominid evolution and ecological change through the African Plio-Pleistocene. Journal of Human Evolution. 1997;32:289–322. DOI: 10.1006/jhev.1996.0106

[26] Eaton S.B. The ancestral human diet: What was it and should it be a paradigm for contemporary nutrition? Proceedings of the Nutrition Society. 2006;65(1):1–6. DOI: 10.1079/PNS2005471

[27] Innis S.M. Dietary lipids in early development: Relevance to obesity, immune and inflammatory disorders. Current Opinion in Endocrinology, Diabetes and Obesity. 2007;14:359–364. DOI: 10.1097/MED.0b013e3282be90b9

[28] Mayer K., Seeger W. Fish oil in critical illness. Current Opinion in Clinical Nutrition & Metabolic Care. 2008;11:121–127. DOI: 10.1097/MCO.0b013e3282f4cdc6

[29] Leonard B.E. Inflammation, depression and dementia: Are they connected? Neurochemical Research. 2007;32:1749–1756. DOI: 10.1007/s11064-007-9385-y

[30] Bourre J.M. Dietary omega-3 fatty acids and psychiatry: Mood, behaviour, stress, depression, dementia and aging. The Journal of Nutrition Health & Aging. 2005;9:31–38.

[31] Hadders-Algra M. Prenatal and early postnatal supplementation with long-chain polyunsaturated fatty acids: Neurodevelopmental considerations. American Journal of Clinical Nutrition. 2011;94(suppl):1874S–1879S. DOI: 10.3945/ajcn.110.001065

[32] Sun G.Y., Sun A.Y. Effect of chronic ethanol administration on phospholipid acyl groups of synaptic plasma membrane fraction isolated from guinea pig brain. Research Communications in Chemical Pathology & Pharmacology. 1979;24:405–408. DOI: 10.1111/j.1530-0277.2001.tb02131.x

[33] Hibbeln J.R., Salem N. Jr. Dietary polyunsaturated fatty acids and depression: When the cholesterol does not satisfy. American Journal of Clinical Nutrition. 1995;62:1–9.

[34] Appleton K.M., Rogers P.J., Ness A.R. Updated systematic review and meta-analysis of the effects of n23 long-chain polyunsaturated fatty acids on depressed mood. American Journal of Clinical Nutrition. 2010;91:757–770. DOI: 10.3945/ajcn.2009.28313.

[35] Albanese E., Dangour A.D., Uauy R., Acosta D., Guerra M., Guerra S.S. et al. Dietary fish and meat intake and dementia in Latin America, China, and India: A 10/66 Dementia Research Group population-based study. American Journal of Clinical Nutrition. 2009;90:392–400. DOI: 10.3945/ajcn.2009.27580

[36] Das U.N. Essential fatty acids: Biochemistry, physiology, and pathology. Biotechnology Journal. 2006;1:420–439. DOI: 10.1002/biot.200600012

[37] Bang H.O, Dyerberg J., Hjoorne N. The composition of food consumed by Greenland Eskimos. Journal of Internal Medicine. 1976;200:69–73. DOI: 10.1111/j.0954-6820.1976.tb08198.x

[38] Rissanen T., Voutilainen S., Nyyssonen K., Lakka T.A., Salonen J.T. Fish oil-derived fatty acids, docosahexaenoic acid and docosapentaenoic acid, and the risk of acute coronary events: The Kuopio Ischaemic heart disease risk factor study. Circulation. 2000;102:2677–2679. DOI: 10.1161/01.CIR.102.22.2677

[39] Oomen C.M., Feskens E.J., Räsänen L., Fidanza F., Nissinen A.M., Menotti A. Fish consumption and coronary heart disease mortality in Finland, Italy, and The Netherlands. American Journal of Epidemiology. 2000;151:999–1006.

[40] Gillum R.F., Mussolino M., Madans J.H. The relation between fish consumption, death from all causes, and incidence of coronary heart disease. The NHANES I Epidemiologic Follow-up Study. Journal of Clinical Epidemiology. 2000;53:237–244. DOI: 10.1016/S0895-4356(99)00149-3

[41] Yuan J.M., Ross R.K., Gao Y.T., Yu M.C. Fish and shellfish consumption in relation to death from myocardial infarction among men in Shanghai, China. American Journal of Epidemiology. 2001;154:809–816. DOI: 10.1093/aje/154.9.809

[42] Hu F.B., Bronner L., Willett W.C., Stampfer M.J., Rexrode K.M., Albert C.M. Fish and omega-3 fatty acid intake and risk of coronary heart disease in women. JAMA. 2002;287:1815–1821. DOI: 10.1001/jama.287.14.1815

[43] Mozaffarian D., Lemaitre R.N., Kuller L.H., Burke G.L., Tracy R.P., Siscovick D.S. Cardiac benefits of fish consumption may depend on the type of fish meal consumed: The cardiovascular health study. Circulation. 2003;107:1372–1377. DOI: 10.1161/01.CIR.0000055315.79177.16

[44] Ascherio A., Rimm E.B., Stampfer M.J., Giovannucci E.L., Willett W.C. Dietary intake of marine n3 fatty acids, fish intake, and the risk of coronary disease among men. The New England Journal of Medicine. 1995;332:977–982. DOI: 10.1056/NEJM199504133321501

[45] Osler M., Andreasen A.H., Hoidrup S. No inverse association between fish consumption and risk of death from all-causes, and incidence of coronary heart disease in middle-aged, Danish adults. Journal Clinical of Epidemiology. 2003;56:274–279. DOI: 10.1016/S0895-4356(02)00600-5

[46] He K., Song Y., Daviglus M.L., Liu K., Van Horn L., Dyer A.R., et al. Accumulated evidence on fish consumption and coronary heart disease mortality: A meta-analysis of cohort studies. Circulation. 2004;109:2705–2711. DOI: 10.1161/01.CIR.0000132503.19410.6B

[47] The International Society for the Study of Fatty Acids and Lipids. ISSFAL Board, 2009. Available from: http://www.issfal.org.uk/Recommendations.htm

[48] Institute of Medicine. 2005. Available from: http://www.iom.edu/Object.File/Master/7/300/0.pdf

[49] Kris-Etherton P.M., Harris W.S., Appel L.J. Fish consumption, fish oil, omega-3 fatty acids, and cardiovascular disease. Circulation. 2002;106:2747–2757. DOI: 10.1161/01.CIR.0000038493.65177.94

[50] Breslow J.L. N-3 Fatty acids and cardiovascular diseases. American Journal of Clinical Nutrition. 2006;83(suppl):1477S–1482S.

[51] Connor W.E. Importance of n-3 fatty acids in health and disease. American Journal of Clinical Nutrition. 2000;71(suppl):171S–175S.

[52] Leaf A, Kang JX, Xiao YF, Billman GE. Clinical prevention of sudden cardiac death by n-3 polyunsaturated fatty acids and mechanism of prevention of arrhythmias by n3 fish oils. Circulation. 2003;107:2646–2652. DOI: 10.1161/01.CIR.0000069566.78305.33

[53] Harris W.S. N-3 Fatty acids and serum lipoproteins: Human studies. American Journal of Clinical Nutrition. 1997;65(suppl):1645S–1654S.

[54] Sacks F.M, Katan M. Randomized clinical trials on the effects of dietary fat and carbohydrate on plasma lipoproteins and cardiovascular disease. The American Journal of Medicine. 2002;113(suppl):13S–24S. DOI: 10.1016/S0002-9343(01)00987-1

[55] Geleijnse J.M., Giltay E.J., Grobbee D.E., Donders A.R., Kok F.J. Blood pressure response to fish oil supplementation: Meta-regression analysis of randomized trials. Journal of Hypertension. 2002;20:1493–1499.

[56] Knapp H.R. Dietary fatty acids in human thrombosis and hemostasis. American Journal of Clinical Nutrition. 1997;65(suppl):1687S–1698S.

[57] Hornstra G. Influence of dietary fatty on arterial thrombosis tendency. The Journal of Nutrition Health and Aging. 2001;5:160–166.

[58] Harris W.S., Rambjor G.S., Windsor S.L., Diederich D. n-3 Fatty acids and urinary excretion of nitric oxide metabolites in humans. American Journal of Clinical Nutrition. 1997;65:459–464.

[59] Goodfellow J, Bellamy M.F, Ramsey M.W, Jones C.J, Lewis M.J. Dietary supplementation with marine omega-3 fatty acids improve systemic large artery endothelial function in subjects with hypercholesterolemia. Journal of the American College of Cardiology. 2000;35:265–270. DOI: 10.1016/S0735-1097(99)00548-3

[60] Calder P.C. Grimble R.F. Polyunsaturated fatty acids, inflammation, and immunity. Lipids. 2001;36:1007–1024. DOI: 10.1038/sj.ejcn.1601478

[61] Roche H.M., Gibney M.J. Effect of long-chain n-3 polyunsaturated fatty acids on fasting. American Journal of Clinical Nutrition. 2000;71(suppl):232S–237S.

[62] Talon L.C., Oliveira E.P., Moreto F., McLellan K.C.P., Burini R.C. Omega-3 fatty acids supplementation decreases metabolic syndrome. Journal of Functional Foods. 2015;19:922–928. DOI: 10.1016/j.jff.2015.01.022

[63] Barbosa D.S., Cecchini R., El Kadri M.Z., Rodríguez M.A., Burini R.C., Dichi I. Decreased oxidative stress in patients with ulcerative colitis supplemented with fish oil omega-3 fatty acid. Nutrition. 2003;19:837–842. DOI: 10.1016/S0899-9007(03)00162-X

[64] Kaushik M., Mozaffarian D., Spiegelman D., Manson J.E., Willett W.C., Hu F.B. Long-chain omega-3 fatty acids, fish intake, and the risk of type 2 diabetes mellitus. American Journal of Clinical Nutrition. 2009;90:613–620. DOI: 10.3945/ajcn.2008.27424.

[65] Undurti N. Das. Do polyunsaturated fatty acids behave like an endogenous "polypill? Medical Hypotheses. 2008;70:430–434. DOI: 10.1016/j.mehy.2007.05.023

[66] Hippisley-Cox J, Coupland C. Effects of combinations of drugs on all mortality in patients with disease: Nested case-control analysis. British Medical Journal. 2005;330:1059–1063. DOI: 10.1136/bmj.330.7499.1059